MODERN INTRODUCTION TO CLASSICAL MECHANICS & CONTROL

MATHEMATICS & ITS APPLICATIONS

Series Editor: Professor G. M. Bell
Chelsea College, University of London

Mathematics and its applications are now awe-inspiring in their scope, variety and depth. Not only is there rapid growth in pure mathematics and its applications to the traditional fields of the physical sciences, engineering and statistics, but new fields of application are emerging in biology, ecology and social organisation. The user of mathematics must assimilate subtle new techniques and also learn to handle the great power of the computer efficiently and economically.

The need of clear, concise and authoritative texts is thus greater than ever and our series will endeavour to supply this need. It aims to be comprehensive and yet flexible. Works surveying recent research will introduce new areas and up-to-date mathematical methods. Undergraduate texts on established topics will stimulate student interest by including applications relevant at the present day. The series will also include selected volumes of lecture notes which will enable certain important topics to be presented earlier than would otherwise be possible.

In all these ways it is hoped to render a valuable service to those who learn, teach, develop and use mathematics.

The Foundation Programme includes:

MODERN INTRODUCTION TO CLASSICAL MECHANICS & CONTROL
David Burghes, University of Newcastle-upon-Tyne
Angela Downs, University of Sheffield

VECTOR & TENSOR METHODS
Frank Chorlton, University of Aston, Birmingham

Lecture notes in
QUEUEING SYSTEMS
Professor Brian Conolly, Chelsea College, University of London

MODERN INTRODUCTION TO CLASSICAL MECHANICS & CONTROL

DAVID N. BURGHES
University of Newcastle upon Tyne

ANGELA M. DOWNS
University of Sheffield

ELLIS HORWOOD LIMITED

Publisher · Chichester

Halstead Press: a division of
JOHN WILEY & SONS INC.
New York · London · Sydney · Toronto

The publisher's colophon is reproduced from James Gillison's drawing of the Ancient Market Cross, Chichester.

First published in 1975 by

ELLIS HORWOOD
Coll House, Westergate, Chichester, Sussex, England

Distributors:
Australia, New Zealand, South-east Asia:
JOHN WILEY & SONS, AUSTRALASIA PTY LIMITED
110 Alexander Street, Crow's Nest, N.S.W. Australia
Canada, Europe, Africa:
JOHN WILEY & SONS LIMITED
Baffins Lane, Chichester, Sussex, England
U.S.A., Mexico, S. America and the rest of the world:
HALSTED PRESS a division of
JOHN WILEY & SONS INC.,
605 Third Avenue, New York, N.Y. 10016, U.S.A.

Library of Congress Cataloging in Publication Data
Burghes, David N., Modern Introduction to Classical Mechanics & Control
1. Mechanics. 2. Mathematics - 1961
I. Downs, Angela M., joint author. II. Title.
QA805.B94 1975 531'.01'51 75-16463
ISBN 0-470-12362-1 (Wiley)
ISBN 85312-039-0 (Ellis Horwood Ltd.)

Set in 10 point Press Roman by Speedpress Repro Services
14 St. John's Precinct, Hebburn, Tyne & Wear

Printed by Unwin Brothers Limited, The Gresham Press, Old Woking, Surrey

Contents

Chapter 8 Stability

Chapter 9 Oscillations and Vibrations

Chapter 10 Variational Calculus

Chapter 11 Optimal Control

[* indicates that this section can be omitted without affecting the understanding of the subsequent sections.]

Preface

The text, examples and exercises in this book have developed from lectures in applied mathematics given over a number of years to first year undergraduates at Newcastle and Sheffield Universities. The book has been written not only for mathematics undergraduates but also for those, for example physicists, chemists and engineers, who also need mathematics.

The book has two major aims. Firstly it develops a modern approach to classical Newtonian mechanics, showing that there are many interesting and stimulating problems. Secondly, it demonstrates that applied mathematics is not merely 'theoretical mechanics', but also includes important topics which are in no way dependent on Newton's laws of motion. In recent years there have been a number of changes in many first year courses in applied mathematics. For example, some teaching centres have replaced traditional mechanics by courses based on mathematical modelling. It is our contention that mechanics, provided that it is presented in an up-to-date manner as in this text, can be a most enjoyable and lively subject. Nevertheless the concept of mathematical modelling and the importance of numerical and computing work should also be stressed. We have included, as well as chapters on Newtonian mechanics, introductory chapters on the stability of solutions of differential equations, variational calculus and optimal control. These three topics were selected because we feel that they can be readily taught to first year students and because they also link well with the mechanics chapters.

Throughout the book we normally use the S.I. system of units and notation. The main details of this system of units are summarized in Chapter 3. We have assumed that the reader has some knowledge of differential equations and vector analysis, although for those who have not met vector differentials we have included appendices which summarize the results needed for this book.

We must finally acknowledge the great help we have received from many people during the preparation of this text. In particular, we thank Professor P. C. Kendall for his many ideas and constant encouragement, and acknowledge our debt to the many students at Sheffield and Newcastle Universities whose reactions, criticisms and work have helped to form this book.

Chapter 1

Introduction

1.1 Modern Applied Mathematics

Modern applied mathematics is concerned with the study of systems, both their *behaviour* and their *control*. The study of their *behaviour* can generally be termed 'classical applied mathematics' and in many cases depends on the application of Newton's laws to the system. In this text we try to emphasize those parts of classical applied mathematics which are applicable to modern problems. We are not interested, for instance, in such mathematical complexities as spheres rolling down smooth inclined planes, which are slipping on rough horizontal surfaces! Unfortunately this type of problem appears still to be the ideal of some applied mathematics examiners. We shall concentrate on such aspects as rocket flight optimization and satellite orbits, which have more relevance today.

The *control* of a system embraces such topics as optimal control theory, simulation and optimization, and does not, directly at least, depend on the application of Newton's laws. In other words, modern applied mathematics is not just modern theoretical mechanics. In fact, theoretical mechanics is just one sub-section of applied mathematics. A simple example which well illustrates the difference between 'classical' and 'control' applied mathematics is that of a rod being balanced vertically at its lowest point on a person's finger as illustrated in Fig. 1.1. Clearly if no action is taken, the rod will fall freely (due to the slight vibration of the finger, which will initiate the motion). The description of the manner in which the rod falls is a problem of classical applied mathematics. But if we consider what could be done to prevent the fall of the rod, that is to stabilize the vertical position of the rod, we have a problem of control. With just the use of the one pivoting finger, there is a simple practical answer to this control problem, which is given in Ch. 11.1.

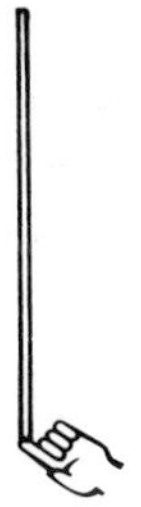

Fig. 1.1 Control of pole

1.2 Model Building

Nearly all applied mathematics, both classical and control, requires the setting up of a model which describes the system under consideration. Usually the applied mathematician has some *observed data* which relate to some property of the system, and his aim is to set up a mathematical model which characterizes the system and its nature. The resulting equations (usually differential equations) are solved, giving some *predicted data.* If the two sets of data are in good agreement, the applied mathematician concludes that he has a good model for the system. If the two sets of data are not in close agreement, he must look back at his model, and decide what important features were neglected in setting up the model. With a new, and probably more complex, model he goes through the same procedure as before, and again compares observed and predicted data. The art of the applied mathematician is to find a model for the system, which incorporates all the important aspects but yet is simple enough for the resulting equations to be solved and such that the predicted and observed data are a close fit. This art is termed *model building.*

As an example consider the motion of a particle lying on a smooth horizontal table and joined to one end of a spring, which is fixed at its other end. The system is illustrated in Fig. 1.2.

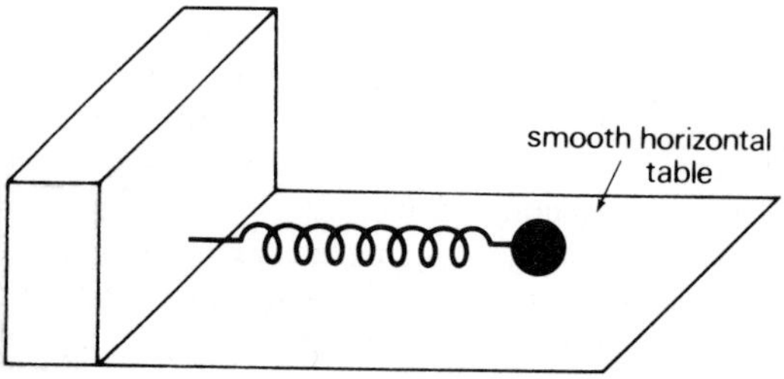

Fig. 1.2 Particle and spring system

We set up a model for the physical system by assuming that

(a) the *tension* in the spring is proportional to the extension of the spring (Hooke's law);

(b) the *force* acting on the particle is equal to the rate of change of its linear momentum. (Newton's laws).

Thus if $x = x(t)$ is the extension of the spring, at any time t, from its equilibrium (i.e. non-extended) position, we see that (a) and (b) can be written as

(a) $T = \lambda x.$

(b) $-T = m\dfrac{d^2x}{dt^2},$

T being the tension, λ the constant of proportionality and m the mass of the particle. We can summarize by saying that the model is given by

$$T = \lambda x; \qquad -T = m\frac{d^2x}{dt^2}.$$

Eliminating T, we obtain the differential equation

$$\boxed{\frac{d^2x}{dt^2} + \frac{\lambda}{m}x = 0\,.} \tag{1.1}$$

which predicts that the motion of the particle is simple harmonic with period of oscillation $2\pi/(\lambda/m)^{\frac{1}{2}}$.

Now if the observed data shows that the particle is moving in S.H.M. with approximate period $2\pi/(\lambda/m)^{\frac{1}{2}}$, we conclude that this model is a good one for the system. But if the observed motion of the particle does not correlate with the predicted results, we must return to our model and see what assumptions have been made. Perhaps there is some significant force (e.g. friction) acting on the particle which we have neglected or maybe the spring does not obey Hooke's law, but obeys a law of the form

$$T = \lambda x + \mu x^3; \tag{1.2}$$

[springs of this type are called 'non-linear'].

The fundamentals of model building are illustrated in the diagram below:

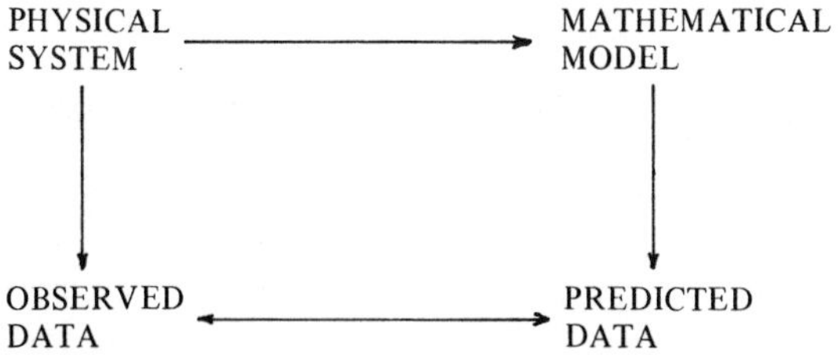

When the observed and predicted data are compatible, the applied mathematician has achieved his aim.

1.3 Numerical and Approximate Solutions

As more and more features are brought into the mathematical model of a particular system, its description and consequently the mathematical equations resulting from this description become more complex. If the equations appear to have no analytic solution (i.e. no solution that can be expressed in terms of known functions), the applied mathematician has at least two other methods open to him; namely, either to try a numerical approach to the solution of the equations, or to try to find an approximate solution to the equations, using an appropriate method if one exists.

For instance, if the mathematical description of the model produces the differential equation

$$\frac{d^2x}{dt^2} + x = \epsilon x^2 \text{ (}\epsilon \text{ small constant)} \tag{1.3}$$

subject to the conditions $x = 1, dx/dt = 0$ at $t = 0$, we shall have difficulty in finding an analytic solution. But there are plenty of numerical techniques (e.g. Runge-Kutta) that can be used to find the solution x, tabulated at various values of t. With the help of modern high speed computers numerical solutions of this type can quickly be found. In fact, the rapid increase in the speed and storage space of modern computers means that many very complex differential equations can now be solved numerically in a very short time. This is extremely important for the applied mathematician for it means that he is able to improve his model whilst still being able to obtain his predicted data. Unfortunately a numerical solution does not always indicate the full significance of a particular term in the equation, and its influence on the solution. Also one has to be very careful that the correct solution is in fact found. This last remark applies particularly to the solutions of partial differential equations.

Methods of approximate solution, on the other hand, can be very useful as they often indicate the importance of particular terms. For instance, in the differential equation above, provided ϵ is a *small* constant, we can attempt to find an approximate solution by assuming that

$$x = x_0(t) + \epsilon x_1(t) + O(\epsilon^2)\,. \tag{1.4}$$

where x_0, x_1 do not depend on ϵ and $O(\epsilon^2)$ means terms of order ϵ^2 and higher order terms.

Substitution in (1.3) gives

$$\frac{d^2x_0}{dt^2} + x_0 + \epsilon\left[\frac{d^2x_1}{dt^2} + x_1 - {x_0}^2\right] + O(\epsilon^2) = 0\,.$$

and equating the coefficient of each power of ϵ to zero, we obtain

$$\frac{d^2x_0}{dt^2} + x_0 = 0\,, \tag{1.5}$$

$$\frac{d^2x_1}{dt^2} + x_1 = {x_0}^2\,. \tag{1.6}$$

Solving (1.5) subject to $x_0(0) = 1$, $dx_0/dt\,(0) = 0$ gives $x_0 = \cos t$, and so (1.6) becomes

$$\begin{aligned}\frac{d^2x_1}{dt^2} + x_1 &= \cos^2 t \\ &= \tfrac{1}{2}(\cos 2t - 1)\,.\end{aligned}$$

Solving (1.6) subject to $x_1(0) = dx_1/dt\,(0) = 0$ gives

$$x_1 = \tfrac{2}{3}\cos t - \tfrac{1}{6}\cos 2t - \tfrac{1}{2}\,.$$

In this way we can build up a solution of the form (1.4), subject to the correct conditions, and provided the series in ϵ converges, we can find the solution correct to any desired accuracy. Using this type of approach we can clearly see the significance of the parameter ϵ for any value of t. This solution, rather than a purely numerically tabulated solution, gives us more insight into the form of the solution and the importance of the parameter ϵ.

On the other hand, we must take care not to underestimate the importance of the modern high speed computer to the applied mathematician. Differential equations which a decade ago were considered far too difficult and time-consuming for either man or a computer of that vintage, can be solved in a few seconds on present day machines. Many problems in different fields of applied mathematics can now be solved, where previously no analytic or approximate solutions were available. One word of warning however. It may be tempting to look immediately to the computer in order to solve the equations of a problem, when in fact the problem possesses an elegant and simple analytic solution. This latter solution can tell us far more about the nature of the problem than any computed solution. So we should look first for any analytic or approximate method before rushing into a computed numerical solution. This book is not primarily concerned with approximations, but in order to show their importance in applied mathematics we have included a number of problems designed to be solved by running a computer programme.

Chapter 2

Kinematics

2.1 Velocity and Acceleration

Kinematics is the study of the 'motion of a particle'. It does not deal with the causes of the motion, which will be discussed in a later chapter, but with the description of this motion.

Suppose a particle, P, moves on a smooth curve, $\mathscr{C}$, in 'three dimensional space'. At any instant of time, the particle's position on this curve can be described by its rectangular Cartesian coordinates $x(t)$, $y(t)$ and $z(t)$ where t denotes time. In vectorial notation, we write

$$\mathbf{r} = x\mathbf{i} + y\mathbf{j} + z\mathbf{k} = (x,y,z)$$

where $\mathbf{r}$ is the position vector of the particle, and $\mathbf{i}$, $\mathbf{j}$ and $\mathbf{k}$ are unit vectors parallel to the rectangular axes. If the particle is in motion from time t_0 to time t_1, we write

$$\boxed{\mathbf{r} = \mathbf{r}(t), \; t_0 \leqslant t \leqslant t_1} \tag{2.1}$$

which indicates that x, y, z are in general functions of t defined on the interval $t_0 \leqslant t \leqslant t_1$. Equation (2.1) is called a parametric equation of the particle's path.

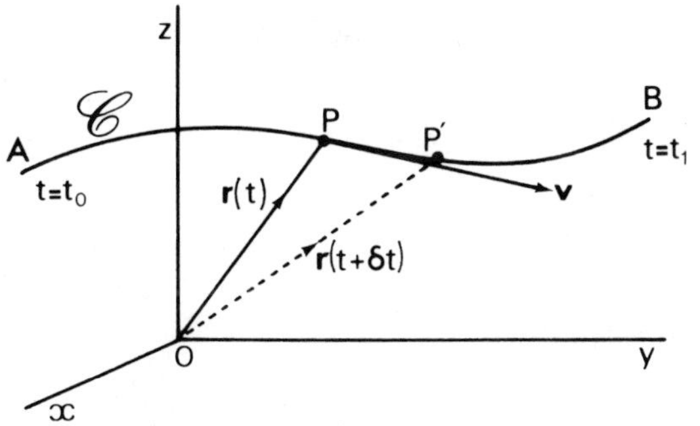

Fig. 2.1 Motion of particle on curve $\mathscr{C}$

A simple example is a car moving uniformly on a circular horizontal track of radius a, which completes a circle in time τ.

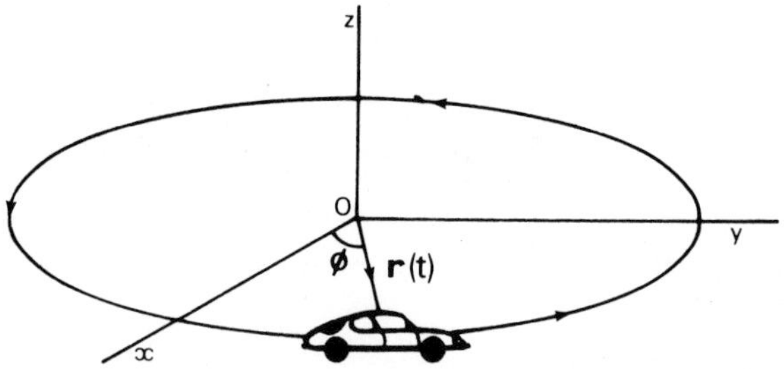

Fig. 2.2 Motion of car on circular track

A description of its path is given by

$$\mathbf{r} = a\cos\phi\,\mathbf{i} + a\sin\phi\,\mathbf{j}$$

and since it traverses a complete circuit in time τ we must have $\phi/2\pi = t/\tau$. Thus the position vector for the particle is given by

$$\mathbf{r} = a\cos\left(\frac{2\pi t}{\tau}\right)\mathbf{i} + a\sin\left(\frac{2\pi t}{\tau}\right)\mathbf{j}.$$

[We have chosen the orientation of the x-axis so that initially, at $t = 0$, the particle is at $(a, 0, 0)$.]

We now define the verms *velocity* and *acceleration*. If in a small time δt, P moves to position P′ on $\mathscr{C}$, where P′ has position vector $\mathbf{r} = \mathbf{r}(t + \delta t)$, as shown in Fig. 2.1, the change in position of P in time δt is given by

$$\mathbf{r}(t + \delta t) - \mathbf{r}(t)$$

Thus the average rate of change of the position of P is

$$\frac{\mathbf{r}(t + \delta t) - \mathbf{r}(t)}{\delta t}$$

and we make the following definition:

DEFINITION:

The *velocity* of P is defined as

$$\mathbf{v} = \lim_{\delta t \to 0}\left\{\frac{\mathbf{r}(t + \delta t) - \mathbf{r}(t)}{\delta t}\right\} \tag{2.2}$$

and by definition of the derivative of a vector function,

$$\boxed{\mathbf{v} = \frac{d\mathbf{r}}{dt}} \quad . \tag{2.3}$$

From this definition, we see that velocity is the instantaneous rate of change of the position vector of the particle, and we note that velocity is a *vector*, directed along the tangent to $\mathscr{C}$ at P, and in the sense of t increasing. The magnitude of the velocity vector is called the *speed*, which is a positive scalar quantity. We denote speed by v, so that

$$v = |\mathbf{v}| . \tag{2.4}$$

The term $d\mathbf{r}/dt$ is given, in terms of rectangular Cartesian components, by

$$\frac{d\mathbf{r}}{dt} = \frac{dx}{dt}\,\mathbf{i} + \frac{dy}{dt}\,\mathbf{j} + \frac{dz}{dt}\,\mathbf{k} .$$

Thus in the example described above we have

$$\mathbf{v} = \frac{d\mathbf{r}}{dt} = -\frac{2\pi a}{\tau}\sin\left(\frac{2\pi t}{\tau}\right)\mathbf{i} + \frac{2\pi a}{\tau}\cos\left(\frac{2\pi t}{\tau}\right)\mathbf{j}$$

and the speed v is given by

$$v = |\mathbf{v}| = \frac{2\pi a}{\tau}\left\{\sin^2\left(\frac{2\pi t}{\tau}\right) + \cos^2\left(\frac{2\pi t}{\tau}\right)\right\}^{1/2}$$

$$= \frac{2\pi a}{\tau} .$$

DEFINITION:

The *acceleration* of P is defined by

$$\boxed{\mathbf{f} = \frac{d\mathbf{v}}{dt}} \quad . \tag{2.5}$$

This means that acceleration is the rate of change of velocity of P, and using (2.3) we see that

$$\mathbf{f} = \frac{d^2\mathbf{r}}{dt^2} . \tag{2.6}$$

In the example above we have

$$\mathbf{f} = -\left(\frac{2\pi a}{\tau}\right)^2\cos\left(\frac{2\pi t}{\tau}\right)\mathbf{i} - \left(\frac{2\pi a}{\tau}\right)^2\sin\left(\frac{2\pi t}{\tau}\right)\mathbf{j} .$$

EXAMPLE:

The position vector of a particle is given by

$$\mathbf{r} = \{a\cos(\omega t),\ b\sin(\omega t), 0\},\quad 0 \leqslant t < \infty$$

where a, b and ω are constants. Show that the particle moves on an elliptic path and find its velocity and acceleration, showing that $\mathbf{f} = -\omega^2\mathbf{r}$.

SOLUTION:

If $\mathbf{r} = (x, y, z)$, we see that the coordinates of the particle P are given by

$$\begin{aligned} x &= a\cos(\omega t), \\ y &= b\sin(\omega t), \\ z &= 0. \end{aligned}$$

Eliminating the parameter t, we find that the locus of P is

$$\frac{x^2}{a^2} + \frac{y^2}{b^2} = 1,\quad z = 0,$$

which is the equation of an ellipse in the plane $z = 0$.

Differentiating $\mathbf{r}$ with respect to t, we see that

$$\begin{aligned} \mathbf{v} &= \{-a\omega\sin(\omega t),\ b\omega\cos(\omega t), 0\} \\ \mathbf{f} &= \{-a\omega^2\cos(\omega t),\ -b\omega^2\sin(\omega t), 0\} \end{aligned}$$

and clearly

$$\mathbf{f} = -\omega^2\mathbf{r}.$$

2.2 Motion in a Straight Line: Constant Acceleration

Suppose a particle is constrained to move along a straight line, as illustrated in Fig. 2.3. Choose a fixed point O as origin, and let x be the distance of P from O. The distance x will be a function of time t; i.e. $x = x(t)$. The position vector of P is given by

$$\mathbf{r} = \{x(t), 0, 0\},\quad t_0 \leqslant t \leqslant t_1,$$

and the velocity and acceleration are given by

$$\begin{aligned} \mathbf{v} &= (\dot{x}, 0, 0) \\ \mathbf{f} &= (\ddot{x}, 0, 0)\,, \end{aligned}$$

where $\cdot$ denotes differentiation with respect to t. It is conventional for one-dimensional motion to write

$$v = \dot{x}, \; f = \ddot{x} \tag{2.7}$$

but we note that this is *not* in accordance with (2.4), which gives

$$v = |\dot{x}| \tag{2.8}$$

In this section we will use the definitions (2.7) for v and f.

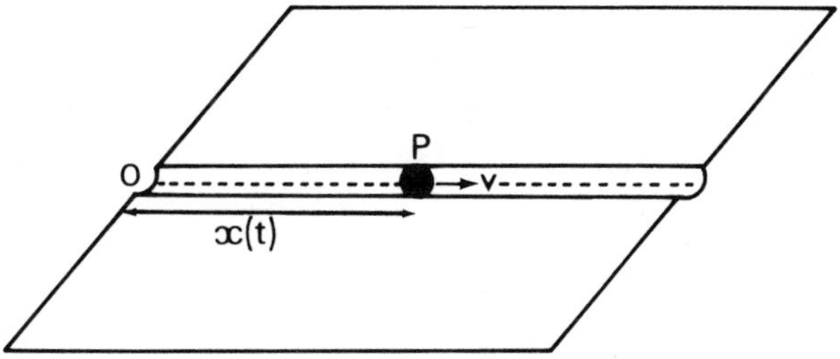

Fig. 2.3 Motion in a straight line

Consequently for one-dimensional motion, we have

$$\boxed{\frac{d^2x}{dt^2} = f} \; . \tag{2.9}$$

When f is a constant, we can integrate (2.9) with respect to time t and determine the velocity; i.e.

$$v = \frac{dx}{dt} = ft + A,$$

where A is an arbitrary constant. If $v = u$ at $t = 0$, $A = u$ and

$$\boxed{v = u + ft} \; . \tag{2.10}$$

Integrating again,

$$\begin{aligned} x &= \int (u + ft)dt + B \\ &= ut + \tfrac{1}{2} ft^2 + B, \end{aligned}$$

and if at $t = 0$, the particle is at the origin O, $B = 0$ and

$$\boxed{x = ut + \tfrac{1}{2} ft^2} \; . \tag{2.11}$$

Equations (2.10) and (2.11) determine the position and velocity of the particle at any time t. The parameter t can in fact be eliminated from (2.10) and (2.11) to give

$$\boxed{v^2 = u^2 + 2fx} \qquad (2.12)$$

which determines the speed at any position x. Note that the three equations (2.10), (2.11) and (2.12) are *not* independent since (2.12) is deduced from (2.10) and (2.11). We also stress that these equations are *only valid for constant acceleration.*

EXAMPLE:

An express train has a maximum speed of 144 km per hour, with maximum acceleration $0{\cdot}25\ \mathrm{m\,s^{-2}}$ and maximum deceleration $0{\cdot}50\ \mathrm{m\,s^{-2}}$. Find the shortest time to travel between two stations 30 km apart, assuming the train stops at both stations.

SOLUTION:

The journey has three distinct stages:

(A) acceleration to maximum speed;

(B) uniform speed;

(C) deceleration to zero speed.

We will determine the distance travelled (and time taken) in stages (A) and (C), from which we will be able to deduce the distance travelled in stage (B).

Acceleration: From (2.12) with $u = 0$, $v = 144 \times 10^3/3600\ \mathrm{m\,s^{-1}} = 40\ \mathrm{m\,s^{-1}}$ and $f = 0{\cdot}25\ \mathrm{m\,s^{-2}}$ we see that the distance travelled in the first stage is

$$x_1 = \frac{40^2}{2 \times 0{\cdot}25} = 3200\ \mathrm{m},$$

and from (2.10), the corresponding time is

$$t_1 = \frac{40}{0{\cdot}25} = 160\ \mathrm{s}.$$

Deceleration: Again, using (2.12), with $u = 40\ \mathrm{m\,s^{-1}}$, $v = 0$, and $f = -0{\cdot}50\ \mathrm{m\,s^{-2}}$ gives the distance travelled in the third stage as

$$x_3 = \frac{40^2}{2 \times 0{\cdot}50} = 1600\ \mathrm{m},$$

and from (2.10) the corresponding time is

$$t_3 = \frac{40}{0{\cdot}5} = 80\text{ s.}$$

Uniform Speed: The distance x_2 travelled in the second stage is given by

$$x_2 = 30000 - (3200 + 1600)$$
$$= 25200\text{ m.}$$

From (2.11), with $u = 40\text{ m s}^{-1}$, $x = 25200\text{ m}$, $f = 0$, we have

$$t_2 = \frac{25200}{40} = 630\text{ s.}$$

Thus the minimum time for the journey is given by

$$t_1 + t_2 + t_3 = 870\text{ s} = 14\tfrac{1}{2}\text{ mins.}$$

Note: The description of the above example can be conveniently illustrated by a velocity-time graph, as shown in Fig. 2.4. The area beneath the curve is given by

$$A = \int v\,dt$$
$$= \int \frac{dx}{dt}\,dt$$
$$= x\,,$$

showing that the area under the graph is the total distance travelled. Also the slope of the curve is the rate of change of speed, giving the acceleration.

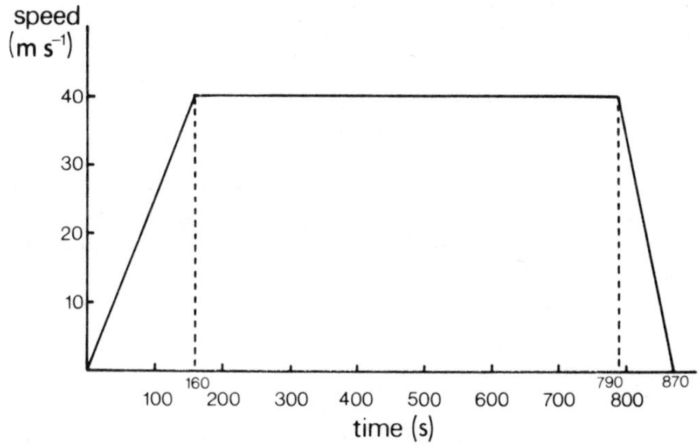

Fig. 2.4 Velocity – time graph

EXAMPLE:

The British Highway Code states that a car travelling at v m.p.h. cannot stop in a distance less than $(v + v^2/20)$ft. The first term represents the 'driver's reaction' and the second the 'braking distance'.

Two cars are travelling in the same direction at 60 m.p.h. at a distance x ft. apart. If

(a) the first car has a head-on collision, which brings it to rest instantaneously,

(b) the first car makes an emergency stop, due to a hazard of which the second car has no prior warning;

find, in each case, the maximum distance x between the cars for which a collision is inevitable. [Assume that the second car is not able to take any emergency action except braking.]

SOLUTION:

(a) In this case, to have any chance of avoiding a collision between the cars, the second car's stopping distance must be less than x;

$$\text{i.e.} \qquad 60 + \frac{(60)^2}{20} < x$$

$$\text{i.e.} \qquad x > 240 \text{ ft.}$$

Hence a collision is inevitable if $x < 240$ ft.

(b) Clearly both cars will have the same 'braking distance', and the driver of the second car will start his emergency stop as soon as he sees the brake lights of the first car. Thus a collision can be avoided only if

$$x > \text{'reaction time'}$$

$$\text{i.e.} \qquad x > 60 \text{ ft.}$$

Hence a collision is inevitable if $x < 60$ ft.

Note: It appears to the authors (both car drivers) that most drivers in practice use distances similar to model (b) above for judging their safe distances between cars. It also appears that some motorway drivers consider that they have reaction times of the order $v/10$ ft!

2.3 Motion in a Plane

We now generalize our results so far by considering the motion of a particle constrained to move in a plane. A particular example would be that of a snooker ball moving on a snooker table. Introducing Cartesian coordinates Oxy, as in Fig. 2.5, we see that the position vector of P is given by

$$\mathbf{r} = (x, y, 0) ,$$

where in general x and y are functions of time t. The velocity and acceleration are then given by

$$\mathbf{v} = (\dot{x}, \dot{y}, 0)\,,$$
$$\mathbf{f} = (\ddot{x}, \ddot{y}, 0)\,.$$

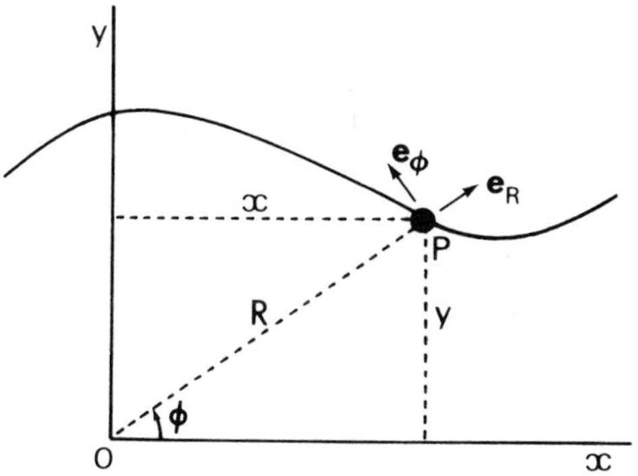

Fig. 2.5 Motion in a plane

In practice, for motion in a plane, it is often more convenient to use plane polar coordinates. Let us express $\mathbf{r}$, $\mathbf{v}$ and $\mathbf{f}$ in terms of plane polar coordinates (R, ϕ), as defined in Fig. 2.5. The position vector $\mathbf{r}$ can now be expressed as

$$\boxed{\mathbf{r} = R\mathbf{e}_R} \tag{2.13}$$

where $\mathbf{e}_R$ is a unit tangent to the R-coordinate line as shown in Fig. 2.5. (For an explanation of coordinate systems, see Appendix 3.) Differentiating (2.13), and noting that $d(\mathbf{e_R})/dt \neq \mathbf{0}$ since its direction is changing, we see that

$$\mathbf{v} = \dot{R}\,\mathbf{e}_R + R\frac{d}{dt}(\mathbf{e}_R)\,.$$

In the usual notation

$$\mathbf{e}_R = \cos\phi\mathbf{i} + \sin\phi\mathbf{j}\,,$$

giving

$$\begin{aligned}\frac{d}{dt}(\mathbf{e}_R) &= -(\sin\phi)\dot{\phi}\mathbf{i} + (\cos\phi)\dot{\phi}\mathbf{j}\\ &= \dot{\phi}(-\sin\phi\,\mathbf{i} + \cos\phi\,\mathbf{j})\\ &= \dot{\phi}\,\mathbf{e}_\phi\,,\end{aligned}$$

where $\mathbf{e}_\phi$ is a unit tangent to the ϕ-coordinate line. Thus

$$\boxed{\mathbf{v} = \dot{R}\,\mathbf{e}_R + R\dot{\phi}\,\mathbf{e}_\phi}\,, \tag{2.14}$$

showing that the velocity of the particle, in general, has components in both $\mathbf{e}_R$, $\mathbf{e}_\phi$ directions, called respectively the *radial* and *transverse* components.

Differentiating (2.14), we see that

$$\mathbf{f} = \ddot{R}\,\mathbf{e}_R + 2\dot{R}\dot{\phi}\,\mathbf{e}_\phi + R\ddot{\phi}\,\mathbf{e}_\phi + R\dot{\phi}\frac{d}{dt}(\mathbf{e}_\phi).$$

It can easily be shown that

$$\frac{d}{dt}(\mathbf{e}_\phi) = -\dot{\phi}\,\mathbf{e}_R\,,$$

giving

$$\boxed{\mathbf{f} = (\ddot{R} - R\dot{\phi}^2)\mathbf{e}_R + (2\dot{R}\dot{\phi} + R\ddot{\phi})\mathbf{e}_\phi} \quad . \tag{2.15}$$

Hence the acceleration has, in general, radial and transverse components.

Returning to the example of a car moving on a circular track, we can clearly write

$$\mathbf{r} = a\,\mathbf{e}_R\,,$$

where $\mathbf{e}_R$ is a unit vector in the outward radial direction, so that in the notation above $R = a$, which is a constant. Thus $\dot{R} = 0$ and (2.14) reduces to

$$\mathbf{v} = a\dot{\phi}\,\mathbf{e}_\phi\,,$$

where $\dot{\phi}$ is the rate of change of the angle swept out. As expected $\mathbf{v}$ has a component only in the tangential direction. The use of polar coordinates for this example has obvious advantage over rectangular coordinates as can be seen by comparing the expression for $\mathbf{v}$ above with that obtained in Ch. 2.1.

It is also useful to define angular velocity. The angular speed is given by $\omega = \dot{\phi}$. For motion in a plane we define angular velocity as

$$\boldsymbol{\omega} = \dot{\phi}\,\mathbf{e}_z\,, \tag{2.16}$$

where $\mathbf{e}_z$ is a unit vector parallel to the z-axis, which is perpendicular to the plane of motion. Angular velocity is a vector quantity, whose magnitude is the angular speed, and whose direction is the axis of rotation, perpendicular to the plane of motion. This definition for the special case of plane motion is consistent with the more general definition of angular velocity for three-dimensional motion, which we do not consider here. In terms of $\boldsymbol{\omega}$, we can write

$$\mathbf{v} = \boldsymbol{\omega} \times \mathbf{r} \tag{2.17}$$

since

$$\boldsymbol{\omega} \times \mathbf{r} = (\dot{\phi}\mathbf{e}_z) \times (a\,\mathbf{e}_R) = a\dot{\phi}(\mathbf{e}_z \times \mathbf{e}_R) = a\dot{\phi}\,\mathbf{e}_\phi\,.$$

For uniform motion of the car, we have seen that $\phi/2\pi = t/\tau$, where τ is the time for one complete cycle. Thus $\omega = \dot{\phi} = 2\pi/\tau$, a constant. Turning now to the acceleration, *for constant* ω, with $R = a$, (2.15) gives

$$\mathbf{f} = -a\,\omega^2\,\mathbf{e}_R = -\frac{v^2}{a}\mathbf{e}_R$$

showing that the acceleration of magnitude v^2/a is directed *inwards* towards the origin. This acceleration is termed the centripetal acceleration. The existence of a centripetal acceleration expresses the fact that the velocity, although of constant magnitude, is changing in direction.

If ω *is not a constant*, we have

$$\mathbf{f} = -a\,\omega^2\,\mathbf{e}_R + a\,\dot{\omega}\,\mathbf{e}_\phi$$

$$\text{i.e.} \qquad \mathbf{f} = -\frac{v^2}{a}\mathbf{e}_R + \dot{v}\,\mathbf{e}_\phi\,, \tag{2.18}$$

showing that the transverse component is due to the non-uniform rotation.

We conclude this section by noting again that if a particle is in uniform circular motion of radius a about a fixed origin, it experiences an acceleration component directed towards the origin, of magnitude v^2/a.

2.4* Three-dimensional Motion

Before considering general three dimensional motion, we first summarize some results from differential geometry (we refer the reader to *Vector Analysis* by Bourne and Kendall, Ch. 3, for a full treatment). Suppose $\mathscr{C}$ is a smooth curve in three-dimensional space, with *parametric equation*

$$\mathbf{r} = \mathbf{r}(t), \quad t_0 \leqslant t \leqslant t_1\,, \tag{2.19}$$

as illustrated in Fig. 2.6. We define *arc length*, s, along this curve by

$$s = \int_{t_0}^{t} (\dot{x}^2 + \dot{y}^2 + \dot{z}^2)^{1/2}\,dt\,. \tag{2.20}$$

The *intrinsic* equation of the curve is defined by

$$\mathbf{r} = \mathbf{r}(s), \quad 0 \leqslant s \leqslant L\,, \tag{2.21}$$

where $L = s(t_1)$, the total length of the curve. In other words, the intrinsic equation of a curve uses s, the arc length, as the parameter instead of time t. It can be shown that the vector $d\mathbf{r}/dt$ is in a direction tangential to the curve (2.19). Thus

$$\hat{\mathbf{T}} = \frac{d\mathbf{r}}{dt}\Big/\left|\frac{d\mathbf{r}}{dt}\right| \tag{2.22}$$

is a unit tangent to $\mathscr{C}$. Also

$$\hat{\mathbf{T}} = \frac{d\mathbf{r}}{dt} \Big/ (\dot{x}^2 + \dot{y}^2 + \dot{z}^2)^{1/2}$$

$$= \frac{d\mathbf{r}}{dt} \Big/ \frac{ds}{dt} \quad \text{using (2.20)},$$

$$\text{i.e.} \quad \hat{\mathbf{T}} = \frac{d\mathbf{r}}{ds} . \tag{2.23}$$

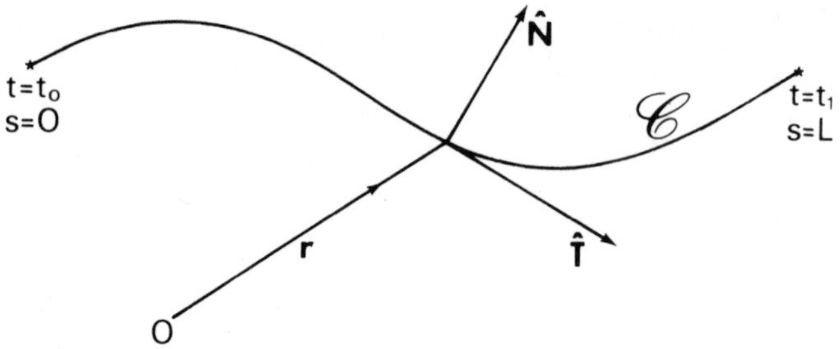

Fig. 2.6 Three-dimensional motion

Now the derivative of a *unit* vector is perpendicular to the vector, so that $d\hat{\mathbf{T}}/ds$ is perpendicular to $\hat{\mathbf{T}}$. We define ρ, the radius of curvature, and $\hat{\mathbf{N}}$, the principal unit normal to $\mathscr{C}$, by

$$\frac{d\hat{\mathbf{T}}}{ds} = \frac{1}{\rho}\,\hat{\mathbf{N}}, \quad \rho \geqslant 0, \tag{2.24}$$

and we note that $\hat{\mathbf{N}}$ is in fact perpendicular to $\hat{\mathbf{T}}$, and so normal to $\mathscr{C}$.

Using the results above, we can now attempt a description of the motion of a particle, P, moving on a smooth curve $\mathscr{C}$, as defined by (2.19), where t is the time. By definition

$$\mathbf{v} = \frac{d\mathbf{r}}{dt} = \frac{d\mathbf{r}}{ds}\frac{ds}{dt},$$

$$\text{i.e.} \quad \boxed{\mathbf{v} = \dot{s}\,\hat{\mathbf{T}}} \tag{2.25}$$

using (2.23). This shows that the velocity of the particle is, as expected, along the tangent vector to its path. Differentiating again with respect to t gives

$$\mathbf{f} = \ddot{s}\,\hat{\mathbf{T}} + \dot{s}^2 \frac{d\hat{\mathbf{T}}}{ds},$$

and using (2.24),

$$\boxed{\mathbf{f} = \ddot{s}\,\hat{\mathbf{T}} + \frac{\dot{s}^2}{\rho}\,\hat{\mathbf{N}}} . \tag{2.26}$$

We note that the acceleration has two components, one in the tangential direction and the other of magnitude v^2/ρ in the direction towards the centre of curvature. This result compares closely with (2.18), the acceleration v^2/ρ representing the more general form of the centripetal acceleration. Clearly (2.18) and (2.26) are equivalent for the special case of circular motion, for which $\hat{\mathbf{N}}$ is in a direction radially inwards and $\hat{\mathbf{T}}$ is in the transverse direction.

2.5 Relative Velocity

If particles at P and Q have velocities $\mathbf{u}, \mathbf{v}$ respectively, the *relative velocity* of Q to P is defined as $\mathbf{v}-\mathbf{u}$. The following examples illustrate the use of relative velocity.

EXAMPLE:

A man cycling east at 8 m.p.h. finds that the wind appears to blow directly from the north. On doubling his speed it appears to blow from the N.E. Find the actual velocity of the wind.

SOLUTION:

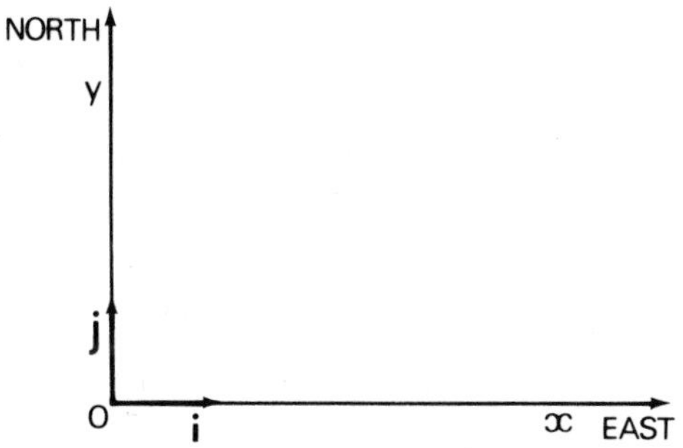

Fig. 2.7 Axes for example

Choose axes as shown in Fig. 2.7. In the first case, the velocity of the man is $\mathbf{u} = 8\mathbf{i}$. Let the actual velocity of the wind be given by

$$\mathbf{v} = v_1\,\mathbf{i} + v_2\,\mathbf{j} \quad ;$$

then the relative velocity of the wind with respect to the man is

$$\mathbf{v} - \mathbf{u} = (v_1 - 8)\mathbf{i} + v_2\,\mathbf{j} \quad .$$

This vector is parallel to $-\mathbf{j}$, giving

$$\nu_1 - 8 = 0, \quad \nu_2 < 0 .$$

Hence

$$\nu_1 = 8 \text{ m.p.h.}$$

In the second case, $\mathbf{u} = 16\mathbf{i}$, and the relative velocity of the wind with respect to the man is

$$(\nu_1 - 16)\,\mathbf{i} + \nu_2\,\mathbf{j} .$$

This must be parallel to $-\mathbf{i} - \mathbf{j}$, so that $\nu_1 - 16 = \nu_2$. Hence $\nu_2 = -8$ and

$$\mathbf{v} = 8\mathbf{i} - 8\mathbf{j} .$$

Thus $\nu = 8\sqrt{2}$ m.p.h., and the wind is from the N.W.

EXAMPLE:

An airport's runway faces north. A light aircraft is accidentally flying south at a height of 3 km and speed 150 km h^{-1} on a straight path which passes over the runway. At the time the light aircraft is 3 km from the runway end, a jet plane takes off from the airport and climbs at a constant angle a to the horizontal at constant speed 600 km h^{-1}. Find the angle α which results in a direct collision.

SOLUTION:

This problem is most easily considered by looking at the motion of the jetplane *relative* to the light aircraft. Introducing axes so that the end of the runway is at the origin, with the x-axis northwards pointing and the y-axis vertically upwards, we see that the light aircraft's velocity is

$$\mathbf{v}_A = -150\,\mathbf{i}$$

and the jetplane's velocity is

$$\mathbf{v}_J = 600 \cos a\,\mathbf{i} + 600 \sin a\,\mathbf{j} .$$

Applying a velocity $-\mathbf{v}_A$ to the whole system, which is equivalent to considering the motion of the jetplane relative to the light aircraft, the light aircraft remains 'at rest' and the jetplane moves with velocity

$$\mathbf{v}_J - \mathbf{v}_A = (600 \cos a + 150)\,\mathbf{i} + 600 \sin a\,\mathbf{j} .$$

The situation is illustrated in Fig. 2.8.

Fig. 28.

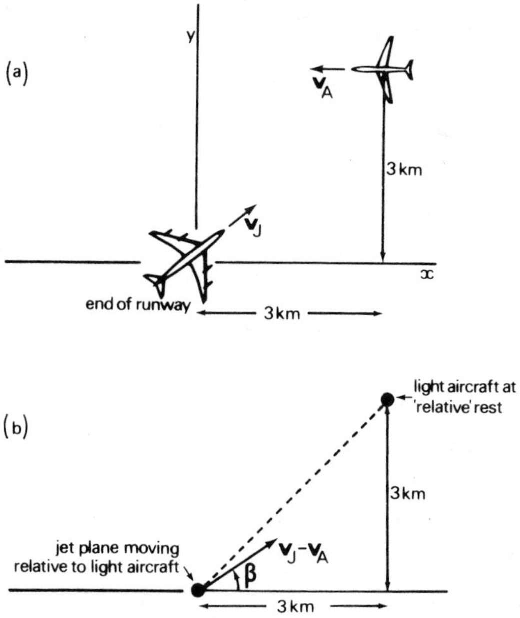

(a) Actual motion; (b) Motion relative to light aircraft

The direction of relative motion with the horizontal is now given by

$$\tan\beta = \frac{600\sin\alpha}{(600\cos\alpha + 150)} = \frac{4\sin\alpha}{(4\cos\alpha + 1)} .$$

For a direct hit we would have $\beta = 45°$, giving

$$4\sin\alpha = 4\cos\alpha + 1 .$$

To solve this equation, we can put $t = \tan\alpha/2$,

to give
$$\frac{8t}{(1+t^2)} = \frac{4(1-t^2)}{(1+t^2)} + 1 ,$$

which reduces to
$$3t^2 + 8t - 5 = 0 .$$

Solving for t, taking the positive root, we obtain $t = 0{\cdot}523$, giving $\alpha = 55°12'$. [In practice one would not want α to be anywhere near this critical value!]

2.6* Pursuit Curves

We finish this chapter with a pursuit problem in kinematics. In this type of problem, one object is constrained to follow another in some specified manner. For instance, consider a guided missile which moves towards its target, an aircraft say, in such a way that the missile's guidance system locks on to the aircraft and the missile moves so that it is always pointing in the direction of the aircraft. We will try to find the missile's path and the 'capture time'.

Introduce coordinate axes Oxy such that initially the missile is at the origin (0, 0) and the aircraft at (a, b). We assume that the aircraft travels parallel to the x-axis with speed v_A. The missile is assumed to travel with constant speed v_M. Let the missile's path be given by

$$\mathbf{r} = (x_M(t), y_M(t), 0), \quad t \geq 0\,, \tag{2.27}$$

where x_M, y_M are unknown functions of time t. The problem is illustrated in Fig. 2.9.

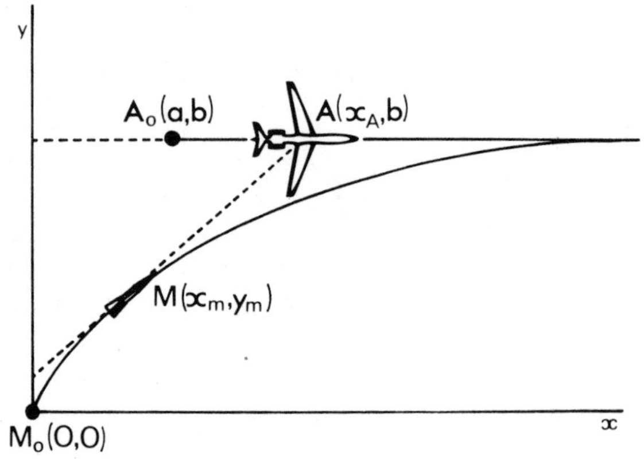

Fig. 2.9 Pursuit curve

If, at time t, the missile is at $M(x_M, y_M)$, the tangent to the path of the missile at M will pass through the position of the aircraft, A, at that time. The equation of this tangent is

$$y - y_M = \left(\frac{dy_M}{dt}\Big/\frac{dx_M}{dt}\right)(x - x_M)\,,$$

and $A(x_A, b)$ lies on this tangent. Now $x_A = a + v_A t$, giving

$$b - y_M = (\dot{y}_M/\dot{x}_M)\,(a + v_A t - x_M)\,.$$

Also, since v_M is the speed of the missile,

$$v_M = (\dot{x}_M^2 + \dot{y}_M^2)^{1/2}.$$

The above two equations constitute a set of simultaneous differential equations which when solved determine the path of the missile, equation (2.27). Dropping the M suffix, and rearranging

$$\dot{x}(b-y) - \dot{y}(a+v_A t - x) = 0\,, \tag{2.28}$$

$$\dot{x}^2 + \dot{y}^2 = v_M^{\,2}\,. \tag{2.29}$$

The boundary conditions on x and y are

$$x(0) = 0, \quad y(0) = 0\,. \tag{2.30}$$

The above non-linear simultaneous differential equations can in fact be solved analytically (see Problem 2.16) but as a useful general method, we will solve the problem graphically. For instance, if

$$a = 5000\,\text{m}, \quad b = 3000\,\text{m}, \quad v_A = 1000\,\text{m}\,\text{s}^{-1}, \quad v_M = 2000\,\text{m}\,\text{s}^{-1}\,,$$

we see that in one second the missile travels 2000 m and the aircraft 1000 m. Let us approximate to the missile's path by assuming that the missile travels for one second *in a straight line* towards the position of the aircraft at the beginning of that second. At the end of each second the missile changes its direction. In this manner we can build up an approximation to the path of the missile by a set of continuous straight lines. The method is illustrated in Fig. 2.10, and the graph shows that the collision takes place after about 5½ seconds. Clearly if we reduced the time interval for the approximation we would find a more accurate path and capture time. But as the time interval is reduced, the graphical method will become too complicated, and the problem becomes suitable for solution on a computer. In this case, the time interval can be reduced in order to give any required degree of accuracy (Problem 2.16).

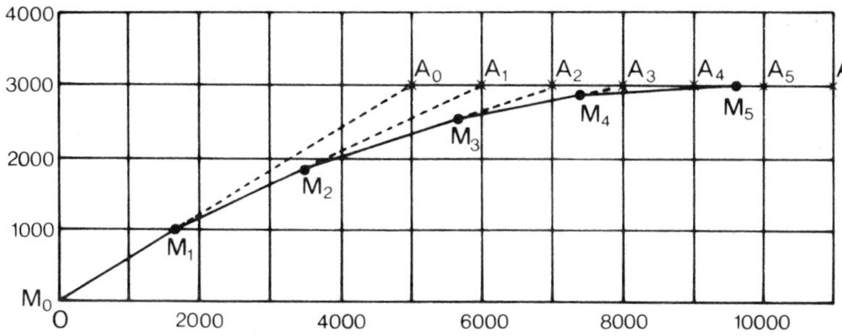

Fig. 2.10 Graphical solution for pursuit problem
[M_i, A_i are positions of the missile and aircraft respectively after the ith second]

The problem described above is relatively straightforward, as it does in fact possess an analytical solution. We finally consider a more general problem in which the aircraft moves on a known path

$$\mathbf{r} = \mathbf{r}_A = (x_A, y_A, z_A)\,, \tag{2.31}$$

and suppose the path of the missile is

$$\mathbf{r} = \mathbf{r}_M = (x_M, y_M, z_M)\,. \tag{2.32}$$

The situation is illustrated in Fig. 2.11.

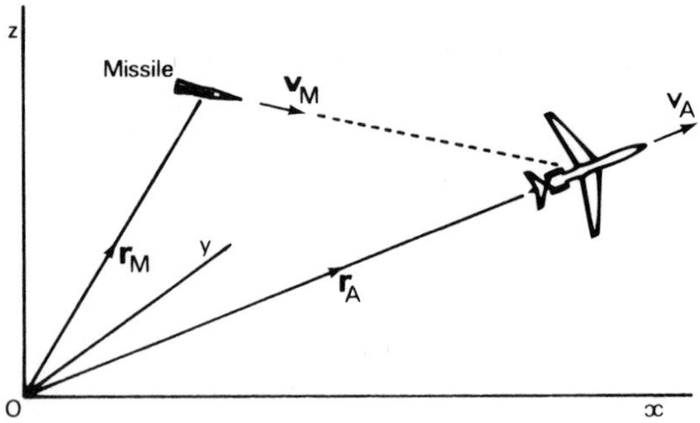

Fig. 2.11 General pursuit curve

The velocity of the missile is always in the direction $\overrightarrow{MA}$, and assuming the missile travels with constant speed v_M,

$$\frac{d\mathbf{r}}{dt} = v_M \frac{(\mathbf{r}_A - \mathbf{r}_M)}{|(\mathbf{r}_A - \mathbf{r}_M)|} . \tag{2.33}$$

Equation (2.33) gives three differential equations for the three components of $\mathbf{r}_M$. For example, in the problem above,

$$\mathbf{r}_M = (x_M, y_M, 0) ,$$

$$\mathbf{r}_A = (a + v_A t, b, 0) ;$$

and substituting into (2.33) gives

$$\frac{d\mathbf{r}_M}{dt} = v_M \frac{(a + v_A t - x_M, b - y_M, 0)}{|(\mathbf{r}_A - \mathbf{r}_M)|} .$$

Taking components gives

$$\frac{dx_M}{dt} = v_M \frac{(a + v_A t - x_M)}{|(\mathbf{r}_A - \mathbf{r}_M)|} ,$$

$$\frac{dy_M}{dt} = v_M \frac{(b - y_M)}{|(\mathbf{r}_A - \mathbf{r}_M)|} .$$

Division of these two equations gives (2.28), and squaring and adding gives (2.29).

Worked Examples

EXAMPLE 2.1:

The position vector of a particle is given by

$$\mathbf{r} = (\cos\theta, \sin\theta \cos\phi, \sin\theta \sin\phi),$$

where $\theta = 2\pi t$ and ϕ is constant. Describe the particle's path.

SOLUTION:

The particle's coordinates are given by

$$x = \cos\theta, \quad y = \sin\theta \cos\phi, \quad z = \sin\theta \sin\phi$$

so that

$$x^2 + y^2 + z^2 = \cos^2\theta + \sin^2\theta\,(\cos^2\phi + \sin^2\phi) = 1.$$

Thus the particle is moving on a sphere, radius 1, centre the origin, We can in fact go further. Reference to Fig. 2.12 indicates the geometrical interpretation of the angles θ and ϕ. Since ϕ is constant and θ is proportional to time t, we clearly have circular motion, the plane of the circle making an angle ϕ with the xy-plane.

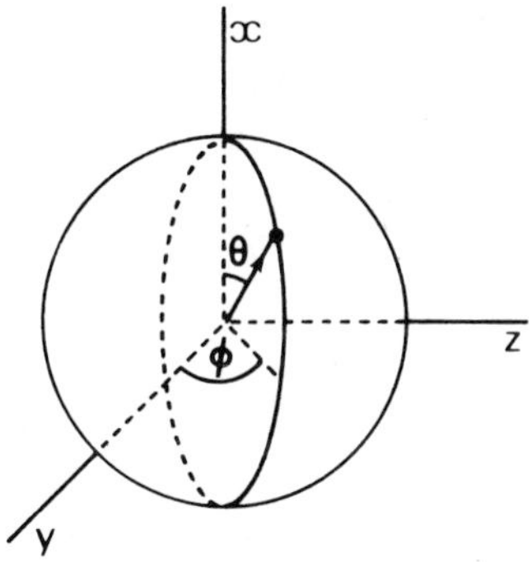

Fig. 2.12 Motion of particle on circle

EXAMPLE 2.2:

Describe the path of a particle whose position vector is

$$\mathbf{r} = \{a\cosh(\omega t), b\sinh(\omega t), 0\}, \quad 0 \leqslant t < \infty, \quad (a, b, \omega \text{ positive constants}).$$

Find its velocity and acceleration.

SOLUTION:

The particle's coordinates are

$$x = a \cosh(\omega t), \quad y = b \sinh(\omega t), \quad z = 0,$$

giving

$$\frac{x^2}{a^2} - \frac{y^2}{b^2} = 1, \quad z = 0.$$

This is the equation of an hyperbola, lying in the xy-plane, centre at the origin, and the particle's path is illustrated in Fig. 2.13.

The particle's velocity is given by

$$\mathbf{v} = \frac{d\mathbf{r}}{dt} = \{a\,\omega \sinh(\omega t),\ b\,\omega \cosh(\omega t), 0\}.$$

and acceleration

$$\mathbf{f} = \frac{d\mathbf{v}}{dt} = \{a\,\omega^2 \cosh(\omega t),\ b\,\omega^2 \sinh(\omega t), 0\}.$$

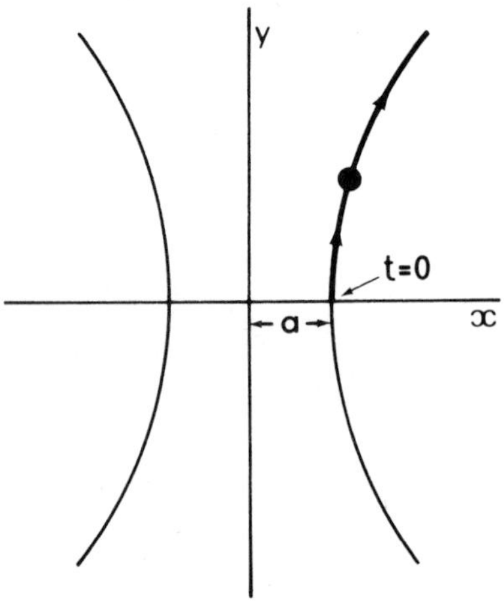

Fig. 2.13 Motion of particle on portion of hyperbola

EXAMPLE 2.3:

A particle, P, is initially a distance a from a fixed point, O, and is given an initial velocity u in the direction OP. The acceleration of the particle for $t \geqslant 0$ is directed towards O and of magnitude G/x^2, where G is a positive constant and x is the distance of the particle from O along the straight line OP produced.

Show that for $u^2 \geqslant 2G/a$, the particle will not return to its initial position. What happens if $u^2 < 2G/a$?

SOLUTION:

Using equation (2.7),

$$\frac{d^2x}{dt^2} = -\frac{G}{x^2}$$

But $d^2x/dt^2 = dv/dt = (dv/dx)/(dx/dt) = v\,dv/dx$, regarding v as a function of x. Thus

$$v\frac{dv}{dx} = -\frac{G}{x^2},$$

which can be integrated to give

$$\int v\,dv = -G\int\frac{1}{x^2}\,dx + A$$

i.e.

$$\tfrac{1}{2}v^2 = \frac{G}{x} + A\,;$$

and applying the initial conditions, $v = u$, $x = a$ at $t = 0$, gives

$$A = \tfrac{1}{2}u^2 - \frac{G}{a}.$$

Thus in the subsequent motion

$$v^2 = u^2 - \frac{2G}{a} + \frac{2G}{x}.$$

The particle will not return to its initial position if it tends to infinity, with positive velocity. Since, as $x \to \infty$,

$$v^2 \to u^2 - \frac{2G}{a}$$

it follows that, provided $u^2 - 2G/a \geqslant 0$, the particle will reach infinity. The limiting velocity, $u = \sqrt{2G/a}$, is called the escape velocity, which we will meet later in the book.

If $u^2 - 2G/a < 0$, the particle will instantaneously come to rest when $x = 2G/\omega^2$ where $\omega^2 = (2G/a) - u^2$. Under the acceleration towards O, the particle will subsequently return, passing through its initial position and approaching O with a velocity tending to infinity.

EXAMPLE 2.4:

The cross-section of a simple piston, P, driving a crankshaft AB is illustrated in Fig. 2.14. Find the displacement of the piston after a time t if the crankshaft is revolving at a constant rate of n revolutions per unit time given that OA $= a$ and AB $= b$, $(b > a)$.

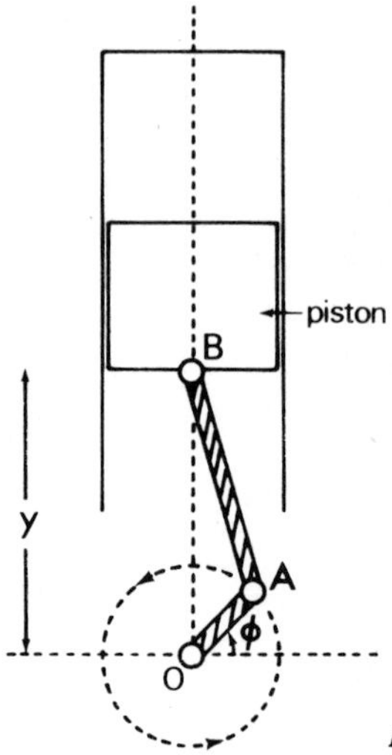

Fig. 2.14 Piston and crankshaft

SOLUTION:

If ω denotes the angular speed of the crankshaft (in radians per unit time) then

$$\frac{d\phi}{dt} = \omega = 2\pi n.$$

Integrating, $\phi = 2\pi nt$, choosing $\phi = 0$ when $t = 0$. To determine y, the displacement of the piston from O, we apply the cosine law to triangle OAB, which gives

$$b^2 = y^2 + a^2 - 2ay\cos(\pi/2 - \phi)\,.$$

Thus

$$y^2 - 2ay\sin\phi - (b^2 - a^2) = 0\,,$$

and solving for y gives

$$y = \tfrac{1}{2}\{2a\sin\phi \pm (4a^2\sin^2\phi + 4b^2 - 4a^2)^{1/2}\}\,.$$

Since at $\phi = \pi/2$, $y = a + b$, we require the positive square root, and

$$y = a\sin\phi + (b^2 - a^2\cos^2\phi)^{1/2}\,.$$

Substituting for ϕ then gives

$$y = a \sin(2\pi nt) + \{b^2 - a^2 \cos^2(2\pi nt)\}^{1/2}.$$

EXAMPLE 2.5:

A car of width d and length l is moving with constant speed u with its nearside a distance a from the edge of a straight road. A pedestrian, walking with speed v in a direction making an angle α with the forward direction of motion of the car, steps onto the road, at a distance b in front of the car.

If neither velocity changes, show that the pedestrian will be hit by the car if

$$\frac{b \sin\alpha + (a+d)\cos\alpha}{(a+d)} \leqslant \frac{u}{v} \leqslant \frac{a\cos\alpha + (b+l)\sin\alpha}{a}.$$

SOLUTION:

One method of solution is to consider motion relative to the car. The relative motion is illustrated in Fig. 2.15.

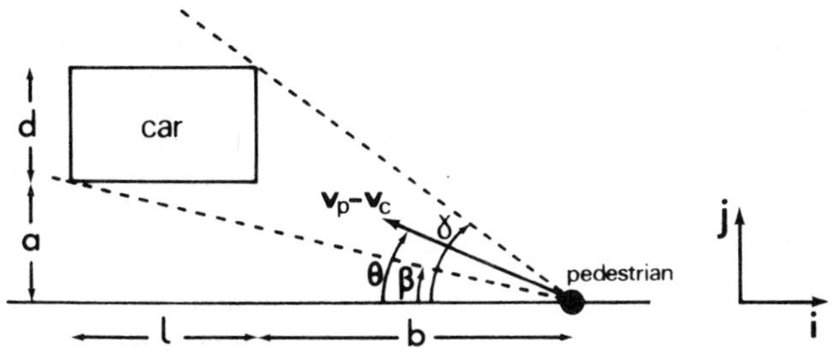

Fig. 2.15 Motion of pedestrian relative to car

For a hit the angle, θ, made by the pedestrian's velocity, $\mathbf{v}_p$, relative to the car's velocity, $\mathbf{v}_c$, i.e. $\mathbf{v}_p - \mathbf{v}_c$, with the backwards direction of motion of the car must be bounded by

$$\beta \leqslant \theta \leqslant \gamma$$

or, alternatively,

$$\tan\beta \leqslant \tan\theta \leqslant \tan\gamma$$

where

$$\tan\beta = \frac{a}{(b+l)}, \quad \tan\gamma = \frac{(a+d)}{b}.$$

But $\mathbf{v}_c = u\mathbf{i}$, and $\mathbf{v}_p = v\cos\alpha\,\mathbf{i} + v\sin\alpha\,\mathbf{j}$, introducing unit vectors **i**, **j** as in Fig. 2.15. Thus

$$\mathbf{v}_p - \mathbf{v}_c = (v\cos\alpha - u)\mathbf{i} + v\sin\alpha\,\mathbf{j}$$

and

$$\tan(\pi - \theta) = \frac{v\sin\alpha}{v\cos\alpha - u}$$

$$\text{i.e.} \qquad \tan\theta = \frac{v\sin\alpha}{u - v\cos\alpha}.$$

Thus for a hit, we require

$$\frac{a}{(b+l)} \leqslant \frac{v\sin\alpha}{(u - v\cos\alpha)} \leqslant \frac{(a+d)}{b}$$

Since, clearly, $u - v\cos\alpha > 0$ for a hit, the first inequality reduces to

$$au - va\cos\alpha \leqslant v(b+l)\sin\alpha$$

$$\text{i.e.} \qquad u \leqslant v\{a\cos\alpha + (b+l)\sin\alpha\}/a$$

and the second inequality to

$$vb\sin\alpha \leqslant u(a+d) - v(a+d)\cos\alpha$$

$$\text{i.e.} \qquad u \geqslant v\{b\sin\alpha + (a+d)\cos\alpha\}/(a+d)$$

Hence if (u/v) lies between the bounds

$$\{b\sin\alpha + (a+d)\cos\alpha\}/(a+d) \leqslant (u/v) \leqslant \{a\cos\alpha + (b+l)\sin\alpha\}/a$$

there will be a hit.

Problems

2.1 The position vector of a particle is given by

$$\mathbf{r} = \mathbf{a} + (\mathbf{b} - \mathbf{a})t, \quad 0 \leqslant t < \infty,$$

where **a** and **b** are constant vectors. Show that the particle moves on a straight line.

2.2 Describe the locus of a particle whose position vector is given by

$$\mathbf{r} = \{ a \sin(\omega t), a \cos(\omega t), bt \},$$

where a, b and ω are constants. Find the velocity and acceleration of the particle and prove that their magnitudes are both constant.

2.3 Prove that the velocity of a particle in three-dimensional space is independent of the fixed origin, O, to which its position vector is referred.
[Hint: Let O$'$ be a second fixed origin, with position vector **a** referred to origin O.]

2.4 A particle is moving on a straight line with constant acceleration f. If x is the distance from some fixed point on the straight line, show that the equation $d^2x/dt^2 = f$ can be written as

$$v\frac{dv}{dx} = f,$$

where v is the particle's speed. If $v = u$ at $x = 0$, integrate this equation and show that

$$v^2 = u^2 + 2fx.$$

2.5 An express train is required to travel between two stations, 42 km apart, in 20 mins. If the train stops at both stations and has maximum acceleration 0·25 m s^{-2} and maximum deceleration 0·50 m s^{-2}, find the least maximum speed which the train must achieve in order to complete the journey on time.

2.6 A car accelerates to 60 m.p.h. in the following way:

1st gear:	$0 \rightarrow 10$ m.p.h.	in	2 secs,
2nd gear:	$10 \rightarrow 20$ m.p.h.	in	3 secs,
3rd gear:	$20 \rightarrow 30$ m.p.h.	in	4 secs,
4th gear:	$30 \rightarrow 60$ m.p.h.	in	15 secs.

Assuming that the car travels with constant acceleration in each gear, find the distance travelled in reaching 60 m.p.h.
[Hint: use velocity-time graph.]

2.7 The details of the gear ratios of a car are listed below:

Gear	Possible speeds (m.p.h.)	Constant acceleration that can be maintained (feet per sec^2)
1	$0 \to 20$	7
2	$10 \to 35$	8
3	$30 \to 45$	9
4	$35 \to 70$	8

Using the data above, at what speeds should the driver change gear in order to reach 60 m.p.h. in the shortest possible time? Find this time.

2.8 Two cars are 100 ft. apart when the first car, travelling at 45 m.p.h. makes an emergency stop. Using the British Highway Code stopping distances (see Ch. 2.2), show that a collision will be unavoidable if the speed of the second car exceeds 55 m.p.h. Assume that the second car has no prior warning of the hazard and is unable to take any emergency action except braking.

2.9 Spherical polar coordinates (r, θ, ϕ) are defined by

$$x = r\cos\phi\sin\theta, \quad y = r\sin\phi\sin\theta, \quad z = r\cos\theta,$$

where $r \geqslant 0$, $0 \leqslant \theta \leqslant \pi$, and $0 \leqslant \phi < 2\pi$. Unit tangents to the coordinate lines are given by

$$\begin{aligned} \mathbf{e}_r &= (\sin\theta\cos\phi,\ \sin\theta\sin\phi,\ \cos\theta), \\ \mathbf{e}_\theta &= (\cos\theta\cos\phi,\ \cos\theta\sin\phi,\ -\sin\theta), \\ \mathbf{e}_\phi &= (-\sin\phi,\ \cos\phi,\ 0). \end{aligned}$$

The position vector $\mathbf{r}$ of a particle in three-dimensional space is given by

$$\mathbf{r} = r\mathbf{e}_r$$

Prove that the velocity and acceleration of the particle are given by

$$\mathbf{v} = \dot{r}\,\mathbf{e}_r + r\dot{\theta}\,\mathbf{e}_\theta + r\sin\theta\,\dot{\phi}\,\mathbf{e}_\phi,$$

$$\begin{aligned} \mathbf{f} = {} & (\ddot{r} - r\dot{\theta}^2 - r\sin^2\theta\,\dot{\phi}^2)\,\mathbf{e}_r + (2\dot{r}\dot{\theta} + r\ddot{\theta} - r\sin\theta\cos\theta\,\dot{\phi}^2)\,\mathbf{e}_\theta \\ & + (2\dot{r}\sin\theta\,\dot{\phi} + 2r\cos\theta\,\dot{\theta}\,\dot{\phi} + r\sin\theta\,\ddot{\phi})\,\mathbf{e}_\phi. \end{aligned}$$

If ϕ = constant, which implies that the particle moves in a plane, show that $\mathbf{v}$ and $\mathbf{f}$ reduce to equations (2.14) and (2.15) with R and ϕ replaced by r and θ. Explain this result.

2.10 A particle moves on the surface of a sphere of radius a. Evaulate the velocity and acceleration components in terms of spherical polar coordinates, origin the centre of the sphere. Deduce that if the particle moves on the rim of a hemisphere with constant angular velocity, then the velocity and acceleration are of the form

$$\mathbf{v} = a\dot{\theta}\,\mathbf{e}_\theta, \quad \mathbf{f} = -a\dot{\theta}^2\mathbf{e}_r .$$

2.11 A particle P, with position vector $\mathbf{r}$, describes a curve $\mathscr{C}$, of radius of curvature ρ. Show that, if $\hat{\mathbf{T}}$, $\hat{\mathbf{N}}$ and s have their usual meanings, the acceleration of P is given by

$$\ddot{\mathbf{r}} = \frac{d}{ds}\left(\tfrac{1}{2}v^2\right)\hat{\mathbf{T}} + \frac{v^2}{\rho}\hat{\mathbf{N}} .$$

2.12 At time t a particle is moving on a curve, with position vector

$$\mathbf{r} = \{\,a(\cos\theta - \sin\theta),\; a(\cos\theta + \sin\theta),\; a\sqrt{2}\,\theta\}$$

where a is a constant, and θ is a function of t. Show that the acceleration at any instant is

$$\ddot{\mathbf{r}} = 2a\ddot{\theta}\,\hat{\mathbf{T}} + a\sqrt{2}\,\dot{\theta}^2\,\hat{\mathbf{N}} .$$

2.13 An aircraft is required to fly from an airfield, A, to another B, 400 km due south of A. In still air the plane can fly at 800 km per hour, but on this particular day there is a constant westerly wind of 20 km per hour. Find the direction the aircraft must navigate in order to reach B directly and find the time taken.

2.14 An aircraft is flying due north at 200 km per hour relative to the ground, and the wind appears to be coming from N 10°E. The plane turns and travels east at the same speed, and the wind appears to come from E 10°S. Determine the speed and direction of the wind.

2.15 Write a computer program for the pursuit problem posed in Ch. 2.6 which determines the 'capture time' in seconds correct to three decimal places. Also determine the position of capture.
[Hint: Start with a prescribed time interval for the approximation, e.g. 0·01 secs, and find the 'capture time'. Keep halving the time interval until the 'capture time' is unaltered to three decimal places.]

2.16 By differentiating (2.28), show that (2.28) and (2.29) can be combined to give

$$\frac{d^2x}{dy^2}\Bigg/\left\{1+\left(\frac{dx}{dy}\right)^2\right\}^{1/2} = v_A/\{v_M(b-y)\}\ .$$

Show that the substitution $u = dx/dy$ and $n = v_M/v_A$, reduces this differential equation to

$$\int \frac{du}{(1+u^2)^{1/2}} = \int \frac{dy}{n\,(b-y)}\ ,$$

with $u = a/b$ at $y = 0$ (i.e. at $t = 0$). Integrate twice in order to show that the capture time is given by

$$T = [a + n\,(a^2+b^2)^{1/2}]/[v_A\,(n^2-1)]\ .$$

Evaluate the capture time, using values as in Ch. 2.6, and compare with your numerical result.

2.17 A cat is running along a straight edge of a garden. A dog, sitting in the garden at a distance b from the edge, sees the cat when it is at its nearest point. The dog immediately chases the cat with twice the cat's speed in such a way that it is always running towards the cat. Using the results of Problem 2.16, find the time that elapses before the cat is caught and show that the cat runs a distance $(2/3)b$ before being caught.

2.18 Consider the problem above, but with the dog at the centre of a circle of radius b, and with the cat running round the circumference of the circle. Use approximate graphical methods to find whether the dog will catch the cat if

(i) their speeds are equal;

(ii) the speed of the dog is less than the speed of the cat.

Chapter 3

Newtonian Mechanics

3.1 Historical Introduction

Throughout the ages man has made many attempts to arrive at an explanation of the motion of bodies. The observation that some effort is needed to lift a heavy body from the ground or to move a body over the surface of the earth led to an intuitive idea of a *force* as something which causes a body to move. It took many centuries, however, before this intuitive idea could be formulated in a quantitative manner.

The Greek philosopher ARISTOTLE (384-322 B.C.) regarded the state of rest as natural for all bodies. In his consequent aim to explain the state of motion, a distinction was made between the motions of celestial and terrestrial bodies. He believed that the heavenly bodies were transported by transparent spheres, each moved by its own soul. The motion of 'heavy' bodies falling freely near the earth's surface was explained by attributing to such bodies a 'nature' of their own. This nature urged them to seek the centre of the universe, which was thought to be in the depths below the earth. Other terrestrial motions, such as along the earth's surface, were termed 'violent', and persisted only so long as a cause persisted. The cause was transmitted by the medium (air, water, etc.) through which the body moved.

There was little advance on this theory until around the 14th century, when attempts were made to abandon the idea of spirits and souls. At this time, the theories of *impetus* were put forward. These suggested that the persistence of violent motion was due to 'impetus' acquired in the process of projection. The impetus gradually weakened unless renewed, thus explaining observed horizontal motions. In the case of a falling body it was assumed that impetus was continuously acquired.

The next important contribution was made by GALILEO (1564-1642), famous for his reputed exploits at Pisa. Galileo did perform some very important experiments at Pisa, but it is most unlikely that he ever dropped cannon balls from the leaning tower as has been supposed. The 'tower experiment' however, if it could be performed 'in vacuo', would lead to some of the conclusions which Galileo reached by less spectacular means. He realised that free fall and horizontal motion were both limiting cases of motion on an inclined plane, and were not essentially different as was previously thought. He carried out a series of experiments, using metal balls on the polished grooved surfaces of wooden planes. Variation of the angle of elevation of the planes from 0° to 90° allowed him to reproduce the two extremes of motion, and all intermediate cases. The concept of acceleration was evolved as a result of these experiments and Galileo showed that accelerations of different bodies down a fixed inclined plane are the same under ideal conditions. (The ideal conditions referred to perfectly smooth surfaces).

A most important result was obtained by means of experiments involving two planes, arranged as in Fig. 3.1. It was found that a ball released from rest at A, after rolling down the first plane, continued up the second plane until it came to rest at the point B, always on the same horizontal level as A. To a certain extent, this showed that only the vertical component of motion was important, and

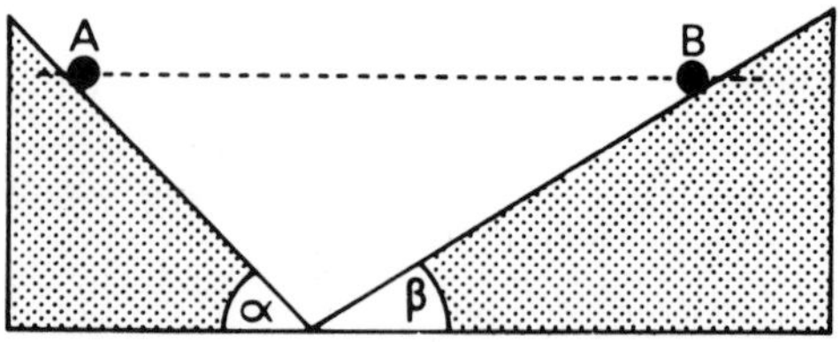

Fig. 3.1 Arrangement of two inclined planes for Galileo's experiment

consideration of the limiting case $\beta = 0$ led to Galileo's most important contribution to dynamics. When $\beta = 0$, clearly the ball was never able to regain its original height and must consequently continue to roll with undiminished speed along the horizontal plane, never coming to rest in the absence of any other forces. This result had previously been obscured by the almost universal presence of forces of resistance in commonly observed phenomena. The conditions of Galileo's experiments were near enough to ideal for an appreciation of the state of affairs in the *absence* of resistance. Aristotle's concept of the natural state of a body could now be extended to:

> 'under ideal conditions on a horizontal plane, rest and uniform motion in a straight line are equally natural conditions'.

Galileo regarded this result as purely local, valid only for small motions at the earth's surface. In his attempt to generalize his result he was no longer able to call upon experimental evidence. He was for instance unaware that, given sufficient velocity, a body would leave the earth's surface and travel off into space. He imagined his horizontal 'plane' encircling the earth along a great circle, and so came to regard uniform circular motion about the centre of the earth as the natural state on the larger scale. This latter conjecture was later found to be incorrect. The present belief stands thus:

> 'rest or uniform motion in a straight line, anywhere in an Euclidean universe†, is the natural state for a body completely free from external influences'.

† Euclidean universe: universe in which there exists a reference system, relative to which all other points in space can be described.

Following Galileo, other scientists and philosophers, including Descartes, Huygens and Hooke, performed further experiments and contributed useful speculations. A substantial amount of experimental evidence was thus assembled, awaiting the arrival of someone with the necessary touch of genius to appreciate its full significance. That genius was found in the person of ISAAC NEWTON (1642-1727), who searched for a fundamental law to explain all the observations. Collision experiments with hard spherical balls had demonstrated that the ratio of the magnitudes of the velocity changes of two colliding spheres depends on two factors; the inverse ratio of their volumes and the material of which the balls are made. This result led Newton to the concept of *mass* as a property of a body, related to its weight, but essentially different. Galileo had recognised *weight* as the force needed to support a body in free fall, or alternatively the force accelerating its unimpeded motion towards the earth. Unfortunately he just failed to realize that, for a given body,

$$\frac{\text{weight}}{\text{acceleration in free fall}} = \frac{\text{force to prevent fall}}{\text{acceleration in free fall}}$$

$$= \frac{\text{force to prevent sliding down inclined plane}}{\text{acceleration down inclined plane}}$$

$$= \text{constant}.$$

Newton termed this constant ratio the *mass* of the body, an intrinsic property of the body and not restricted solely to the case of motion under gravity. In Fig. 3.2, W and W′ denote forces preventing motion and f and f′ denote accelerations in unrestricted motion.

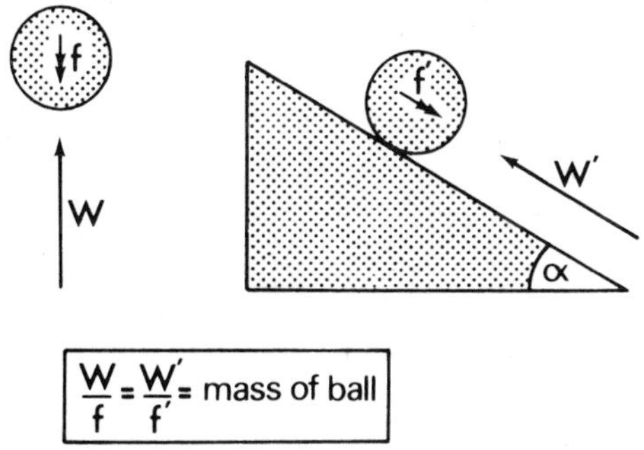

Fig. 3.2 The concepts of mass and weight

We now proceed to consideration of Newton's laws, which were postulated as a result of the combination of experimental evidence and a great deal of intuition. We note the use of the word 'postulated', for the laws which follow cannot be deduced by any rigorous argument. Their validity needs to be tested by experimental verification of deductions made by assuming the laws to be true.

3.2 Newton's Laws of Motion

Before stating Newton's laws we give two definitions, also due to Newton.

Definition 1: Quantity of matter, or *mass,* of a body is the product of its *density* and its volume.

Definition 2: Quantity of *momentum* of a body is the product of its mass and its velocity.

NEWTON'S LAWS OF MOTION:

Law 1: Every body continues in its state of rest or uniform motion in a straight line unless compelled to do otherwise by forces impressed upon it.

Law 2: The rate of change of momentum of a body is proportional to the impressed force and is effective in the direction of the force.

Law 3: An action is always opposed by an equal reaction; or, the mutual actions of two bodies are always equal in magnitude and act in opposing directions.

INTERPRETATION OF THE LAWS:

We now examine briefly the consequences of each of the definitions and laws:

Definition 1 cannot define both mass and density. We will consider it to define density as mass per unit volume, where mass has yet to be defined. Denoting mass by m, density by ρ and volume by V, definition 1 becomes

$$m = \rho V . \tag{3.1}$$

Definition 2 defines momentum provided we can at some stage define mass. Since velocity is a vector, it follows that momentum is a vector. Denoting velocity by $\mathbf{v}$ and momentum by $\mathbf{p}$, we have

$$\mathbf{p} = m\mathbf{v} . \tag{3.2}$$

Law 1 introduces the concept of a force as the cause of non-uniform motion. This is a purely qualitative law expressing a cause and effect relationship.

Law 2 gives a quantitative definition of force as a precise vector quantity which produces a specific change in the state of motion of a body. Denoting force by **F**, law 2 may be written

$$\frac{d}{dt}(m\mathbf{v}) = \lambda\mathbf{F} ,$$

where t denotes time and λ is a constant of proportionality. We are free to choose suitable units for the new physical quantity **F**. It is convenient to do this in such a way that $\lambda = 1$. With this choice of units, the mathematical formulation of law 2 is

$$\boxed{\frac{d}{dt}(m\mathbf{v}) = \mathbf{F} .} \qquad (3.3)$$

This vector equation is called the *equation of motion* of the body. For a body of *constant mass*, we can write

$$\boxed{m\frac{d\mathbf{v}}{dt} = \mathbf{F} .} \qquad (3.4)$$

In this special case of Newton's second law, we see that force is the product of mass and acceleration, but note that this is *not* the general form of the law.

Law 3, the essence of Newton's laws, is a statement concerning the real physical world. It provides us with a *practical* definition of mass, without which the previous definitions and laws are incomplete.

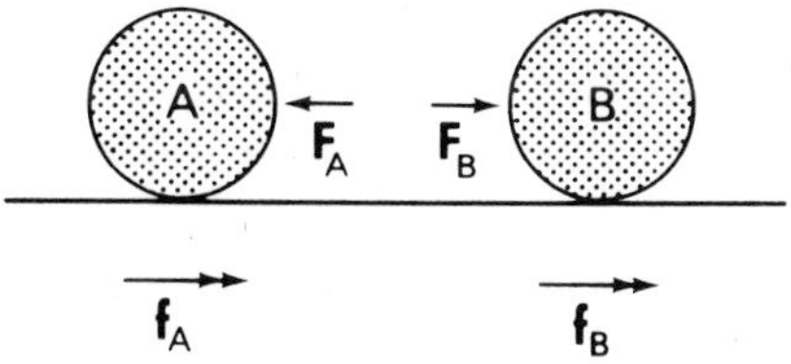

Fig. 3.3 Collision between two hard spheres

Consider an isolated system of two bodies, denoted by A and B. Let $\mathbf{F}_A$ be the force exerted on A by B and let $\mathbf{F}_B$ be the force exerted on B by A. As an example of such a system we can visualize the act of collision between two hard spheres. In Fig. 3.3 the spheres are drawn slightly separated for the sake of clarity. At any instant during the collision (while the spheres are actually in contact), denote by $\mathbf{p}_A$, $\mathbf{p}_B$ the momenta of the respective spheres. Application of law 3 gives

$$\mathbf{F}_A = -\mathbf{F}_B ,$$

and from law 2,

$$\frac{d\mathbf{p}_A}{dt} = -\frac{d\mathbf{p}_B}{dt} .$$

Using definition 2,

$$m_A\frac{d\mathbf{v}_A}{dt} = -m_B\frac{d\mathbf{v}_B}{dt} ,$$

where m_A, m_B are the respective masses and $\mathbf{v}_A$, $\mathbf{v}_B$ the respective velocities of A and B,

$$\text{i.e.}\quad m_A\mathbf{f}_A = m_B(-\mathbf{f}_B) ,$$

where $\mathbf{f}_A$, $\mathbf{f}_B$ are the respective accelerations of A and B. Thus we have

$$\frac{m_A}{m_B} = \frac{|\mathbf{f}_B|}{|\mathbf{f}_A|} . \tag{3.5}$$

We can now adopt one particle, say m_A, as the basic *unit mass* and determine the mass of any other body by comparing the accelerations produced in a collision experiment. We have thus achieved a practical definition of mass according to (3.5); 'the ratio of masses is the inverse ratio of the accelerations produced by equal forces'.

3.3 S.I. Units

We have already met several examples of different systems of units in Ch. 2, where, for instance, we have used feet, miles and metres for the unit of length. In what follows, we adopt the S.I. system of units (Systéme International d'Unités). This system has been adopted by most countries as the recommended system of units for scientific publications. For a complete exposition of the system we refer the reader to 'Symbols, Signs and Abbreviations', published by the Royal Society in 1969.

The basic S.I. units are tabulated below:

PHYSICAL QUANTITY	NAME OF S.I. UNIT	SYMBOL FOR S.I. UNIT
Length	metre	m
Mass	kilogram	kg
Time	second	s
Temperature	kelvin	K
Electric Current	ampere	A
Luminous Intensity	candela	cd

Each unit has a basic definition which is given in the reference cited above.

Derived S.I. units are listed below:

PHYSICAL QUANTITY	NAME OF S.I. UNIT	SYMBOL	DEFINITION OF S.I. UNIT
Energy	Joule	J	$kg\,m^2\,s^{-2}$
Force	Newton	N	$kg\,m\,s^{-2} = J\,m^{-1}$
Power	Watt	W	$kg\,m^2\,s^{-3} = J\,s^{-1}$
Electric Charge	Coulomb	C	$A\,s$
Electric Potential Difference	Volt	V	$kg\,m^2\,s^{-3}\,A^{-1} = J\,A^{-1}\,s^{-1}$
Electric Resistance	Ohm	Ω	$kg\,m^2\,s^{-3}\,A^{-2} = V\,A^{-1}$
Electric Capacitance	Farad	F	$A^2\,s^4\,kg^{-1}\,m^{-2} = A\,s\,V^{-1}$
Area	Square metre	—	m^2
Volume	Cubic metre	—	m^3
Density	Kilogram per unit cubic metre	—	$kg\,m^{-3}$
Velocity	Metre per second	—	$m\,s^{-1}$
Acceleration	Metre per second squared	—	$m\,s^{-2}$
Pressure	Newton per square metre	—	$N\,m^{-2}$
Electric Field	Volt per metre	—	$V\,m^{-1}$
Magnetic Field	Ampere per metre	—	$A\,m^{-1}$

In order to write decimal fractions or multiples we use the following prefixes:

MULTIPLE/FRACTION	PREFIX	SYMBOL
10^6	mega	M
10^3	kilo	k
10^2	hecto	h
10	deka	da
10^{-1}	deci	d
10^{-2}	centi	c
10^{-3}	milli	m
10^{-6}	micro	μ

When writing S.I. units, the symbol remains unaltered in the plural, and is not followed by a full stop, except when it occurs at the end of a sentence:

e.g. 3 cm but *not* 3 cms or 3 c.m. or 3 c.ms.

A product or quotient of two units is written in the form

$$\text{kg s}^2 \quad \text{or} \quad \text{kg s}^{-2},$$

with a space between the units; whereas prefixes for units should be written with *no* space between the prefix and unit; thus:

$$10^{-6}\text{m} = \mu\text{m}$$

Thus 5 m s means 5 metre seconds, whereas 5 ms means 5×10^{-3} seconds. Also, for example,

$$\text{cm}^2 = (0.01\text{m})^2$$

3.4 Applications to one-dimensional motion

The simplest application which we can consider is to the motion of a particle, P, moving in a straight line. Choose axes such that the motion is along the x-axis of a rectangular Cartesian coordinate system (Fig. 3.4). Then P has position vector $\mathbf{r} = (x, 0, 0)$, velocity $\mathbf{v} = (\dot{x}, 0, 0)$ and acceleration $\mathbf{f} = (\ddot{x}, 0, 0)$ at time t.

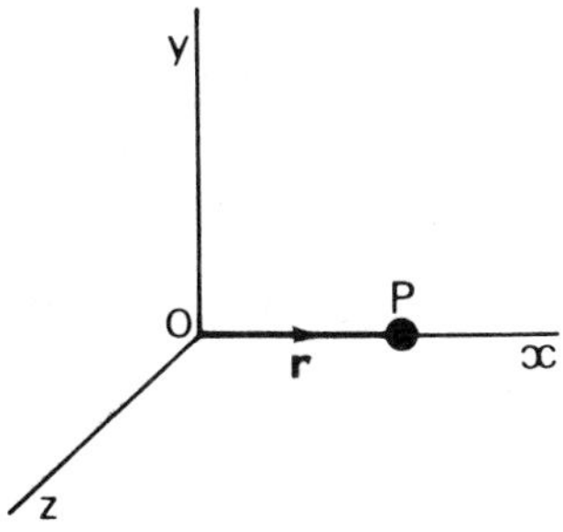

Fig. 3.4 One dimensional motion: Particle moves along x-axis

Suppose P has constant mass m and is acted upon by a force **F**. By Newton's second law, (3.4),

$$\mathbf{F} = m\mathbf{f} = m(\ddot{x}, 0, 0)\,,$$

showing that the force must be parallel to Ox in order that the particle moves only along the x-axis. Since all the vector quantities in which we are interested have components only in the x-direction, it is usual to work with the scalar quantities x, $\dot{x}$, $\ddot{x}$ and F. We shall be concerned with variable forces F which we assume to depend on the position and speed of P, and also perhaps explicitly on time t; thus

$$F = F(x, \dot{x}, t)\,,$$

and writing $F(x, \dot{x}, t) = mf(x, \dot{x}, t)$, Newton's second law may now be written as

$$\boxed{\ddot{x} = f(x, \dot{x}, t)}\,. \tag{3.6}$$

The solution of the second order differential equation (3.6), for *any* given function f, is a fairly formidable task. If we restrict ourselves to seeking analytic solutions [i.e. solutions expressing $x(t)$ in terms of familiar functions], we are able to deal only with a very small class of functions f. Once we have expressed a problem in the form (3.6) we can, however, proceed by a numerical method to complete the solution for $x(t)$.

Even when we restrict ourselves by considering only forces with no dependence on velocity, then existence of a 'simple' solution is the exception rather than the rule. In fact, the general solution of the equation

$$\ddot{x} = f(x, t) \tag{3.7}$$

is still a subject of current research in differential equations. Having issued a warning against expecting simple solutions as a matter of course, it is nevertheless interesting and worthwhile to look for simple solutions before rushing prematurely into a numerical solution on a computer. Analytic solutions can often shed light on the mechanisms operating in a physical situation which would in all probability be obscured by a purely numerical solution. As we are not dealing with specific numerical methods in this text we will be concerned mostly with investigating problems which do possess analytic solutions. At the same time we should remember that once a problem is expressed in the form (3.6), together with suitable boundary conditions, it is essentially soluble, numerically at least.

3.5 Vertical Motion Under Gravity

Our first physical problem is to determine the height reached when a ball is thrown vertically upwards at a known initial speed. Alternatively, can we estimate the initial speed if we can measure the height?

The first step is to set up a suitable mathematical model. There are many factors which could be taken into account concerning the nature of the forces acting on the ball. If we incorporate every factor possible in our model, this apparently simple problem will begin to look very complex. The essence of a good mathematical model is that it extracts the really important features of the real situation, so that the mathematics is kept reasonably simple, at the same time taking care not to 'throw the baby out with the bathwater!'

It seems reasonable to propose a model for the present situation based on the following assumptions:

(i) The ball can be represented as a 'point particle' of mass m.
(ii) The ball is subject to a constant downward force, equal to its weight, due to gravity†.
(iii) To a first approximation, no other forces act on the ball. This means we are neglecting any resistance to motion offered by the surrounding air and also the effect of winds.
(iv) The ball is thrown directly upwards with a known speed u.

We should note that, in a different context, it may be quite inappropriate to make some or all of assumptions (i), (ii) and (iii) above. For our present purpose, however, they represent a reasonable approximation.

We now consider three approaches to the solution of our problem. We will show that the problem can be solved by integration of the equation of motion, by a direct method called imbedding, and also by using a conservation law.

† This is the result deduced by Galileo and is justified in Ch. 7.1 for near earth motions.

3.5.1 DIRECT INTEGRATION OF EQUATION OF MOTION

This is the most straightforward method available. The situation at a time t after the ball was thrown is represented in Fig. 3.5 below.

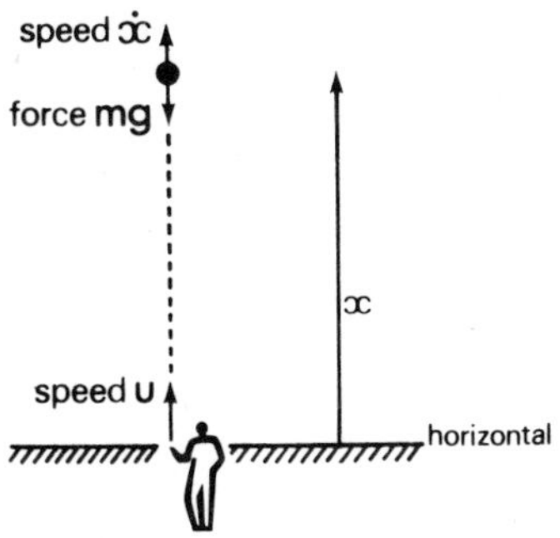

Fig. 3.5 Vertical motion of a ball

Denoting by g the known downward acceleration due to gravity, the downward force acting on the ball is given by

$$F = mg.$$

Measuring the upward displacement by x, the upward acceleration of the ball is $\ddot{x}$. Thus the equation of motion is

$$m\ddot{x} = -mg. \tag{3.8}$$

This is a special case of (3.6) in which $f(x, \dot{x}, t) = -g$, a constant. Two conditions are required to determine the two arbitrary constants in the solution of the second order differential equation (3.8). These are the *initial conditions:*

$$x(0) = 0, \dot{x}(0) = u.$$

We may now proceed to solve (3.8) by direct integration with respect to t, giving

$$\dot{x} = -gt + A,$$

where A is a constant of integration. The condition $\dot{x}(0) = u$ gives $A = u$, whence we have

$$\dot{x} = u - gt. \tag{3.9}$$

Further integration, with $x(0) = 0$ yields

$$x(t) = ut - \tfrac{1}{2}gt^2. \tag{3.10}$$

Returning now to our original question, we observe that the greatest height is attained when the upward velocity ceases; i.e. when $\dot{x} = 0$. From (3.9), this occurs when $t = u/g$, and the greatest height, h, is obtained from (3.10) as

$$h = x(u/g) = u^2/g - u^2/2g$$

$$\text{i.e.} \boxed{h = u^2/2g}\,. \tag{3.11}$$

3.5.2 Method of Imbedding

We now consider an alternative approach which aims to obtain the result directly, without reference to unwanted intermediate results [i.e. (3.9) and (3.10)]. The technique of *imbedding* can be usefully applied in many situations, and is so named because it sets out to imbed the given problem as one particular case within a whole class of similar problems. This simple example illustrates the way in which this is achieved.

It is clear that the greatest height reached as a result of projection with speed u will depend on u. Thus we write

$$h = h(u).$$

Consider the situation at times $t = 0$ and $t = \Delta t$, as illustrated in Fig. 3.6, where Δt is a small time interval. In a time Δt the ball travels a distance

$$P_0P_1 \approx u\Delta t$$

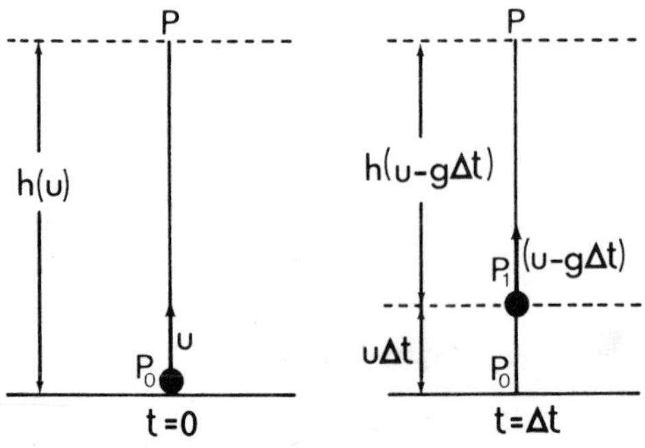

Fig. 3.6 Vertical projection: method of imbedding

and its speed decreases by an amount $g\Delta t$. The speed at P_1 is thus $u - g\Delta t$.

We now pose a second problem equivalent to the original problem: 'what height will the ball reach if projected from P_1 with speed $(u - g\Delta t)$?' Clearly the ball must still reach the point P. Since, from the definition of the function h, we have $P_1P = h(u - g\Delta t)$, it follows that

$$h(u) = u\Delta t + h(u - g\Delta t).$$

Expanding $h(u - g\Delta t)$ by Taylor's theorem,

$$h(u) = u\Delta t + h(u) - g\Delta t\, h'(u) + O(\Delta t^2) ,$$

where dash denotes differentiation with respect to u,

giving $$u = gh'(u) + O(\Delta t).$$

Letting $\Delta t \to 0$ gives $$h'(u) = u/g. \tag{3.12}$$

Since a ball projected with zero speed ($u = 0$) would reach zero height ($h = 0$), the appropriate boundary condition is

$$h(0) = 0. \tag{3.13}$$

Integrating (3.12) and applying (3.13) gives

$$h = u^2/2g,$$

which is in agreement with (3.11).

It should be noted here that we have implicity assumed, in the above analysis, that the function $h(u)$ satisfies the conditions for its Taylor series to converge. Although this was obviously justified in this case, it was not possible to ascertain this beforehand since h was an unknown function.

3.5.3 CONSERVATION LAW

We now return to the equation of motion (3.8), but proceed in a different manner. Multiplying both sides of (3.8) by $\dot{x}$ gives

$$\dot{x}\ddot{x} = -g\dot{x},$$

i.e. $$\frac{d}{dt}\left(\tfrac{1}{2}\dot{x}^2\right) = -g\dot{x} .$$

Integrating with respect to t, we obtain

$$\tfrac{1}{2}\dot{x}^2 + gx = C , \tag{3.14}$$

where C is a constant. This means that the quantity $(\frac{1}{2}\dot{x}^2 + gx)$ remains constant throughout the motion. This is an example of a *conservation law.* The value of the constant may be determined from the initial conditions. In the present example, with $x(0) = 0, \dot{x}(0) = u$, the constant has value $\frac{1}{2}u^2$. The conservation law now becomes

$$\dot{x}^2 + 2gx = u^2 , \tag{3.15}$$

and again we see that the greatest height attained (when $\dot{x} = 0$) is $h = u^2/2g$.

The particular conservation law derived here may be recognised as the *energy equation,* which will be discussed in Ch. 6.6. It has enabled us to arrive more easily at the solution of our problem than by direct integration as in Ch. 3.5.1. (A generalization of this method is considered in Worked Example 4 at the end of this chapter.)

3.6 Vertical Motion under Gravity with Resistance

Having obtained the formula $h = u^2/2g$ or alternatively $u = \sqrt{2gh}$, experimental results might indicate some discrepancy with the theoretical result. For instance, for a given u, $h = u^2/2g$ will probably over-estimate the maximum height, particularly if u is large. Why is this? One plausible reason would be the omission of air resistance effects in our mathematical model in Ch. 3.5.

Air resistance will tend to reduce the ball's speed. It takes the form of a force in a direction *opposing* the direction of motion, which is usually assumed to be proportional to some power of the speed. For example we will assume that this force is given by mkv in a direction opposing motion, where v is the particle's speed and k is a positive constant. The equation of motion is now given by

$$\boxed{m\ddot{x} = -mg - mk\dot{x}} , \tag{3.16}$$

which can be rewritten as

$$\frac{dv}{dt} + kv = -g .$$

This first order differential equation has solution

$$v = A\mathrm{e}^{-kt} - (g/k)$$

where the arbitrary constant, A, is determined from the initial condition $v(0) = u$, giving

$$u = A - (g/k) .$$

Thus

$$v = \frac{dx}{dt} = [u + (g/k)]\,e^{-kt} - g/k \qquad (3.17)$$

and integrating again

$$x = -\frac{1}{k}[u + (g/k)]\,e^{-kt} - \frac{gt}{k} + B .$$

Since $x(0) = 0$, the arbitrary constant B is given by

$$0 = -\frac{1}{k}[u + (g/k)] + B ,$$

so that

$$x = \frac{1}{k}[u + (g/k)](1 - e^{-kt}) - \frac{gt}{k} . \qquad (3.18)$$

The maximum height is achieved when $v = 0$, which from (3.17) occurs at time t given by

$$[u + (g/k)]\,e^{-kt} = g/k$$

or

$$t = \frac{1}{k}\log\left(1 + \frac{uk}{g}\right) .$$

Substituting this time in (3.18) gives

$$\boxed{h = \frac{u}{k} - \frac{g}{k^2}\log\left(1 + \frac{uk}{g}\right)} . \qquad (3.19)$$

The value of k is usually taken to be small, and we can expand h in terms of small k.

Since, for small α, $\log(1 + \alpha) = \alpha - \frac{\alpha^2}{2} + \frac{\alpha^3}{3} - \ldots ,$

we have

$$h = \frac{u}{k} - \frac{g}{k^2}\left\{\frac{uk}{g} - \frac{1}{2}\left(\frac{uk}{g}\right)^2 + \frac{1}{3}\left(\frac{uk}{g}\right)^3 - \ldots\right\}$$

and simplyfying,

$$\boxed{h = \frac{1}{2}\frac{u^2}{g} - \frac{1}{3}\frac{u^3k}{g^2} + \text{terms involving higher order powers of } k} . \qquad (3.20)$$

The first term is independent of k, and identical to the result (3.11) obtained when resistance is neglected. That is, it is the limiting value of h as $k \to 0$. The second term, which will be small, represents the first correction to (3.11) for small air resistance.

We have of course only taken one special form for the air resistance. Other models for the resistance will lead to varying expressions for h. Comparisons between experimental and theoretical data will indicate which is the most suitable model to use.

Fig. 3.7 shows a sketch of ballistic data, plotting the resistive force against the ballistic missile's speed. For small speed, v, a v^2 dependence is indicated, whereas for large values of v, the curve becomes straight, indicating a linear dependence. It should also be noted that the resistive force will not only depend on the body's speed but also on the shape of the body. The missile referred to in Fig. 3.7 has a pointed nose, whereas for *slow* motion of a sphere through a viscous fluid, the resistive force is given by

$$R = 6\pi a\eta v$$

where a is the sphere's radius and η a measure of the viscosity of the fluid.

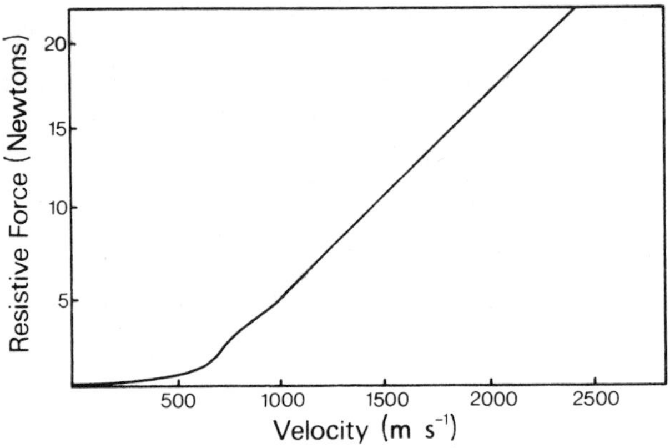

Fig. 3.7 Experimental data from a pointed-nose missile

3.7 Springs and Elastic Strings

Springs and elastic strings have very many useful practical applications, so it is important that we understand how they behave. An elastic string or spring is one which deforms when subjected to a load, resuming its original shape when the load is removed. The elastic string can be extended by the application of a force, whilst the coil spring can be extended or compressed. It is found that elastic strings and springs behave similarly when extended. Experiments have shown that, over an appreciable range, the extension of most strings and springs subjected to a tensile force is directly proportional to the force applied. This is *Hooke's law.* From now on we shall refer to the spring only, remembering that the same results apply for *extended* elastic strings.

Although most springs obey Hooke's law, it is certainly not exactly true for all springs, and usually completely invalid for large extensions. A spring which obeys Hooke's law is called a *linear spring.* The linear spring model provides a good approximation in most situations, provided we are not dealing with excessive loads and extensions. In mathematical notation we may express Hooke's law as

$$\boxed{T = \lambda x/a}\,, \tag{3.21}$$

where T is the applied force (equal and opposite to the force of tension produced in the spring), a is the natural length of the spring when unloaded, x is the extension of the spring beyond its natural length and λ is a constant known as the *modulus of elasticity* of the spring†. Under compression the linear spring is assumed to obey the same law, with x taking negative values. Clearly for the elastic string, (3.21) is valid only for $x \geqslant 0$, as the string cannot support forces of compression, but may become slack in the absence of tension.

HORIZONTAL MOTION OF A LINEAR SPRING

Consider the motion of a bob of mass m attached to one end of a linear spring, of modulus λ and natural length a. The spring is fixed at the other end and lies on a smooth horizontal plane (see Fig. 3.8).

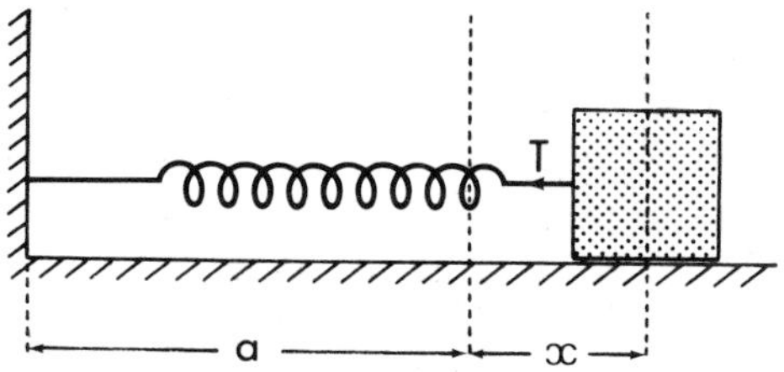

Fig. 3.8 Horizontal linear spring

We assume that the mass of the spring is neligible compared with m and that the frictional resistance due to the horizontal surface supporting the bob may also be neglected. The bob is displaced through a distance b from its equilibrium position, along the length of the spring, and is then released from rest.

Let the displacement from equilibrium at time t be x, and the tension in the spring be T. By Hooke's law,

$$T = \lambda x/a\ .$$

† In engineering it is sometimes useful to work with Young's modulus E, where $\lambda = AE$, and A is the cross-sectional area of the material of the spring. Young's modulus depends only on the material, whereas λ depends also on the geometrical properties of the particular spring.

The *equation of motion* of the bob is

$$m\ddot{x} = -T$$

$$\text{i.e.} \quad \boxed{\ddot{x} = -\lambda x/ma} \; . \tag{3.22}$$

Since (3.22) is a differential equation of the form $\ddot{x} = f(x)$, we may obtain a first integral by multiplying by $\dot{x}$ and integrating (c.f. Ch. 3.5.3). This gives

$$\tfrac{1}{2}\dot{x}^2 = -\frac{\lambda}{2ma}x^2 + A \; ,$$

where A is the constant of integration. Applying the initial conditions

$$x(0) = b, \; \dot{x}(0) = 0 \; , \tag{3.23}$$

gives $A = \lambda b^2/2ma$, whence $\dot{x}^2 = \dfrac{\lambda}{ma}(b^2 - x^2)$. (3.24)

From (3.24), we note that the greatest speed of the bob is obtained when $x = 0$; i.e. as the bob passes through its equilibrium position. Proceeding to solve (3.24) for $x(t)$, we have

$$\int \frac{dx}{(b^2 - x^2)^{1/2}} = \int \sqrt{\lambda/ma}\, dt + B \; ,$$

where B is the constant of integration. Integration yields

$$\sin^{-1}(x/b) = (\sqrt{\lambda/ma})t + B \; .$$

The conditions (3.23) give $B = \sin^{-1} 1 = \pi/2$, so that finally we obtain

$$\boxed{x = b \cos (\sqrt{\lambda/ma})t} \; . \tag{3.25}$$

We see that the motion of the bob consists of an oscillation about equilibrium such that $-b \leqslant x \leqslant b$. The time taken for a complete oscillation is τ, where

$$(\sqrt{\lambda/ma})\tau = 2\pi \; ;$$

for at this time, $x = b$ again, for the first time. We say that the *period of oscillation* of the bob is

$$\boxed{\tau = 2\pi\sqrt{ma/\lambda}} \; . \tag{3.26}$$

Oscillations are discussed further in Ch. 9.

3.8* Frames of Reference

For Newton's laws to have any real meaning we must be able to assess motion against the background of some suitable *frame of reference,* which is assumed to be at rest. Newton's laws pre-suppose that there exists, somewhere in the universe, a state of *absolute rest.* But is there any reason why such a reference frame should exist? It is, in fact, sufficient to seek a frame of reference which is non-accelerating, for it has already been established that rest and uniform rectilinear motion are essentially equivalent states. Thus, if the laws hold in a frame at rest, they will also hold in any frame moving with uniform velocity relative to the frame at rest. Such a non-accelerating frame of reference is called an *inertial frame.* Newton postulated that his laws were valid relative to an inertial frame, and we would not expect them to be exactly true in a non-inertial frame. Let us look at some possible reference frames.

3.8.1 Terrestrial Frame

Consider a frame of reference, with origin at a point O on the surface of the earth, and axes along the vertical and horizontal as shown in Fig. 3.9.

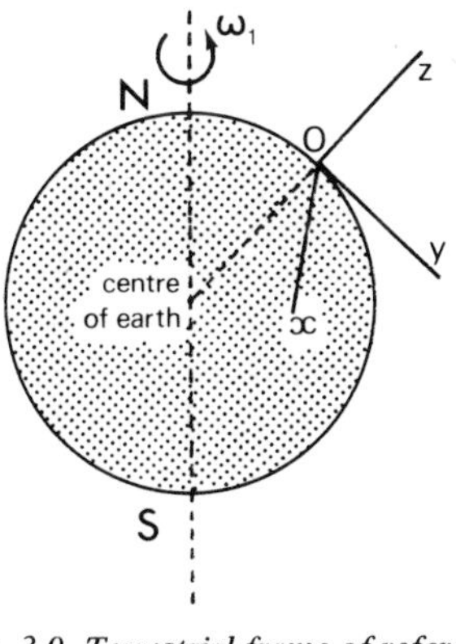

Fig. 3.9 Terrestrial frame of reference

The point O at rest on the earth's surface experiences a centripetal acceleration due to the rotation of the earth about its polar axis. This acceleration will vary in magnitude according to the latitude of the point. At a point on the equator, the centripetal acceleration [equation (2.18)] is given by

$$f_1 = \omega_1^2 R_e$$

where ω_1 is the angular speed of the rotating earth and R_e is the earth's equitorial radius. Substituting appropriate values ($\omega_1 = 2\pi$ radians per day $\approx 7{\cdot}3 \times 10^{-5}\,s^{-1}$; $R_e \approx 6{\cdot}4 \times 10^6\,m$) we obtain

$$f_1 \approx 0.034\ m\ s^{-2}\ . \tag{3.27}$$

In addition this frame experiences a further acceleration due to the earth's motion about the sun. In this case $f_2 = \omega_2^2 R_s$, where $\omega_2 = 2\pi$ radians per year $\approx 2 \times 10^{-7}\ s^{-1}$ and $R_s \approx 1.5 \times 10^{11}$m (the sun-earth distance).

Thus
$$f_2 \approx 0.006\ m\ s^{-2}\ , \tag{3.28}$$

which is an order of magnitude smaller than f_1. In comparison with the acceleration due to gravity ($\approx$9.8 m s^{-2} at the earth's surface) both accelerations, f_1 and f_2, are small. For this reason, we can regard this frame as an approximately inertial frame for local phenomena, with little resultant error in our calculations.

3.8.2 Sidereal Frame

A frame fixed in the sun would be accelerated due to the sun's rotation about its axis. To eliminate this, we construct a frame with its origin at the centre of the sun and axes determined by the directions of certain fixed (distant) stars. Such a sidereal frame is still not strictly inertial as the sun itself describes an orbit about the centre of the galaxy. The acceleration due to this motion is thought to be of the order of 3×10^{-6} m s^{-2}, which is considerably smaller than the corresponding terrestrial acceleration.

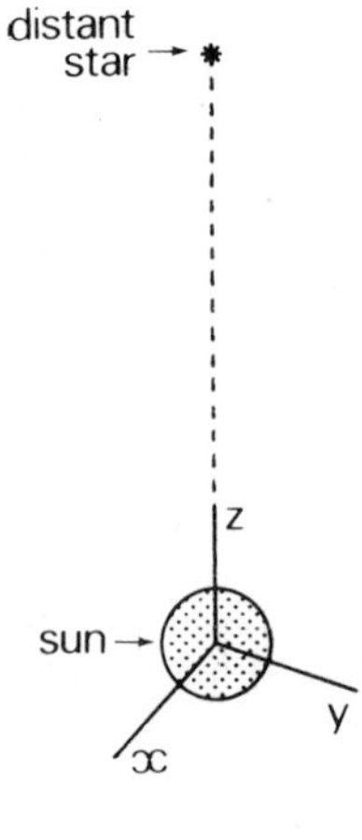

Fig. 3.10

Sidereal frame of reference

It is also probable that the whole galaxy has an acceleration, which would, however be an order of magnitude less than the acceleration of the sun.

To a very good approximation we can take the sidereal frame, defined by the fixed stars, to be inertial; whereas, except for motions near the surface of the earth, a frame fixed in the earth is non-inertial.

We have seen above that we are not able at present to describe a completely inertial frame of reference. It follows that we know of no frame with respect to which Newton's laws are absolutely true. They do however appear to represent an immensely valuable approximation to the truth, which has proved adequate for nearly all physical situations for over 200 years. Moreover, they are still widely applicable today in a great variety of situations. The chief difficulty arises when we are concerned with very high velocities, approaching the speed of light. The resolution of this difficulty was achieved by Einstein's theory of relativity, in which Newton's laws were modified to cope with situations of this nature.

With the above reservations regarding relativistic effects Newton's laws are applicable to all known types of forces, including gravitational, electrostatic, magnetic and nuclear forces.

Worked Examples

EXAMPLE 3.1:

A particle of mass m is sliding on a horizontal table. The resistive force is modelled by $\mathbf{F} = -km\mathbf{v}$, where k is a positive constant and $\mathbf{v}$ is the particle's velocity. If initially the particle is given a prescribed speed u in a given direction, show that the particle cannot travel further than a distance u/k from the initial point.

SOLUTION:

Clearly the particle will move on a straight line in the given initial direction. Let x denote its distance from the initial point along this line, so that the equation of motion

$$m\frac{d\mathbf{v}}{dt} = -km\mathbf{v}$$

reduces to the scalar equation

$$\frac{dv}{dt} = -kv\ ,$$

where v is the particle's speed along the straight line.

Solving for v,
$$v = A e^{-kt},$$

where the arbitrary constant, A, is determined from $v = u$ at $t = 0$.

Thus $A = u$, and
$$v = \frac{dx}{dt} = u e^{-kt}.$$

Integrating again,
$$x = -\frac{u}{k} e^{-kt} + B,$$

where $0 = -\frac{u}{k} + B$. Thus

$$x = \frac{u}{k}(1 - e^{-kt}).$$

Now the particle will come to rest when $v = 0$, and we see that this only occurs as $t \to \infty$. When $t \to \infty$, we note that $x \to u/k$, so that the particle will come to rest at a distance approaching u/k from its initial point.

[Note that in this model the particle travels a finite distance in an infinite time.]

EXAMPLE 3.2

A particle of mass m slides down an inclined plane, subject to a constant vertically downward force of magnitude mg. The motion is resisted by a force mkv^2, k being a positive constant and v the particle's speed. Show that the time required to move a distance d after starting from rest is

$$t = \frac{\cosh^{-1}(e^{kd})}{(kg\sin\theta)^{1/2}},$$

where θ is the angle of inclination of the plane.

SOLUTION:

Fig. 3.11 shows a sketch of the situation.

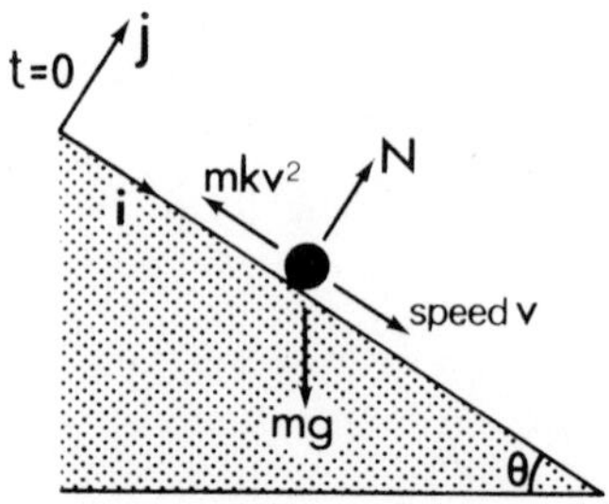

Fig. 3.11 Motion down an inclined plane

The equation of motion of the particle, introducing axes as shown in the diagram, is

$$m\ddot{\mathbf{r}} = -mkv^2\mathbf{i} + mg\sin\theta\,\mathbf{i} - mg\cos\theta\,\mathbf{j} + N\mathbf{j}\ .$$

Here N is the normal reaction exerted by the wedge on the particle. Since there is motion only down the plane,

$$N = mg\cos\theta$$

and the equation of motion is

$$\ddot{\mathbf{r}} = (g\sin\theta - kv^2)\mathbf{i}\ .$$

Now $\mathbf{r} = x\mathbf{i}$, say, so that x is the distance travelled by the particle down the plane. Thus

$$\frac{d^2x}{dt^2} = g\sin\theta - kv^2$$

or

$$\frac{dv}{dt} = g\sin\theta - kv^2\ .$$

We can integrate this differential equation by writing it as

$$\int\frac{dv}{(g\sin\theta - kv^2)} = \int dt$$

or, writing $\lambda^2 = g\sin\theta/k$,

$$\int\frac{dv}{(\lambda^2 - v^2)} = k\int dt$$

i.e.

$$\frac{1}{2\lambda}\int\left[\frac{1}{(\lambda - v)} + \frac{1}{(\lambda + v)}\right]dv = kt + A\ ,$$

and integrating the left hand side,

$$\log\left(\frac{\lambda + v}{\lambda - v}\right) = 2\lambda(kt + A)\ .$$

The constant of integration, A, is determined from the condition $v(0) = 0$, which gives $A = 0$. Thus

$$\frac{\lambda + v}{\lambda - v} = e^{2\lambda kt}\ ,$$

whence
$$v = \lambda \frac{(e^{2\lambda kt} - 1)}{(e^{2\lambda kt} + 1)} = \lambda \tanh(\lambda kt) .$$

Integrating again,

$$x = \frac{1}{k} \log \cosh(\lambda kt) ,$$

the constant of integration being zero since $x(0) = 0$.

When $x = d$,
$$e^{kd} = \cosh(\lambda kt)$$

so that the time to travel a distance d is given by

$$t = \frac{\cosh^{-1}(e^{kd})}{(\lambda k)} = \frac{\cosh^{-1}(e^{kd})}{(kg \sin\theta)^{1/2}}$$

EXAMPLE 3.3

A ball is thrown vertically upwards, and it is possible to measure both the maximum height reached, h, and the time from projection to maximum height, τ. Develop a simple mathematical model which will determine a theoretical relationship between h and τ.

Is the model in good agreement with the following experimental data? (Assume $g = 9.8 \text{ m s}^{-2}$.)

h(m)	τ(s)
7.6	1.2
15.7	1.7
23.9	2.1

If not, develop a more realistic model and again compare the theoretical and experimental data.

SOLUTION:

The simplest model would be that used in Ch. 3.5, where the only force acting is due to gravity. The equation of motion is then given by (3.8), i.e.

$$\frac{d^2x}{dt^2} = -g ;$$

and integrating,

$$\frac{dx}{dt} = -gt + u ,$$

where u is the (unknown) initial speed. Integrating again

$$x = -\tfrac{1}{2}gt^2 + ut ,$$

since initially $x = 0$.

The time to maximum height is given by $dx/dt = 0$, giving $\tau = u/g$. The maximum height is then given by

$$h = -\tfrac{1}{2}g\tau^2 + u\tau ,$$

and eliminating u gives

$$\boxed{h = \tfrac{1}{2}g\tau^2} . \qquad (3.29)$$

Using the same values of τ as in the experimental data the theoretical formula (3.29) gives the following table.

h(m)	τ(s)
7.06	1.2
14.16	1.7
21.61	2.1

The results do show a reasonable agreement, although for each case the theoretical results underestimate the height reached. Clearly the neglect of a resistive force could well be the reason for this discrepancy.

So consider a second model which includes a resistive force of magnitude *mkv* opposed to the direction of motion. This is the model described in Ch. 3.6, and using the results of that section, we see that the maximum height achieved is given by (3.19); i.e.

$$h = \frac{u}{k} - \frac{g}{k^2}\log\left(1 + \frac{uk}{g}\right)$$

where

$$\tau = \frac{1}{k}\log\left(1 + \frac{uk}{g}\right).$$

Eliminating u results in

$$h = \frac{g}{k^2}(e^{k\tau} - 1) - \frac{g\tau}{k},$$

and assuming k small,

$$h \approx \frac{g}{k^2}\left[\left(1 + k\tau + \tfrac{1}{2}k^2\tau^2 + \tfrac{1}{6}k^3\tau^3 + \ldots\right) - 1\right] - \frac{g\tau}{k}.$$

Simplifying,

$$\boxed{h \approx \tfrac{1}{2}g\tau^2\left(1 + \tfrac{1}{3}k\tau\right)}. \qquad (3.30)$$

[As expected, as $k \to 0$, (3.30) becomes (3.29), where resistive forces were neglected.]

We now have a second theoretical prediction for h, although we still have to determine a suitable value for the constant k. We can use the experimental results to do this, for rearranging (3.30) gives

$$k = \frac{3}{\tau}\left(\frac{2h}{g\tau^2} - 1\right).$$

Using the three sets of data gives values of k as 0.193, 0.192 and 0.151 respectively. Taking the average value, $k = 0.179$, we consider the theoretical prediction

$$h \approx \tfrac{1}{2}g\tau^2(1 + 0.06\tau),$$

which gives the table below:

h(m)	τ(s)
7.56	1.2
15.61	1.7
24.33	2.1

The agreement for this model is much better than for the first model, although to have complete confidence in the model we would want to check it with more experimental data.

EXAMPLE 3.4

A particle of mass m moves in one direction subject to a force $f(x)$ in the direction of motion, where x measures its distance along the straight line from its initial point. If v denotes the particle's velocity along the straight line, show that

$$\tfrac{1}{2}mv^2 - \int_0^x f(x)dx = \tfrac{1}{2}mu^2 ,$$

where u is the initial speed.

Find, in an integral form, an expression for the time taken, τ, to move a distance d along the straight line.

SOLUTION:

The equation of motion of the particle is $m\dfrac{d^2x}{dt^2} = f(x)$

and multiplying the equation by $\dfrac{dx}{dt}$ gives $m\dfrac{d^2x}{dt^2}\dfrac{dx}{dt} = f(x)\dfrac{dx}{dt}$.

Thus
$$m\frac{d}{dt}\left[\tfrac{1}{2}\left(\frac{dx}{dt}\right)^2\right] = f(x)\frac{dx}{dt} ,$$

which can be integrated from $t = 0$ to t to give

$$\left[\tfrac{1}{2}m\left(\frac{dx}{dt}\right)^2\right]_{t=0}^{t=t} = \int_0^t f(x)\frac{dx}{dt}dt$$

$$\text{i.e. } \tfrac{1}{2}mv^2 - \tfrac{1}{2}mu^2 = \int_0^x f(x)\,dx ,$$

since at $t = 0$, $x = 0$ and $v = u$. This gives

$$\tfrac{1}{2}mv^2 - \int_0^x f(x)dx = \tfrac{1}{2}mu^2 .$$

[We call this equation a first integral of the equation of motion. It does in fact represent *conservation of energy*, as we will see in Ch. 6.6, since the left hand side has the constant value $\tfrac{1}{2}mu^2$ throughout the motion. Each term of the left hand side will in general vary as the motion progresses, but the sum of $\tfrac{1}{2}mv^2$ and $\{-\int_0^x f(x)dx\}$ will remain constant, i.e. the quantity $\{\tfrac{1}{2}mv^2 - \int_0^x f(x)dx\}$ is conserved].

Now
$$\tfrac{1}{2}m\left(\frac{dx}{dt}\right)^2 = \int_0^x f(x)dx + \tfrac{1}{2}mu^2 ,$$

so defining
$$F(x) = \left[\frac{2}{m}\int_0^x f(x)dx + u^2\right]^{1/2}$$

we have
$$\frac{dx}{dt} = F(x) \,,$$

which can be integrated from $t = 0$ to $t = \tau$

to give
$$\int_0^d \frac{dx}{F(x)} = \int_0^\tau dt \,.$$

Thus
$$\tau = \int_0^d \frac{dx}{F(x)} \,.$$

Problems

3.1 A particle is projected vertically upwards under constant gravity with speed 30 m s^{-1}. One second later another particle is projected vertically upwards from the same initial point. Find the initial speed of the second particle so that the particles collide just as the first particle reaches its highest point ($g \approx 9.8$ m s^{-2}).

3.2 A stone is observed to fall vertically past a 2 m high window in 0·1 s. Neglecting all forces except constant gravity, determine the height above the bottom of the window from which the stone was dropped ($g \approx 9.8$ m s^{-2}).

3.3 A particle is projected vertically downwards under a constant gravitational force of magnitude g per unit mass. Show that, if the air resistance produces a force of magnitude kv per unit mass opposing the motion (where k is a positive constant) then no matter what initial speed is given to the particle, it's speed will eventually tend to g/k (called the *terminal speed*).

What is the terminal speed if the resistive force is of magnitude kv^2 per unit mass?

3.4 A ball is projected vertically upwards under gravity with initial velocity u, the resistance of the air producing a retardation kv^n, where v is the speed of the ball and k and n are positive constants. Using the method of imbedding, show that the maximum height reached, $f(u)$, is given by the integral

$$f(u) = \int_0^u \frac{udu}{(g + ku^n)}$$

3.5 In problem 3.4, show that

$$\text{(i) if } k = 0, \text{ then } f(u) = \frac{u^2}{2g}\,;$$

$$\text{(ii) if } n = 1, \text{ then } f(u) = \frac{1}{k}\left\{u - \frac{g}{k}\log\left(1 + \frac{ku}{g}\right)\right\};$$

$$\text{(iii) if } n = 2, \text{ then } f(u) = \frac{1}{2k}\log\left(1 + \frac{ku^2}{g}\right).$$

In (ii) and (iii), expand the function f in a power series in k, and in each case show that as $k \to 0$, $f \to u^2/2g$.

3.6 A particle of mass m is projected vertically upwards under constant gravity g. The force due to air resistance is $2amv + b^2mv^2$, where v is the particle's speed and a and b are constants. Using imbedding techniques, find integral expressions for h, the greatest height reached, and τ, the time taken to reach its greatest height, in terms of the speed of projection, u.

Hence deduce that

$$b^2h + a\tau = \tfrac{1}{2}\log\left(1 + \frac{2au}{g} + \frac{b^2u^2}{g}\right).$$

3.7 A particle is projected vertically upwards under constant gravity g in a medium for which the resistance per unit mass is kv^n, where v is the particle's speed and k and n are positive constants. If the speed of projection is u, show that the particle returns to its point of projection with speed u_1, such that

$$\text{(i) if } n = 1, \text{ then } u_1 + u = (g/k)\log\{(g + ku)/(g - ku_1)\}\,;$$

$$\text{(ii) if } n = 2, \text{ then } u_1 = gu^2/(g + ku^2)\,.$$

3.8 Develop a mathematical model for the experimental data given in Worked Example 3.3, assuming that the resistive force is proportional to the square of the particle's speed. Does the model present a reasonable explanation for the data?

3.9 A falling raindrop is spherical in shape and has constant density. As the raindrop passes through a cloud, it gains mass at a rate proportional to its cross-sectional area.

If the raindrop enters the cloud at time $t = 0$ with initial radius r_0 and initial speed v_0, and assuming no forces act on the raindrop except gravity,

(i) show that the radius of the raindrop increases linearly with time, and

(ii) neglecting both r_0 and v_0, show that the raindrop's speed increases linearly with time, when in the cloud.

3.10 A spherical falling raindrop *leaves* a cloud with initial radius r_0 and negligible speed. As it passes through the atmosphere, it increases its mass at a rate proportional to the product of its surface area and speed v. Assuming its density is constant, show that its radius, r, increases linearly with x, its distance below the cloud.

If r_0 is very small show that the acceleration of the raindrop is approximately $g - 3v^2/x$. Show also, when r_0 is very small, that the equation of motion of the raindrop can be written in the form

$$\frac{dp}{dx} + \frac{6p}{x} = 2g ,$$

where $p = v^2$. Verify that this equation has solution $p = 2gx/7$ and show that the acceleration of the raindrop is $g/7$.

Chapter 4

Rocket Flight Performance

4.1 Rockets through the Ages

Although the exact origin of the rocket is not clear, we know that a simple version was in use, mainly for signalling purposes, long before the Christian era. The freely rising rocket was first used as a weapon of warfare in the 13th century by the Chinese. By the end of that century it had reached Britain, where it was used by the English in the Scottish War in 1327. The use of rockets in warfare continued for about one century, until it was overshadowed by the development of the gun.

During the Napoleonic Wars, Sir William Congreve resumed experiments on rockets and by 1806 had developed a product so successful that it was employed to attack the city of Boulogne by firing rockets from small boats of the British Fleet. The results were encouraging and the rocket again found its place in warfare.

Congreve's first rockets were made with a tail stick tangentially attached (as in the modern day firework rocket) but as these were hopelessly inaccurate, the stick was subsequently situated centrally. William Hale, an American, made a further improvement by using a stickless variety and stabilizing its spin by liberating gas from tangential portholes.

The modern rocket is principally due to Dr. Robert H. Goddard of Clark University in America, who was interested in experiments at extremely high altitudes. It was clear that he could not attain such high altitudes by firing projectiles from a gun, and so he experimented with rockets, using a powered propellant and a tapered nozzle. Many of his features and proposals were later used by German scientists in the development of the V-2 rocket.

The Germans, Russians and British secretly began to develop rocket weapons several years before World War II started. Clearly the German V-2 rocket was the most successful effort. The rocket weighed 12 tons, reaching heights of 60-70 miles with a range of 200 miles. It is hardly surprising that at the end of the war the Americans captured all the V2s they could find, sending them back to the U.S. They were also fortunate in being able to persuade many of the leading German scientists, in particular Werner von Braun, to continue their work in America. The Russians did not do quite so well, but were still able to collect enough scientists and equipment to enable them to reconstruct similar rockets.

Development continued in both America and Russia in the post war years. At first the importance of orbiting satellites was clearly not appreciated by American politicians, but with the launching of Sputnik 1 on 4th October 1957, and the American Explorer 1 on 31st January 1958, the space race was really on in earnest. This meant that both nations, and to a lesser extent Britain, spent much of their wealth on rocket research.

Rocket performance improved rapidly and Fig. 4.1 illustrates the size of the V-2 rocket in comparison with the Thor missile (British development), Atlas Launcher (used for Mariner satellites to Mars) and the Saturn V Launcher (used for Apollo moonshots). The Saturn V launcher is a 3-stage rocket. It is shown later in this chapter that multistage rockets are certainly needed to launch satellites into Earth or inter-planetary orbits. At present solid and liquid fuels are used as the rocket propellant, but future long range exploration will require nuclear core rockets in order to achieve the necessary speeds.

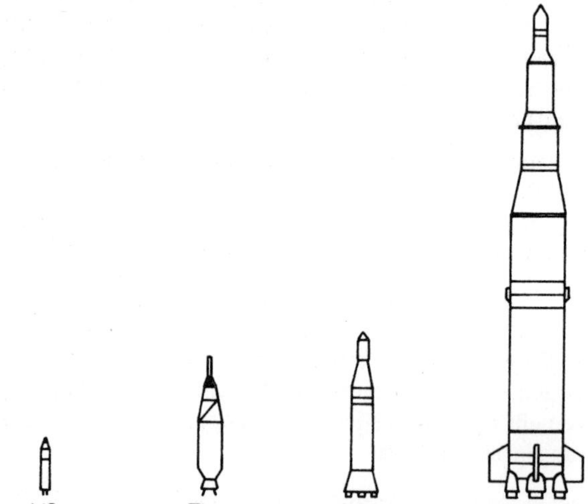

Fig. 4.1 Comparison of Rocket sizes

4.2 Equation of Motion for Variable Mass

In order to formulate the equation of motion for a rocket, we have to allow for the fact that the mass of the rocket is decreasing continually, as its fuel is burnt. We consider first the general problem of the motion of a body of variable mass.

Consider a body of mass $m = m(t)$, a function of time, moving with velocity $\mathbf{v}$. Suppose that in a small time δt the body coalesces with a small mass δm, which is moving with velocity $\mathbf{u}$ (see Fig. 4.2). At time $t + \delta t$ the body has mass $m + \delta m$ and velocity $\mathbf{v} + \delta \mathbf{v}$. It is assumed that the coalition imparts no rotation to the body.

Change of Lin. Momentum = $(m+\delta m)(v+\delta v) - \{mv + \delta m\, u\} = mv + \delta m\, v + m\,\delta v + \delta m\,\delta v - mv - \delta m\, u$
$= \delta m(v+\delta v - u) + m\,\delta v$

F = Rate of change of Lin momentum $= (v-u)\frac{dm}{dt} + m\frac{dv}{dt} = \frac{dP}{dt}$

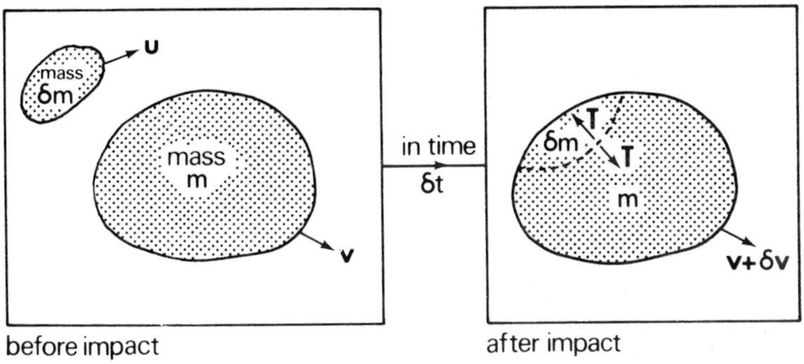

Fig. 4.2 Coalition of bodies

By Newton's third law, equal and oppositely directed forces, $\pm\mathbf{T}$ say, will be exerted by each body on the other. The equation of motion of the small mass is

$$\mathbf{T} = \text{rate of change of linear momentum}$$

$$= \lim_{\delta t \to 0} \left\{ \frac{\delta m(\mathbf{v} + \delta\mathbf{v}) - \delta m\,\mathbf{u}}{\delta t} \right\} \iff \lim_{\delta t \to 0} \left\{ \frac{\delta m}{\delta t}(v + \delta v - u) \right\}$$

$$= (\mathbf{v} - \mathbf{u})\frac{dm}{dt} \tag{4.1}$$

to the first order. The change in linear momentum of the main body in time δt is

$$m(\mathbf{v} + \delta\mathbf{v}) - m\,\mathbf{v} = m\,\delta\mathbf{v}\,.$$

If $\mathbf{F}$ is the applied external force, the equation of motion for the body is

$$\mathbf{F} - \mathbf{T} = \lim_{\delta t \to 0} \left(\frac{m\,\delta\mathbf{v}}{\delta t} \right) = m\frac{d\mathbf{v}}{dt}$$

giving

$$\boxed{\mathbf{F} = (\mathbf{v} - \mathbf{u})\frac{dm}{dt} + m\frac{d\mathbf{v}}{dt}}\,. \tag{4.2}$$

It should be noted that $\mathbf{F} \neq d(m\mathbf{v})/dt$ with the notation defined above as the mass δm has velocity $\mathbf{u}$ before impact. Clearly if $\mathbf{u} = \mathbf{0}$, (4.2) reduces to the familiar equation

$$\mathbf{F} = \frac{d}{dt}(m\,\mathbf{v})\,. \tag{4.3}$$

4.3 Flight Performance of a Single Stage Rocket

Our simple model for a one-stage rocket is a cylinder, closed at one end and supporting a payload (e.g. satellite). Burnt fuel, called the propellant, is ejected from the other end. The model is illustrated in Fig. 4.3.

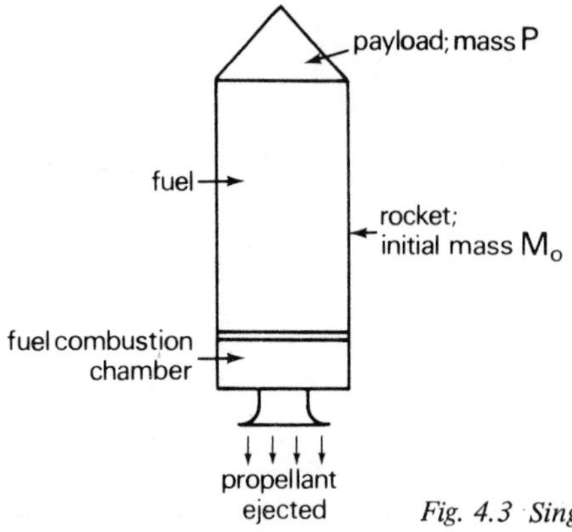

Fig. 4.3 Single stage rocket model

Let the exhaust speed of the propellant, *relative to the rocket* be c. Using the notation in Ch. 4.2, where $\mathbf{v}$ is the velocity of the rocket and $\mathbf{u}$ is the velocity of the fuel,

$$\mathbf{u} - \mathbf{v} = -c\,\mathbf{i} \tag{4.4}$$

$\mathbf{i}$ being a unit vector in the direction of motion of the rocket.

Substituting in (4.2) and considering the motion to be in one direction, $\mathbf{i}$, only gives

$$F = M\frac{dv}{dt} + c\frac{dM}{dt}, \tag{4.5}$$

M being the total rocket and payload mass and F the external force in the $\mathbf{i}$ direction. To understand the characteristics of rocketry we neglect the external force F, which means the neglect of such influences as gravitation and atmospheric resistance. At first this might seem a rather dubious approximation, but the external force is certainly small compared with the thrust term, $c\,(dM/dt)$, in (4.5). We will later consider (4.5) with the gravitational term present.

The total mass, M, is of the form

$$M = P + M_f(t) + M_c \tag{4.6}$$

where P is the mass of the payload, M_f the fuel and M_c the casing. M_f is the only contribution to M that varies with time. Initially let $M_f + M_c = M_0$, and suppose $M_f(0) = \epsilon M_0$ so that $M_c = (1-\epsilon)M_0$. The parameter ϵ is the ratio of the initial fuel supply to the initial rocket mass, and $(1-\epsilon)$ is called the structural factor. Clearly $0 < \epsilon < 1$ and we would like ϵ to be as near to unity as possible. Typical practical values of ϵ are in the range $0{\cdot}7 < \epsilon < 0{\cdot}9$.

At any time t, after the fuel is ignited,

$$M = P + M_f(t) + (1-\epsilon)M_0 \tag{4.7}$$

where $M_f(0) = \epsilon M_0$. Assuming that the fuel is burnt and ejected at a constant rate, k say,

$$\frac{dM}{dt} \overset{*}{=} \frac{dM_f(t)}{dt} = -k, \tag{4.8}$$

* since $(P + (1-\varepsilon)M_0)$ is constant.

the minus sign showing that the rocket is losing mass. Integrating this equation from time $t = 0$ until time t gives

$$M_f(t) = -kt + \epsilon M_o . \tag{4.9}$$

If all the fuel is burnt by time $t = \tau$, so that $M_f(\tau) = 0$, equation (4.9) gives

$$\tau = \epsilon M_o/k . \tag{4.10}$$

Substituting (4.9) in (4.7), the rocket mass at any time t, $0 \leqslant t \leqslant \tau$, is given by

$1 = P + (-kt + \epsilon M_o) + (1-\epsilon) M_o$

$$M(t) = P + M_o - kt , \tag{4.11}$$

and the equation of motion (4.5) becomes

$$(P + M_o - kt) \frac{dv}{dt} = ck. \tag{4.12}$$

Let the rocket have initial velocity v_o (normally zero). Integrating the differential equation (4.12) shows that the velocity of the rocket at any time t, $0 \leqslant t \leqslant \tau$, is

$$v = v_o - c \log \left\{ 1 - \frac{kt}{(M_o + P)} \right\} .$$

The increment in velocity, Δv, gained by the rocket by the time all the fuel has been burnt is thus

$$\boxed{\Delta v = -c \log \left\{ 1 - \frac{\epsilon M_o^*}{(M_o + P)} \right\} .}$$

$*$ since $\tau = \dfrac{\epsilon M_o}{k} \therefore k\tau = \epsilon M_o$ (4.13)

For the case of no payload (i.e. $P = 0$), (4.13) reduces to

$$\Delta v = -c \log (1 - \epsilon) \tag{4.14}$$

and is independent of the initial rocket mass M_o.

In practice the amount of fuel burnt and ejected is carefully programmed. The fuel would not necessarily be consumed in one continuous burn as described in the above theory.

From (4.13), it is clear that the efficiency of the rocket, that is the optimum final velocity given to the payload, depends on

(i) $1 - \epsilon$, the structural factor,

(ii) c, the exhaust speed,

(iii) P/M_o, sometimes known as the mass ratio.

The structural coefficient depends on the design and materials used in the rocket and the mass ratio is fixed by the payload mass (normally predetermined) and the rocket mass, which is related to the overall size and cost of the rocket. In the next section, variations in the exhaust speed are considered.

4.4   Exhaust Speed Parameters

We define the thrust, T, of a rocket as the rate of change of momentum of propellant fuel on burning. Using (4.1), (4.4) and (4.8), the magnitude of the thrust is

(4.1) $T = (r-u)\frac{dm}{dt}$.

(4.4) $(u-v) = -ci$

(4.8) $\frac{dM}{dt} = \frac{d M_{f}(t)}{dt} = -\kappa$.

$$T = ck. \qquad (4.15)$$

The specific impulse, I, is defined as the thrust produced per unit *weight* of fuel consumed per second, i.e.

$$I = T/gk. \qquad (4.16)$$

Eliminating k from (4.15) and (4.16) gives

$$I = c/g. \qquad (4.17)$$

The parameter I is used to compare the exhaust speeds of different types of propellant fuels. Specific impulses for solid, liquid and nuclear rockets are compared and contrasted in Table 4.1 below.

The advantages of solid fuel rockets are simple construction, ease of handling and relatively cheap cost, but they have the disadvantage of low specific impulse and difficulties in controlling the thrust. On the other hand, thrusts due to liquid fuels can easily be controlled but the design is complex and the cost high. Alternatively, nuclear core rockets are attractive as they will give such an enormous energy release compared with chemical reactions. The energy is released in the form of heat and must be transferred to a propellant in order to produce high velocity exhaust gases. This is one of the many problems concerned with the nuclear rocket. Others are the control of the thrust level, heat removal after cut-off, and adequate radiation shielding for the payload. These and many more difficulties must be overcome before a nuclear rocket becomes a practical proposition.

Table 4.1 – Comparison of Specific Impulses

ROCKET MOTOR	PROPELLANT	CHARACTERISTICS	OPTIMUM SPECIFIC IMPULSES
Solid Fuel	Polyurethane/Aluminium/ Ammonium Perchlorate	Relatively few difficulties	250
Solid Fuel	Lithium Hydride/ Lithium Fluoroamide	Not yet fully developed	290
Liquid Fuel	Hydrogen Peroxide/ Kerosene	Requires scrupulously clean storage tanks and there are fire and explosive hazards	265
Liquid Fuel	Liquid Oxygen/ Liquid Hydrogen	Severe fire and explosive risk and causes severe cold burns	385
Liquid Fuel	Liquid Fluorine/ Liquid Hydrogen	Liquid Fluorine reacts vigorously with most substances, frequently with immediate ignition!	425
Nuclear Core	Liquid Hydrogen	Not yet developed	1000-3000

4.5 Effect of Gravity

Assuming a constant gravitational force as the only external force, the equation of motion for a one-stage rocket, moving in one direction, is

$$(P + M_o - kt)\frac{dv}{dt} = ck - (P + M_o - kt)g\,, \tag{4.18}$$

where g is the constant acceleration per unit mass due to the gravitational force. Now, from (4.10), the total burn time is $\tau = \epsilon M_o/k$. Integrating (4.18) from 0 to τ gives

$$\Delta v = -c \log\left\{1 - \frac{\epsilon M_o}{(M_o + P)}\right\} - \frac{g\,\epsilon M_o}{k}\,. \tag{4.19}$$

Using typical values, $\epsilon = 0{\cdot}8$, $P/M_o = 1/100$, $I = 300$ s, the first term in (4.19) is approximately 4·64 km s^{-1}, and taking $M_o = 10^5$ kg, $k = 5 \times 10^3$ kg s^{-1}, the second term is approximately 0·16 km s^{-1}. This shows that the reduction in terminal velocity due to gravity is a relatively small effect. The results obtained with the neglect of gravity can thus be regarded as a reasonable approximation and the characteristics of rocket flight indicated in Ch. 4.4 are valid.

It is also of interest to find the distance travelled while the fuel is burnt. Integrating (4.18) from 0 to t gives

$$\frac{dx}{dt} = -c \log\left\{1 - \frac{kt}{(M_o + P)}\right\} - gt.$$

Integrating again from 0 to t, using integration by parts,

$$x = -\frac{1}{2}gt^2 - c\left[t \log\left\{1 - \frac{kt}{(M_o + P)}\right\} + \frac{k}{(M_o + P)}\int_o^t \frac{tdt}{\left\{1 - \dfrac{kt}{(M_o + P)}\right\}}\right]$$

$$= -\frac{1}{2}gt^2 + ct - c\,[t - (M_o + P)/k] \log\left\{1 - \frac{kt}{(M_o + P)}\right\}.$$

In time $\tau = \epsilon M_o/k$, the height reached, h, is given by

$$h = -\frac{g\epsilon^2 M_o^2}{2k^2} + \frac{c\epsilon M_o}{k} + \frac{c}{k}\,[M_o(1 - \epsilon) + P] \log\left\{1 - \frac{M_o}{(M_o + P)}\right\} \tag{4.20}$$

Using the values above gives $h \approx 30$ km, which is small in comparison with the radius of the Earth, 6400 km.

4.6 Two Stage Rocket

In Ch. 7.5 it is shown that to place a satellite in a circular orbit, height a above the Earth's surface, the satellite must be given a velocity of magnitude

$$v = \left(\frac{\gamma M_E}{a + R_E}\right)^{1/2} . \tag{4.21}$$

Here γ is a constant, called the gravitational constant, M_E is the mass of the Earth and R_E the radius of the Earth. For a typical circular orbit, 100 km above the Earth,

$$v \approx 7{\cdot}8\,\mathrm{km\,s^{-1}} . \tag{4.22}$$

Now the terminal velocity attained by a one-stage rocket, using the assumptions of Ch. 4.3, is given by equation (4.13). Taking as typical values, $\epsilon = 0{\cdot}8$, $P/M_o = 1/100$, $I = 300\,\mathrm{s}$, this gives

$$v \approx 4{\cdot}6\,\mathrm{km\,s^{-1}} . \tag{4.23}$$

This indicates that a one-stage rocket would not be adequate to place a satellite in a typical orbit round the Earth, and this is why multi-stage rockets have been developed.

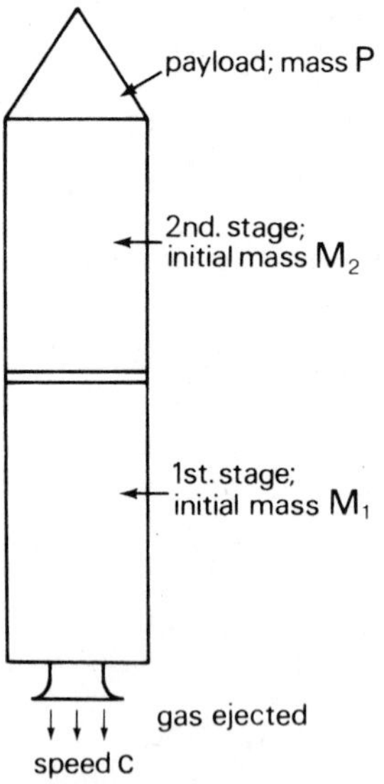

Fig. 4.4 Two stage rocket model

The model we use for a two-stage rocket is illustrated in Fig. 4.4. When all the fuel in the first stage has been burnt, the rocket casing and instruments for this stage are detached and the fuel of the second stage ignited. Suppose that the first stage has mass M_1, with initial fuel of mass ϵM_1, and the second stage is of mass M_2, with initial fuel of mass ϵM_2. By applying the theory of a one-stage rocket, the increment in velocity attained by the first stage of the rocket is given, using (4.13), by

$$\nu_1 = -c \log \left\{ 1 - \frac{\epsilon M_1}{(M_1 + M_2 + P)} \right\}.$$

The exhaust speed of the ejected gas in each stage is assumed to be c. The time required for the second stage fuel to burn is $\tau' = \epsilon M_2/k$. Integrating the equation of motion

$$(P + M_2 - kt)\frac{dv}{dt} = ck$$

between 0 and τ' gives

$$\int_{v_1}^{v_2} dv = \int_0^{\tau'} \frac{ck\,dt}{(M_2 + P - kt)}$$

i.e.

$$\nu_2 - \nu_1 = -c \log \left\{ 1 - \frac{\epsilon M_2}{(M_2 + P)} \right\},$$

and so the final velocity attained by a two-stage rocket is

$$\nu_2 = -c \log \left\{ 1 - \frac{\epsilon M_1}{(M_1 + M_2 + P)} \right\} - c \log \left\{ 1 - \frac{\epsilon M_2}{(M_2 + P)} \right\}. \tag{4.24}$$

If $M_1 = M_2 = 50P$ (so that $P/(M_1 + M_2) = 1/100$), $\epsilon = 0{\cdot}8$ and $I = 300$ s, then

$$\nu \approx 6{\cdot}0 \text{ km s}^{-1}. \tag{4.25}$$

This is certainly a good improvement over the one-stage rocket, but is still not adequate to put the payload into an Earth orbit. In the above two-stage rocket we have taken $M_1 : M_2 = 1 : 1$. However a more astute choice of this ratio might produce a considerable increase in the final velocity. This is a straightforward maximization problem, common to many branches of applied mathematics. Writing $M_o = M_1 + M_2$,

$$\nu_2 = -c \log \left\{ 1 - \frac{\epsilon (M_o - M_2)}{(M_o + P)} \right\} - c \log \left\{ 1 - \frac{\epsilon M_2}{(M_2 + P)} \right\}.$$

If we now assume that the total initial mass, M_o, is fixed we see that ν_2 is effectively a function of M_2 alone. For maximum or minimum values of ν_2, $d\nu_2/dM_2 = 0$. This condition gives

$$\frac{\epsilon/(M_o + P)}{[1 - \epsilon (M_o - M_2)/(M_o + P)]} = \frac{\epsilon P/(M_2 + P)^2}{[1 - \epsilon M_2/(M_2 + P)]}$$

and simplifying, $$(1-\epsilon)(M_2{}^2+2PM_2-PM_o) = 0.$$

Since $\epsilon \neq 1$, $$M_2{}^2 + 2PM_2 - PM_o = 0. \tag{4.26}$$

Solving the quadratic equation for M_2 gives

$$M_2 = -P \pm (P^2+PM_o)^{1/2}.$$

and choosing the appropriate root,

$$\frac{M_2}{M_o} = -\frac{P}{M_o} + \left(\frac{P^2}{M_o{}^2} + \frac{P}{M_o}\right)^{1/2}.$$

Assuming the ratio P/M_o is *small,*

$$\frac{M_2}{M_o} = -\frac{P}{M_o} + \left(\frac{P}{M_o}\right)^{1/2}\left[1+\frac{1}{2}\frac{P}{M_o}+\cdots\right]$$

$$= \left(\frac{P}{M_o}\right)^{1/2} - \left(\frac{P}{M_o}\right) + \frac{1}{2}\left(\frac{P}{M_o}\right)^{3/2} + \cdots,$$

and the first approximation is

$$\frac{M_2}{M_o} \approx \left(\frac{P}{M_o}\right)^{1/2}. \tag{4.27}$$

Taking the value $P/M_o = 1/100$ gives $M_2/M_o \approx 1/10$ and so $M_1/M_o \approx 9/10$ and $M_1 : M_2 \approx 9 : 1$. The optimum choice of the ratio $M_1 : M_2$ shows that the first stage must be constructed very much larger than the second stage.

Putting $P/M_o = \alpha$, where α is small, and substituting $M_2/M_o = \alpha^{1/2} - \alpha + O(\alpha^{3/2})$ in (4.24) gives

$$v_2 = -c\log\left\{1 - \frac{\epsilon[1-\alpha^{1/2}+\alpha+O(\alpha^{3/2})]}{1+\alpha}\right\}$$

$$-c\log\left\{1 - \frac{\epsilon[\alpha^{1/2}-\alpha+O(\alpha^{3/2})]}{\alpha^{1/2}+O(\alpha^{3/2})}\right\}$$

$$= -c\log\{1-\epsilon[1-\alpha^{1/2}+O(\alpha^{3/2})]\}$$

$$-c\log\left\{1-\frac{\epsilon[1-\alpha^{1/2}+O(\alpha)]}{1+O(\alpha)}\right\}$$

$$= -c\log\{1-\epsilon[1-\alpha^{1/2}+O(\alpha^{3/2})]\} - c\log\{1-\epsilon[1-\alpha^{1/2}+O(\alpha)]\}.$$

Hence we obtain

$$v_2 \approx -2c \log \{1-\epsilon(1-\alpha^{1/2})\} \quad . \tag{4.28}$$

If $P/M_o = 1/100$, $\epsilon = 0{\cdot}8$, $I = 300\,\text{s}$, then (4.28) gives

$$v \approx 7{\cdot}5\,\text{km}\,\text{s}^{-1} \ . \tag{4.29}$$

This is a considerable improvement on previous results and further staging should further increase the terminal velocity but at the expense of increasing the overall cost of the rocket.

4.7 Optimization of a multi-stage Rocket

The real problem that has in practice to be solved is to program the rocket thrust in order to give the payload a specific velocity at a specific position. This enables the payload to be placed in a definite orbit with the required velocity. In this section the individual masses of a multi-stage rocket of total mass M_o are chosen in order to give the payload, of fixed mass P, a definite velocity, say v_T, with a minimum value of the total mass M_o.

The model is illustrated in Fig. 4.5, the ith stage having initial total mass M_i and containing fuel $\epsilon_i M_i (1 \leqslant i \leqslant n)$. The exhaust speed of the ith stage is assumed to be $c_i (1 \leqslant i \leqslant n)$. This optimization problem is more complex than the previous one since we wish to choose the individual masses $M_i (1 \leqslant i \leqslant n)$ so as to minimize $\sum_{i=1}^{n} M_i = M_o$ but subject to the *constraint* that the final velocity obtained is v_T.

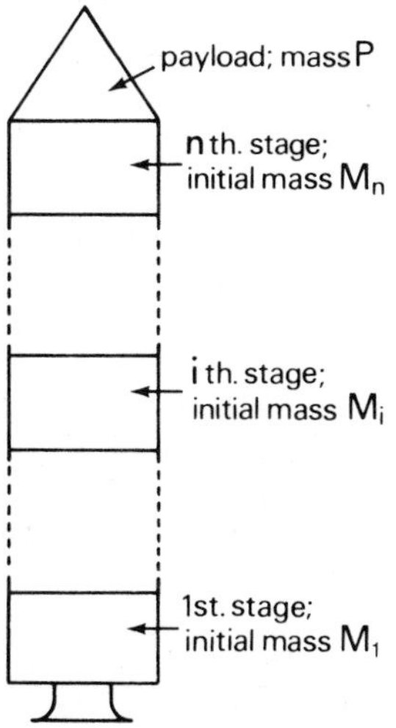

Fig. 4.5 Multi-stage rocket

Lagrange multiplier techniques (see Appendix 4) must be used to solve this problem as we wish to minimize a function of the n variables $M_1, M_2, \ldots, M_n$, namely

$$M_0 = M_1 + M_2 + \ldots + M_n \tag{4.30}$$

subject to the constraint that the final velocity is v_T, and clearly the final velocity is a function of the $M_i (i = 1, 2, \ldots, n)$. The velocity increase during the ith stage burn is given by

$$v_i = -c_i \log\left\{1 - \frac{\epsilon_i M_i}{(M_i + M_{i+1} + \ldots + M_n + P)}\right\}$$

$$= c_i \log\left\{\frac{M_i + \ldots + M_n + P}{(1-\epsilon_i)M_i + M_{i+1} + \ldots + M_n + P)}\right\}$$

and so the constraint is of the form

$$v_T - \sum_{i=1}^{n} c_i \log\left\{\frac{M_i + \ldots. + M_n + P}{(1-\epsilon_i)M_i + M_{i+1} + \ldots + M_n + P}\right\} = 0. \tag{4.31}$$

Introducing the Lagrange multiplier λ, we consider the function

$$f = M_1 + M_2 + \ldots + M_n$$

$$+ \lambda\left[v_T - \sum_{i=1}^{n} c_i \log\left\{\frac{M_i + \ldots + M_n + P}{(1-\epsilon_i)M_i + M_{i+1} + \ldots + M_n + P}\right\}\right] \tag{4.32}$$

and we look for minima of this function by solving $\partial f/\partial M_i = 0 (i = 1, 2, \ldots, n)$.

This leads to some very complicated analysis, and the optimization problem can be simplified by using new variables $\mu_i (i = 1, 2, \ldots, n)$ defined from (4.31), so that

$$v_T - \sum_{i=1}^{n} c_i \log \mu_i = 0. \tag{4.33}$$

Thus

$$\mu_i = \frac{M_i + \ldots. + M_n + P}{(1-\epsilon_i)M_i + M_{i+1} + \ldots + M_n + P} \tag{4.34}$$

and the constraint has the simple form (4.33). We must now express the function M_0 in terms of the μ_i $(i = 1, 2, \ldots, n)$, noting that

$$\frac{M_i + M_{i+1} + \ldots + M_n + P}{M_{i+1} + \ldots + M_n + P} =$$

$$= \frac{\epsilon_i(M_i + \ldots + M_n + P)}{(1-\epsilon_i)M_i + M_{i+1} + \ldots + M_n + P - (1-\epsilon_i)(M_i + M_{i+1} + \ldots + M_n + P)}$$

$$= \frac{\epsilon_i \mu_i}{[1-(1-\epsilon_i)\mu_i]} . \tag{4.35}$$

Hence

$$\frac{M_o + P}{P} = \left[\frac{M_1 + M_2 + \ldots + M_n + P}{M_2 + \ldots + M_n + P}\right]\left[\frac{M_2 + \ldots + M_n + P}{M_3 + \ldots + M_n + P}\right] \ldots$$

$$\ldots \left[\frac{M_{n-1} + M_n + P}{M_n + P}\right]\left[\frac{M_n + P}{P}\right]$$

$$= \prod_{i=1}^{n} \left[\frac{M_i + M_{i+1} + \ldots + M_n + P}{M_{i+1} + \ldots. + M_n + P}\right]$$

$$= \prod_{i=1}^{n} \frac{\epsilon_i \mu_i}{[1-(1-\epsilon_i)\mu_i]} , \text{ using (4.35).}$$

Minimizing M_o is equivalent to minimizing $(M_o + P)/P$, since P is constant; and also equivalent to minimizing $\log [(M_o + P)/P]$. But

$$\log [(M_o + P)/P] = \sum_{i=1}^{n} \left\{ \log \mu_i + \log \epsilon_i - \log [1 - \mu_i(1-\epsilon_i)] \right\} \tag{4.36}$$

and adding λ times the constraint (4.33) to (4.36) gives

$$\log [(M_o + P)/P] = \sum_{i=1}^{n} \left\{ \log \mu_i + \log \epsilon_i - \log [1 - \mu_i(1-\epsilon_i)] \right\}$$

$$+ \lambda \left[\sum_{i=1}^{n} c_i \log \mu_i \right] - \lambda v_T$$

$$= \sum_{i=1}^{n} \left\{ (1 + \lambda c_i)\log \mu_i + \log \epsilon_i - \log [1 - \mu_i(1-\epsilon_i)] \right\} - \lambda v_T ,$$

λ being the Lagrange multiplier. Differentiating this expression with respect to μ_i and equating to zero leads to the values of μ_i, which minimize M_o, in terms of λ. Thus

$$\frac{\partial}{\partial \mu_k} \{ \log [(M_o + P)/P] \} = \frac{1 + \lambda c_k}{\mu_k} + \frac{(1-\epsilon_k)}{[1 - \mu_k(1-\epsilon_k)]} = 0 ,$$

and solving for μ_k gives

$$\mu_k = \frac{1 + \lambda c_k}{\lambda c_k(1-\epsilon_k)} . \tag{4.37}$$

To find the value of λ, substitute the values of μ_k into (4.31) giving

$$v_T = \sum_{i=1}^{n} c_i \log\left[\frac{(1+\lambda c_i)}{\lambda c_i(1-\epsilon_i)}\right] . \tag{4.38}$$

This equation must be solved for λ, and (4.37) then gives the value of μ_k. In general computational techniques have to be used to solve (4.38) for λ, but in a simplified model with equal exhaust speeds (i.e. $c_1 = c_2 = \ldots = c_n = c$, say) and equal structural factors (i.e. $\epsilon_1 = \epsilon_2 = \ldots = \epsilon_n = \epsilon$, say), equation (4.37) reduces to

$$\mu_1 = \mu_2 = \ldots = \mu_n = \frac{1 + \lambda c}{\lambda c(1-\epsilon)} = \mu \text{, say;} \tag{4.39}$$

and (4.33) gives $\qquad v_T = nc \log \mu$

or $\qquad \mu = e^{v_T/nc}$.

Solving (4.39) for λ gives $\qquad \lambda = 1/[(1-\epsilon)e^{v_T/nc} - 1]c \qquad (4.40)$

and subsituting for λ shows that $\dfrac{M_0 + P}{P} = \left[\dfrac{\epsilon\mu}{1-\mu(1-\epsilon)}\right]^n$

or

$$M_0 = \left\{\frac{\epsilon^n e^{v_T/c}}{[1-e^{v_T/nc}(1-\epsilon)]^n} - 1\right\} P . \tag{4.41}$$

Using the typical values for the parameters, $v_T = 7{\cdot}8\ \mathrm{km\,s^{-1}}$, $I = 300$ s, $\epsilon = 0{\cdot}8$ in (4.41) gives the following results:

(i) $n = 1$: M_0, evaluated from (4.41), is negative, indicating that it is impossible to obtain the velocity v_T with a one-stage rocket;

(ii) $n = 2$: $M_0 \approx 147P$;

(iii) $n = 3$: $M_0 \approx 52P$;

(iv) $n = 4$: $M_0 \approx 40P$;

(v) $n = 5$: $M_0 \approx 36P$;

(vi) As $n \to \infty$, $M_0 \to (13{\cdot}2)P$.

These results are illustrated in Fig. 4.6, and indicate why NASA uses 3-stage rockets. The decrease in the rocket mass M_o from a 2-stage to a 3-stage rocket is considerable, but from 3 to 4-stage there is only a relatively small improvement. Also the cost will increase considerably as the number of stages increases.

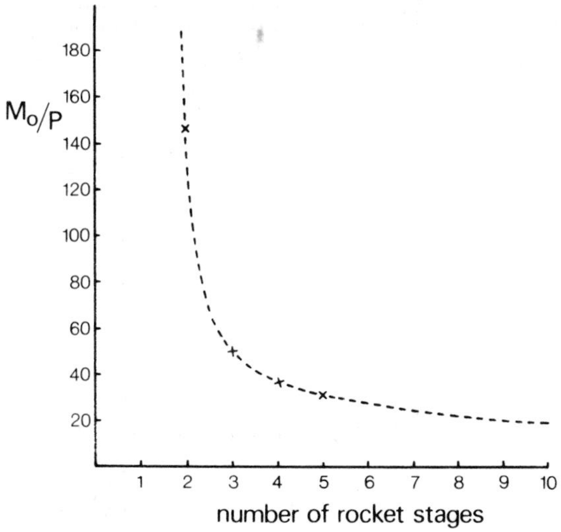

Fig. 4.6 Minimum rocket mass for multi-stage rockets in order to achieve a terminal speed of 7·8 km s^{-1}

To find the ratio of the masses in the simplified model, we must solve (4.34) for M_i in terms of μ_i. Writing

$$\beta = \frac{\mu - 1}{1 - \mu(1-\epsilon)},$$

$$M_n = \beta P, \; M_{n-1} = \beta(\beta+1)P, \quad M_{n-2} = \beta(\beta+1)^2 P, \; \ldots\ldots ,$$

$$M_1 = \beta(\beta+1)^{n-1}P .$$

Thus

$$\frac{M_i}{M_{i+1}} = 1 + \beta = \frac{\epsilon\mu}{[1-\mu(1-\epsilon)]}$$

and for

(i) $n = 2$: $M_1 : M_2 \approx 12:1$;

(ii) $n = 3$: $M_1 : M_2 : M_3 \approx 13{\cdot}7 : 3{\cdot}7 : 1{\cdot}0$.

Worked Examples

EXAMPLE 4.1:

Show that if the payload mass P is very small compared with the initial rocket mass M_o and all external forces are neglected, then the increment in speed given to the payload when all fuel has been burnt is given approximately by

$$\Delta v = -c\log(1-\epsilon) - \frac{c\epsilon(P/M_o)}{(1-\epsilon)},$$

$(1-\epsilon)$ being the structural factor.

SOLUTION:

The increment in speed given to the payload is given by (4.13) as

$$\Delta v = -c\log\left[1 - \frac{\epsilon M_o}{(M_o+P)}\right].$$

If P is small in comparison with M_o,

$$\begin{aligned}
\Delta v &= -c\log\left[1-\epsilon(1+P/M_o)^{-1}\right] \\
&= -c\log\left\{1-\epsilon\left[1-\frac{P}{M_o}+\left(\frac{P}{M_o}\right)^2 - \ldots\right]\right\} \\
&\approx -c\log\left[1-\epsilon+\frac{\epsilon P}{M_o}\right] \\
&= -c\log\left\{(1-\epsilon)\left[1+\frac{\epsilon(P/M_o)}{(1-\epsilon)}\right]\right\} \\
&= -c\log(1-\epsilon) - c\log\left[1+\frac{\epsilon(P/M_o)}{(1-\epsilon)}\right] \\
&= -c\log(1-\epsilon) - c\left[\frac{\epsilon(P/M_o)}{(1-\epsilon)} - \frac{(P/M_o)^2}{2(1-\epsilon)^2} + \ldots\right]
\end{aligned}$$

i.e.

$$\Delta v \approx -c\log(1-\epsilon) - \frac{c\epsilon(P/M_o)}{(1-\epsilon)}.$$

Note that this result indicates that if P can be neglected in comparison with M_o, the final speed, $-c\log(1-\epsilon)$, is independent of the initial rocket mass M_o.

EXAMPLE 4.2:

A rocket is programmed to burn and eject its propellant at the variable rate $\alpha k e^{-kt}$, where k and α are positive constants. The rocket is launched vertically from rest. Neglecting all external forces except gravity, show that the final speed given to the payload, of mass P, when all the fuel has been burnt is

$$v = -c\log\left[1 - \frac{\epsilon M_o}{(M_o + P)}\right] + \frac{g}{k}\log\left[1 - \frac{\epsilon M_o}{\alpha}\right].$$

Here M_o is the initial rocket mass, excluding the payload, and the initial fuel mass is ϵM_o.

SOLUTION:

Since the rate of change of expelled propellant mass is minus the rate of change of the rocket mass, we have

$$\frac{dM}{dt} = -\alpha k e^{-kt},$$

where M is the total rocket mass including the payload. Integrating this equation gives

$$M = \alpha e^{-kt} + A,$$

where, at $t = 0$, $M = M_o + P$, so that $A = M_o + P - \alpha$.

Thus $$M(t) = M_o + P - \alpha(1 - e^{-kt}).$$

From (4.5), the equation of motion is given by

$$\frac{dv}{dt} = \frac{c\alpha k e^{-kt}}{[M_o + P + \alpha(1 - e^{-kt})]} - g.$$

Integrating,

$$v = -c\log[M_o + P - \alpha(1 - e^{-kt})] - gt + B,$$

where, initially,

$$0 = -c\log(M_o + P) + B.$$

Thus $$v(t) = -c\log\left[1 - \frac{\alpha(1 - e^{-kt})}{(M_o + P)}\right] - gt.$$

When all the fuel has been burnt, $M = (1-\epsilon)M_o + P$, so that the time, τ say, at which this occurs is given by

$$(1-\epsilon)M_o + P = M_o + P - \alpha(1-e^{-k\tau})$$

or

$$\alpha(1-e^{-k\tau}) = \epsilon M_o .$$

Thus

$$\tau = -\frac{1}{k}\log(1-\epsilon M_o/\alpha),$$

and substituting into $v(t)$ gives the final speed

$$v(\tau) = -c\log\left[1-\frac{\epsilon M_o}{(M_o+P)}\right]+\frac{g}{k}\log\left(1-\frac{\epsilon M_o}{\alpha}\right)$$

EXAMPLE 4.3:

Show that, in Worked Example 4.2, if all external forces are neglected, the final speed given to the payload is identical with that given by (4.13). Is this final speed independent of the fuel expenditure programme?

SOLUTION:

The neglect of the only external force in Worked Example 4.2 is adjusted for by putting $g = 0$ in the final equation for $v(\tau)$. Thus

$$v(\tau) = -c\log\left[1-\frac{\epsilon M_o}{(M_o+P)}\right]$$

and this is identical with (4.13).

This is to be expected as (4.13) can be derived from (4.5) directly, without reference to the form of the rate of change of mass; for we have

$$0 = M\frac{dv}{dt} + c\frac{dM}{dt},$$

so that

$$\int dv = -c\int(1/M)\,dM + \text{constant}.$$

This gives

$$v = -c\log M + \text{constant},$$

and application of the initial and final mass values results in (4.13).

Thus, under the approximation that all external forces are neglected, (4.13) is true for any fuel expenditure programme, provided the fuel is consumed in one continuous burn.

EXAMPLE 4.4:

A rocket consists of two stages and a payload of mass P. The first stage is of mass $0{\cdot}9M_o$ which contains fuel of mass $0{\cdot}9\epsilon M_o$, and the second stage is of mass $0{\cdot}1M_o$, containing fuel of mass $0{\cdot}1\epsilon M_o$. Show that, when $\epsilon = 0{\cdot}8$, a payload of mass $0{\cdot}07M_o$ can be given a final speed of approximately $1{\cdot}75c$, where c is the constant propellant exhaust speed relative to the rocket.

SOLUTION:

The final speed given to the payload by the two stage rocket is determined by equation (4.24) with $M_1 = 0{\cdot}9M_o$, $M_2 = 0{\cdot}1M_o$, $P = 0{\cdot}07M_o$ and $\epsilon = 0{\cdot}8$; i.e.

$$v = -c\log\left[1 - \frac{(0{\cdot}8)(0{\cdot}9)}{(1 + 0{\cdot}07)}\right] - c\log\left[1 - \frac{(0{\cdot}8)(0{\cdot}1)}{(0{\cdot}1 + 0{\cdot}07)}\right]$$

$$= -c\log\left(\frac{0{\cdot}35}{1{\cdot}07}\right) - c\log\left(\frac{0{\cdot}09}{0{\cdot}17}\right).$$

Thus $v \approx 1{\cdot}75c$.

EXAMPLE 4.5:

A rocket consists of a payload of mass P propelled by a two stage rocket. The first stage is of mass M_1 and has initial fuel mass $\epsilon_1 M_1$ and the second stage is of mass M_2 with initial fuel mass $\epsilon_2 M_2$. The propellant exhaust speeds relative to the rocket are given by c_1 and c_2 respectively.

If the total initial mass, $M_o = M_1 + M_2$, is fixed, show that the ratio M_2/M_o for maximum final speed is given by the positive root α of the quadratic equation

$$a\alpha^2 + b\beta\alpha - c\beta = 0,$$

where $\alpha = M_2/M_o$, $\beta = P/M_o$, terms of order β^2 have been neglected and the constants a, b and c are defined by

$$a = c_1\epsilon_1(1-\epsilon_2),$$

$$b = c_1\epsilon_1(2-\epsilon_2) - c_2\epsilon_1\epsilon_2,$$

$$c = c_2\epsilon_2(1-\epsilon_1).$$

For small β, deduce that

$$\alpha \approx \left[\frac{c_2\epsilon_2(1-\epsilon_1)}{c_1\epsilon_1(1-\epsilon_2)}\right]^{1/2} \beta^{1/2}$$

and show that the maximum final velocity is given approximately by

$$v \approx -c_1\log(1-\epsilon_1) - c_2\log(1-\epsilon_2) - 2\left[\frac{c_1c_2\epsilon_1\epsilon_2}{(1-\epsilon_1)(1-\epsilon_2)}\right]^{1/2}\beta^{1/2}.$$

SOLUTION:

The speed attained by the first stage is given by

$$v_1 = -c_1\log\left[1-\frac{\epsilon_1M_1}{M_1+M_2+P}\right]$$

and the increment in speed during the second stage is

$$v_2 = -c_2\log\left[1-\frac{\epsilon_2M_2}{M_2+P}\right].$$

Thus the final speed given to the payload is

$$v = -c_1\log\left[1-\frac{\epsilon_1M_1}{M_1+M_2+P}\right] - c_2\log\left[1-\frac{\epsilon_2M_2}{M_2+P}\right].$$

Furthermore, since $M_1+M_2 = M_0$, where M_0 is fixed, we may express v as a function of M_2 only, giving

$$v = -c_1\log\left[1-\frac{\epsilon_1(M_0-M_2)}{M_0+P}\right] - c_2\log\left[1-\frac{\epsilon_2M_2}{M_2+P}\right].$$

Writing $M_2/M_0 = \alpha$, $P/M_0 = \beta$ gives

$$v = -c_1\log\left[1-\epsilon_1\frac{(1-\alpha)}{(1+\beta)}\right] - c_2\log\left[1-\frac{\epsilon_2\alpha}{(\alpha+\beta)}\right]. \qquad (4.42)$$

For maximum final speed v we require

$$dv/dM_2 = 0; \quad \text{i.e. } dv/d\alpha = 0.$$

This condition gives

$$-\frac{c_1\epsilon_1/(1+\beta)}{[1-\epsilon_1(1-\alpha)/(1+\beta)]} - \frac{c_2(-\epsilon_2\beta)/(\alpha+\beta)^2}{[1-\epsilon_2\alpha/(\alpha+\beta)]} = 0\,,$$

which simplifies to

$$c_1\epsilon_1(\alpha+\beta)(\alpha+\beta-\epsilon_2\alpha) = c_2\epsilon_2\beta(1+\beta-\epsilon_1+\epsilon_1\alpha)\,.$$

Finally, neglecting terms involving β^2 and defining a, b and c as above, we obtain the condition

$$a\alpha^2 + b\beta\alpha - c\beta = 0\,.$$

Solving for α,

$$\begin{aligned}\alpha &= \frac{-b\beta + (b^2\beta^2 + 4ac\beta)^{1/2}}{2a}\\ &= \frac{-b\beta + (4ac\beta)^{1/2}(1+(b^2/4ac)\beta)^{1/2}}{2a}\\ &= \frac{-b\beta}{2a} + \left(\frac{c}{a}\right)^{1/2}\beta^{1/2}\left(1+\frac{b^2}{8ac}\beta + \ldots\right)\\ &= \left(\frac{c}{a}\right)^{1/2}\beta^{1/2} - \frac{b}{2a}\beta + O(\beta^{3/2})\,.\end{aligned}$$

Thus, for small β,

$$\alpha \approx \left(\frac{c}{a}\right)^{1/2}\beta^{1/2} = \left[\frac{c_2\epsilon_2(1-\epsilon_1)}{c_1\epsilon_1(1-\epsilon_2)}\right]^{1/2}\beta^{1/2}\,.$$

Obtaining the corresponding final velocity requires a careful approach to the approximations involved. If we let $\alpha = k\beta^{1/2} + O(\beta)$, where $k = (c/a)^{1/2}$,substitution into (4.42) gives the maximum final speed as

$$v = -c_1\log\left\{1-\epsilon_1\frac{[1-k\beta^{1/2}+O(\beta)]}{(1+\beta)}\right\} - c_2\log\left\{1-\frac{\epsilon_2[k\beta^{1/2}+O(\beta)]}{k\beta^{1/2}+\beta+O(\beta)}\right\}.$$

It is clear that we need to retain terms of order β in the final term of the above expression, as these will become terms in $\beta^{1/2}$ when we divide numerator and denominator by $\beta^{1/2}$. It is therefore necessary to consider the exact form of the term of order β in the expansion for a, and hence we must take

$$\alpha = k\beta^{1/2} + \lambda\beta + O(\beta^{3/2})\,,$$

where $\lambda = -b/2a$.

Substitution into (4.42) now gives

$$\nu = -c_1 \log \left\{ 1 - \epsilon_1 \frac{[1 - k\beta^{1/2} + O(\beta)]}{1+\beta} \right\} - c_2 \log \left\{ 1 - \frac{\epsilon_2 [k\beta^{1/2} + \lambda\beta + O(\beta^{3/2})]}{k\beta^{1/2} + (1+\lambda)\beta + O(\beta^{3/2})} \right\}$$

$$= -c_1 \log \{1 - \epsilon_1 [1 - k\beta^{1/2} + O(\beta)]\} - c_2 \log \{1 - \epsilon_2 [1 + (\lambda/k)\beta^{1/2} + O(\beta)] \; [1 + ((1+\lambda)/k)\beta^{1/2} + O(\beta)]^{-1}\}$$

$$= -c_1 \log \{1 - \epsilon_1 [1 - k\beta^{1/2} + O(\beta)] - c_2 \log \{1 - \epsilon_2 [1 + (\lambda/k)\beta^{1/2} + O(\beta)] \; [1 - ((1+\lambda)/k)\beta^{1/2} + O(\beta)]\}$$

$$= -c_1 \log \{1 - \epsilon_1 [1 - k\beta^{1/2} + O(\beta)]\} - c_2 \log \{1 - \epsilon_2 [1 - (\beta^{1/2}/k) + O(\beta)]\}$$

$$= -c_1 \log \left\{ (1-\epsilon_1) \left[1 + \frac{\epsilon_1 k\beta^{1/2} + O(\beta)}{(1-\epsilon_1)} \right] \right\} - c_2 \log \left\{ (1-\epsilon_2) \left[1 + \frac{\epsilon_2 \beta^{1/2}/k + O(\beta)}{(1-\epsilon_2)} \right] \right\}$$

$$= -c_1 \log(1-\epsilon_1) - c_2 \log(1-\epsilon_2) - c_1 \log \left[1 + \frac{\epsilon_1 k\beta^{1/2} + O(\beta)}{(1-\epsilon_1)} \right] - c_2 \log \left[1 + \frac{\epsilon_2 \beta^{1/2}/k + O(\beta)}{(1-\epsilon_2)} \right]$$

$$= -c_1 \log(1-\epsilon_1) - c_2 \log(1-\epsilon_2) - \frac{c_1 \epsilon_1 k\beta^{1/2}}{(1-\epsilon_1)} - \frac{c_2 \epsilon_2 \beta^{1/2}}{k(1-\epsilon_2)} + O(\beta) \, .$$

Thus we finally obtain, neglecting terms of order β, and substituting for k,

$$\nu \approx -c_1 \log(1-\epsilon_1) - c_2 \log(1-\epsilon_2) - 2\left[\frac{c_1 c_2 \epsilon_1 \epsilon_2}{(1-\epsilon_1)(1-\epsilon_2)} \right]^{1/2} \beta^{1/2} \, .$$

Problems

4.1 A single stage rocket of mass M_0 and structural factor $(1-\epsilon)$ carries a payload of mass P. If $\epsilon = 0{\cdot}8$, neglecting external forces show that the maximum payload which can be given a final speed of $1{\cdot}3c$ is $0{\cdot}1M_0$, c being the exhaust speed.

4.2 Before firing, a single stage rocket has total mass M_0, which comprises the casing, instruments etc, which have mass M_c, and the fuel. The fuel is programmed to burn and be ejected at a variable rate which is such that the total mass of the rocket $m(t)$ at any time t, during which the fuel is being burnt, is given by

$$m(t) = M_0 \exp(-Kt/M_0)$$

where K is a constant.

The rocket is launched vertically from rest. Neglecting all external forces except gravity, show that the height H attained at the instant the fuel is fully consumed is

$$H = \frac{M_0^2}{2K^2}\left(\frac{cK}{M_0} - g\right)\left(\log \frac{M_0}{M_c}\right)^2,$$

c being the exhaust speed relative to the rocket.

4.3 A single stage rocket of mass M_0 and structural factor $(1-\epsilon)$ carries a payload of mass P. Suppose it is technically possible to discard the casing continuously at a constant rate with zero speed relative to the rocket whilst the fuel is burning so that *no* casing remains when all the fuel has been burnt. If the fuel is burnt at a constant rate k, show that the casing must be discarded at the rate $(1-\epsilon)k/\epsilon$.

Neglecting all external forces, find the rocket's final speed if $\epsilon = 0{\cdot}8$, $P = M_0/100$ and $I = 300\,\text{s}$.

4.4 The stages of a two stage rocket have initial masses M_1 and M_2 respectively and carry a payload of mass P. Both stages have equal structural factors and equal relative exhaust speeds. If the rocket mass, $M_1 + M_2$, is fixed, show that the condition for maximum final speed is

$$M_2^2 + PM_2 = PM_1 .$$

Find the optimum ratio M_1/M_2 when $P/(M_1 + M_2) = 0{\cdot}1$.

4.5 A rocket consists of a payload of mass P propelled by two stages of masses M_1 (first stage) and M_2 (second stage), both with structural factor $(1-\epsilon)$. The exhaust speed of the first stage is c_1, and of the second stage c_2.

If the initial total mass $M_1 + M_2$ is fixed, show that the ratio $M_2/(M_1+M_2)$ for a maximum final velocity is given by

$$(c_2\alpha/c_1)^{1/2} + O(a)$$

where $\alpha = P/(M_1 + M_2)$ is assumed very small.

Also show that this maximum velocity is

$$-(c_1 + c_2)\log(1-\epsilon) - 2\epsilon(c_1 c_2 \alpha)^{1/2}/(1-\epsilon) + O(a).$$

4.6 A rocket consists of a satellite of fixed mass P, propelled by three rocket stages of mass $M_i (i = 1, 2, 3)$ with structural factor $(1-\epsilon_i)\ (i = 1, 2, 3)$. The exhaust speed of each stage is c.

If the satellite requires a final speed v_T, show that the minimum total rocket mass, $M_o + P$ (where $M_o = \sum_{i=1}^{3} M_i$), that can attain this speed is given by

$$M_o + P = \frac{\epsilon_1 \epsilon_2 \epsilon_3 P \exp(v_T/c)}{[1 - b\exp(v_T/3c)]^3},$$

where $b^3 = \prod_{i=1}^{3}(1-\epsilon_i)$.

When $v_T = 9{\cdot}0 \times 10^3\,\mathrm{m\,s^{-1}}$ and $c = 3{\cdot}0 \times 10^3\,\mathrm{m\,s^{-1}}$, compare the values of M_o/P in the cases:

(i) $\epsilon_1 = 0{\cdot}9,\ \epsilon_2 = 0{\cdot}8,\ \epsilon_3 = 0{\cdot}7$;

(ii) $\epsilon_1 = \epsilon_2 = \epsilon_3 = 0{\cdot}8$.

Chapter 5

Fields of Force

We now consider the application of the laws of motion to some important physical problems. We confine our attention to forces of a gravitational or electromagnetic nature.

5.1 The Gravitational Field

5.1.1 Free Motion Under Gravity

Consider a particle P of mass m projected from an origin O with initial velocity $\mathbf{u}$ in a uniform gravitational field†. Choose axes Ox, Oy such that Oy is vertically upwards and $\mathbf{u}$ lies in the xy-plane (Fig. 5.1). The force on P is

$$m\mathbf{g} \;=\; -\,mg\mathbf{j}\;. \tag{5.1}$$

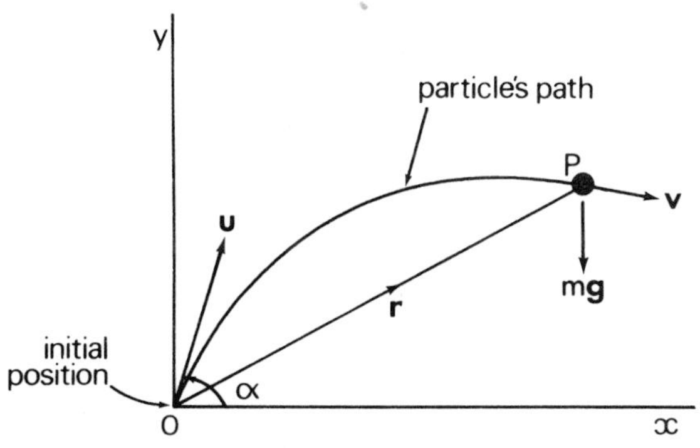

Fig. 5.1 Free motion under gravity

Assuming an inertial frame of reference, the subsequent motion of the particle must be confined to the xy-plane, since the initial velocity vector $\mathbf{u}$ and the force $m\mathbf{g}$ both lie in this plane. When dealing with large scale motions, a measurable deviation would occur due to the non-inertial nature of the frame of

† The uniform field approximation, valid for near-earth motions, is justified in Ch. 7.1.

reference fixed on the earth. This can be incorporated by using the theory of rotating axes which takes into account the earth's rotation. We shall here, however, assume the deviation to be negligible, so that the motion is essentially two-dimensional.

Let $\mathbf{r}(t)$ be the position vector of P at a time t after projection.

The equation of motion of P is

$$\ddot{\mathbf{r}} = \mathbf{g} \,, \tag{5.2}$$

where $\mathbf{r} = (x,\ y,\ 0)$. Integrating (5.2) with respect to t,

$$\dot{\mathbf{r}} = \mathbf{g}t + \mathbf{a} \,,$$

where $\mathbf{a}$ is a constant vector. Since $\dot{\mathbf{r}}(0) = \mathbf{u}$,

$$\dot{\mathbf{r}} = \mathbf{u} + \mathbf{g}t \,.$$

Integrating again,
$$\mathbf{r} = \mathbf{u}t + \tfrac{1}{2}\mathbf{g}t^2 + \mathbf{b} \,,$$

where $\mathbf{b}$ is a constant vector. Since $\mathbf{r}(0) = \mathbf{0}$, we see that $\mathbf{b} = \mathbf{0}$ and

$$\mathbf{r} = \mathbf{u}t + \tfrac{1}{2}\mathbf{g}t^2 \,. \tag{5.3}$$

Writing $\mathbf{u} = (u\cos\alpha, u\sin\alpha, 0)$ and $\mathbf{g} = (0, -g, 0)$, and taking components gives

$$\boxed{x = (u\cos\alpha)t,\ y = (u\sin\alpha)t - \tfrac{1}{2}gt^2,\ z = 0} \,. \tag{5.4}$$

From (5.4) we can deduce some elementary properties of *projectile* motion:

(a) THE TIME OF FLIGHT, t_f, is the time taken for the particle to return to its initial height. Putting $y = 0$ in (5.4), we obtain

$$t_f = (2u\sin\alpha)/g$$

(b) THE RANGE, R, is the horizontal distance covered during the time of flight. Thus evaluating x at $t = t_f$,

$$R = (u^2\sin 2\alpha)/g \,. \tag{5.5}$$

The maximum range for a given speed of projection u is obtained by maximizing R with respect to α. This gives $\sin 2\alpha = 1$, or $\alpha = \pi/4$, the maximum range being u^2/g.

(c) THE GREATEST HEIGHT reached, h, is obtained by putting $\dot{y} = 0$. This gives $t = (u \sin \alpha)/g = ½t_f$, and

$$h = (u^2 \sin^2 \alpha)/2g \ . \tag{5.6}$$

(d) THE EQUATION OF THE TRAJECTORY is determined by eliminating t from (5.4). This gives

$$\boxed{y = x \tan \alpha - (gx^2 \sec^2 \alpha)/2u^2} \ . \tag{5.7}$$

After some analysis it can be shown that (5.7) can be written as

$$(x - R/2)^2 = (R^2/4)\,(1 - y/h) \ .$$

This is the equation of a parabola, illustrated in Fig. 5.2.

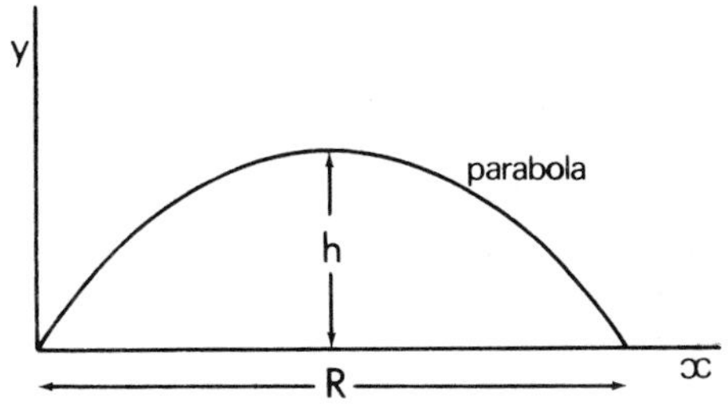

Fig. 5.2 Trajectory of particle

(e) THE PARABOLA OF SAFETY. Suppose we have a gun which can fire projectiles with a given speed u in any chosen direction. We require to determine the extent of the region of space which is accessible.

A point Q with coordinates $(X,Y,0)$ is accessible if and only if there exists a trajectory from O which passes through Q. The coordinates of Q must therefore satisfy (5.7) for some real value of α. This condition gives

$$Y = X \tan \alpha - (gX^2 \sec^2 \alpha)/2u^2 \ .$$

Writing $\sec^2 \alpha = 1 + \tan^2 \alpha$ and rearranging, we obtain a quadratic equation for $\tan \alpha$:

$$\tan^2 \alpha - \frac{2u^2}{gX} \tan \alpha + \left(\frac{2u^2 Y}{gX^2} + 1 \right) = 0 \tag{5.8}$$

Thus the point $Q(X, Y, 0)$ can be reached by projection (in the vertical plane containing OQ) at an angle α to the horizontal, provided (5.8) has at least one real root. The point is inaccessible if (5.8) has complex roots. The locus of points Q which can just be reached is found by examining the critical case of equal roots of (5.8). The condition for equal roots of a quadratic equation gives

$$\frac{u^4}{g^2 X^2} = \frac{2u^2 Y}{gX^2} + 1 ,$$

$$\text{i.e. } X^2 = \frac{2u^2}{g}\left(\frac{u^2}{2g} - Y\right) .$$

The accessible region in the xy-plane is thus bounded by the parabola

$$\boxed{x^2 = \frac{2u^2}{g}\left(\frac{u^2}{2g} - y\right)} \quad . \tag{5.9}$$

Each point Q_1 on this parabola can just be reached by projection from O at a unique angle α, determined by solving (5.8). Points Q_2, situated within the bounding parabola, are such that there are two possible angles of projection. Points Q_3, outside the parabola, cannot be reached at all, and (5.8) has no real solutions. The parabola (5.9) is called the *parabola of safety*. All trajectories from O must touch the parabola of safety at some point. For, substituting (5.4) in (5.9) and solving the quadratic equation for t gives $t_p = u/(g \sin \alpha)$, which is the time when the projectile meets the parabola of safety. We note that

$$t_p - \tfrac{1}{2} t_f = \frac{u}{g \sin \alpha}(1 - \sin^2 \alpha) = \frac{u \cos^2 \alpha}{g \sin \alpha} .$$

Since $0 \leqslant \alpha \leqslant \pi$, it follows that $t_p \geqslant \frac{1}{2} t_f$. Thus the trajectory touches the parabola of safety after it has attained its greatest height. The above results are depicted in Fig. 5.3.

The complete region of space accessible from O (with projection speed u) is obtained by rotating the parabola of safety about the y-axis. This generates the paraboloid of safety, which contains all trajectories for all possible directions of projection with given speed u from O.

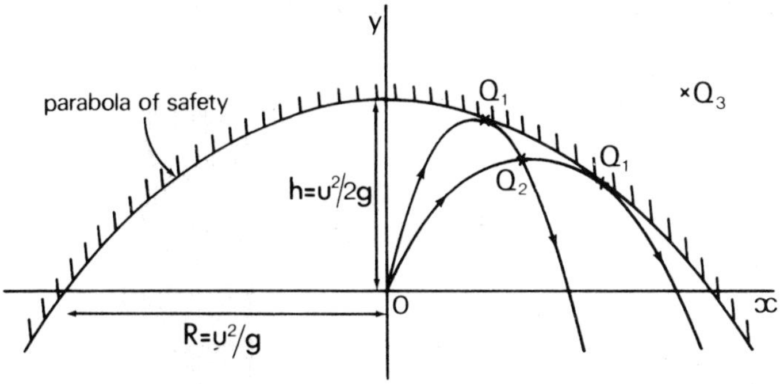

Fig. 5.3 The parabola of safety and typical trajectories

The theory above relates to the particle moving under a constant gravitational force only, whereas in practice air resistance is bound to have some influence on the motion. Also, if the distances travelled are appreciably large, the effect of the earth's rotation, the curved surface of the earth and the change in gravity with height (see Ch. 7.1) might all be significant factors. The most important feature neglected above is certainly air resistance and its effect on the results obtained above is next considered.

5.1.2 Resisted Projectile Motion: Resistance Proportional to Speed

Consider now the problem of a particle P projected in a medium which exerts a resistance $-mk\mathbf{v}$, where $\mathbf{v}$ is the particle's velocity and k is a positive constant. The equation of motion is

$$\frac{d\mathbf{v}}{dt} = \mathbf{g} - k\mathbf{v} \; . \tag{5.10}$$

Multiplying by the integrating factor e^{kt} and integrating gives

$$\mathbf{v} = (\mathbf{u} - \mathbf{g}/k)e^{-kt} + \mathbf{g}/k \; , \tag{5.11}$$

where we have used the initial condition $\mathbf{v}(0) = \mathbf{u}$. Integrating again, with $\mathbf{r}(0) = \mathbf{0}$, gives

$$\mathbf{r} = \frac{1}{k}(\mathbf{u} - \mathbf{g}/k)(1 - e^{-kt}) + \mathbf{g}t/k \; . \tag{5.12}$$

Taking components (with axes as defined above), we obtain

$$\boxed{\begin{aligned} x &= \frac{u\cos\alpha}{k}(1 - e^{-kt}) \\ y &= \frac{1}{k}\left(u\sin\alpha + \frac{g}{k}\right)(1 - e^{-kt}) - \frac{gt}{k} \\ z &= 0 \end{aligned}} \quad . \tag{5.13}$$

From (5.11) we see that, as $t \to \infty$,

$$\mathbf{v} \to \mathbf{g}/k = -g\mathbf{j}/k \ ;$$

i.e. the particle tends to fall vertically with constant speed g/k. This speed is called the *terminal speed* and represents the achievement of a balance between the opposing effects of gravity and the resistance. As $\mathbf{v} \to \mathbf{g}/k$, (5.10) shows that the acceleration tends to zero. Thus the effect of resistance is to produce a limiting value to the downward speed, which increases indefinitely in the absence of resistance.

Since the limiting velocity is vertically downwards, there must also be a limit to the horizontal distance which can be covered. From (5.13) we obtain

$$x \to (u\cos\alpha)/k \quad \text{as } t \to \infty \ .$$

The trajectory thus has an asymptote at $x = (u\cos\alpha)/k$. This is illustrated in Fig. 5.4.

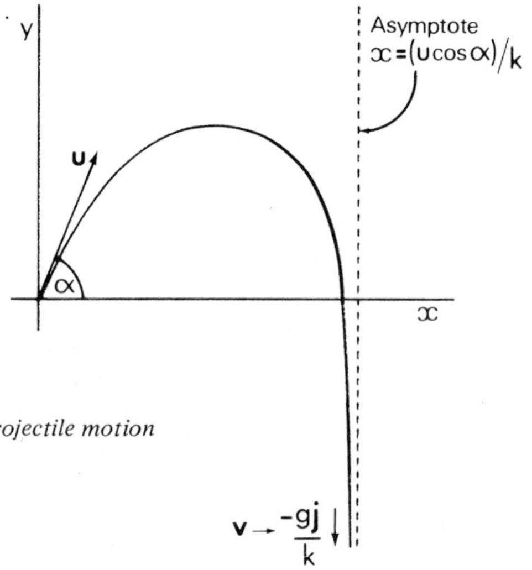

Fig. 5.4 Resisted projectile motion

Note: Whenever a problem is generalized by the inclusion of an effect previously neglected, it is a good idea to check that the original result is still obtained when the magnitude of the extra effect tends to zero in the revised results. In the above problem, we must consider the limit as $k \to 0$.

Since
$$e^{-kt} = 1 - kt + \frac{k^2t^2}{2} - \dots ,$$

so that
$$1 - e^{-kt} = kt - \frac{k^2t^2}{2} + O(k^3) ,$$

we have, from (5.13),

$$x = \frac{u \cos \alpha}{k}\left(kt - \frac{k^2t^2}{2} + O(k^3)\right) ,$$

$$y = \frac{1}{k}(u \sin \alpha + g/k)\left(kt - \frac{k^2t^2}{2} + O(k^3)\right) - \frac{gt}{k} .$$

Letting $k \to 0$ gives

$$x = (u \cos \alpha)t$$

$$y = (u \sin \alpha)t - \tfrac{1}{2}gt^2 ,$$

which is the result for an unresisted projectile.

5.1.3 RESISTED PROJECTILE MOTION: THE GENERAL PROBLEM

More generally, let us suppose the resistance per unit mass is some general function of velocity and position, $\mathbf{R} = \mathbf{R}(\mathbf{v}, \mathbf{r})$. Consider also the variation of the gravitational force with position, so that $\mathbf{g} = \mathbf{g}(\mathbf{r})$. The equation of motion is

$$\ddot{\mathbf{r}} = \mathbf{g}(\mathbf{r}) + \mathbf{R}(\mathbf{v},\mathbf{r}) . \qquad (5.14)$$

This situation may occur if we wish to investigate large scale motions in the atmosphere. We know that g decreases with height according to an inverse square law. As we shall see later (Ch. 7.1) this variation is small throughout the atmosphere and is usually neglected. The resistance however will depend on the density of the atmosphere, as well as on the particle's speed.

Fig. 5.5 illustrates the variation of density with height above the earth. H is a scale height, and n_a is the density at ground level.

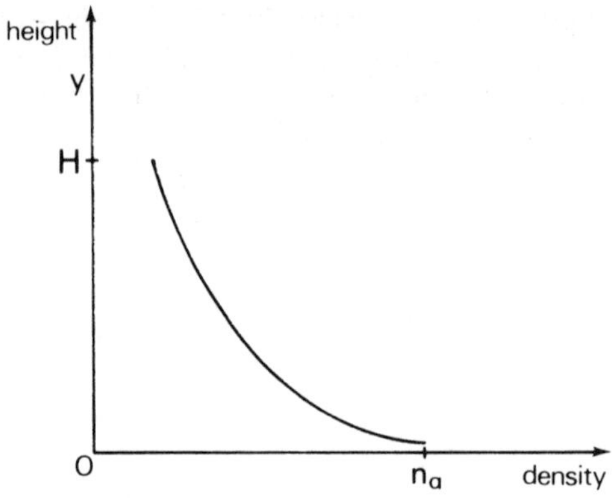

Fig. 5.5 Variation of density with height

To a good approximation, the density decreases exponentially with height, and we can take as a possible form for **R**,

$$\mathbf{R} = -\lambda v^2 n_a \exp(-y/H)\,\hat{\mathbf{v}}\ , \tag{5.15}$$

where λ is a constant, and we have assumed dependence on the square of the speed and on air density. Writing the unit velocity vector $\hat{\mathbf{v}} = \mathbf{v}/v = \dot{\mathbf{r}}/v$, the equation of motion (5.14), assuming constant gravity, becomes

$$\ddot{\mathbf{r}} = \mathbf{g} - \lambda v n_a \exp(-y/H)\,\dot{\mathbf{r}}\,.$$

Taking components,

$$\left.\begin{aligned} \ddot{x} &= -\lambda n_a v \dot{x} \exp(-y/H) \\ \ddot{y} &= -g - \lambda n_a v \dot{y} \exp(-y/H) \end{aligned}\right\} \tag{5.16}$$

where $v^2 = \dot{x}^2 + \dot{y}^2$.

In this way the problem has been reduced to solving a set of two simultaneous second order differential equations. We are unable to proceed further by simple analytical methods, but given appropriate initial conditions we can use numerical techniques.

5.2* Electric and Magnetic Fields

In this section we further illustrate the application of Newton's laws by considering some problems involving forces of an electrical origin. In this book we will only state the nature of the forces produced by electric and magnetic fields. We refer readers to *Electromagnetic Theory* by V. C. A. Ferraro (Athlone Press, 1954) for a complete description of electric and magnetic fields.

These vector fields can be regarded as mental constructions, employed to express the electrical properties of space. The existence of an electric field in a region is observable when a particle with a resultant positive or negative electrical charge is present. A magnetic field may be detected only when the charged particle is in motion. These facts are expressed by the *Lorentz force,* which is described below.

5.2.1 FORCE ON A CHARGED PARTICLE

We assume, without further discussion, that we may define two vector functions, **E** and **B**, where **E** is the electric field vector and **B** the magnetic induction vector in some region of space. Consider a charged particle Q with charge q, moving with velocity **v** under the influence of the above fields. The quantity q, which may be positive or negative, is a measure of the electrical state of the particle. By convention the charge on an *electron* is defined to be negative and of magnitude e. Thus we write $q = -e$ for an *electron,* $q = +e$ for a *proton* where e is a constant. The total force acting on Q is the *Lorentz force,* **F**, given by

$$\boxed{\mathbf{F} = q[\mathbf{E} + (\mathbf{v} \times \mathbf{B})]} \quad . \qquad (5.17)$$

Here **F** is measured in Newtons, provided the other units are as follows:

q	coulombs (denoted by C)
E	NC^{-1}
B	$NC^{-1}\ m^{-1}\ s$
v	$m\,s^{-1}$.

We have introduced a new unit, the coulomb. In terms of basic S.I. units

$$1\,C = 1\,A\,s$$

and the charge on a proton is given by $e = 1.6 \times 10^{-19}$ C. Equation (5.17) expresses the fact that a charged particle *at rest* does *not* experience a magnetic force, whereas a *moving* charged particle is affected by both a magnetic force $q(\mathbf{v} \times \mathbf{B})$ and an electrostatic force $q\mathbf{E}$. Note that the magnetic force acts in a direction perpendicular to both the magnetic induction vector and the velocity of the particle, whereas the electrostatic force is parallel to **E**. Initially we will confine our attention to electric and magnetic fields which are both uniform (in space) and constant (in time).

5.2.2 THE MOTION OF A CHARGED PARTICLE IN A UNIFORM ELECTRIC FIELD

Consider a particle of mass m and charge q, initially at rest in a uniform, constant electric field **E**. Using (5.17), the equation of motion is

$$\boxed{\ddot{\mathbf{r}} = \frac{q}{m}\mathbf{E}} \quad . \tag{5.18}$$

Since **E**, q and m are all constants, this equation is analogous to the equation of motion of a particle in a constant gravitational field [cf (5.2)].

The trajectory of a charged particle in a constant electric field is thus equivalent to that of a particle moving freely under gravity. If released from rest, the particle will move with acceleration $|qE/m|$ in a direction parallel or anti-parallel to the electric field, depending on the sign of the charge q. A particle projected at an angle to the electric field will move in a parabolic path, as illustrated in Fig. 5.6. We note from (5.18) that the acceleration of a charged particle in an electric field depends on its mass. An electron is therefore accelerated more rapidly than a proton, which has greater mass.

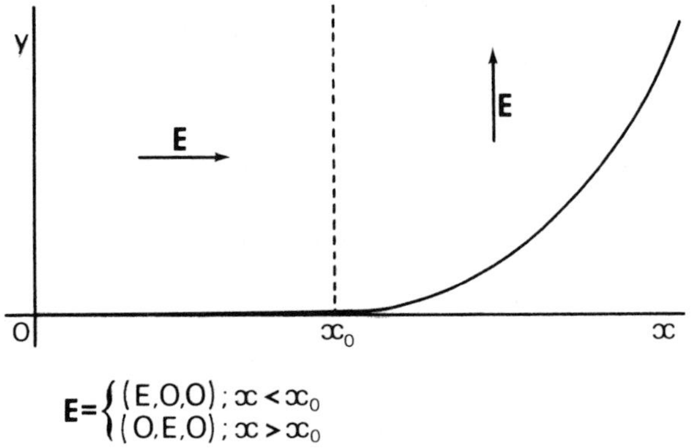

Fig. 5.6 Trajectory of a proton released from rest at O *in the electric field* **E**

5.2.3 THE NEGLECT OF GRAVITATIONAL FORCES

In the previous section we considered the charged particle to be influenced solely by the electric field. In any terrestrial application, the particle is also subject to the usual gravitational force. The above analysis will only be valid if we are justified in neglecting gravity. In order to reassure ourselves on this point, consider the two forces (electrical and gravitational) acting on a *proton* in an electric field **E** at the earth's surface. A proton has mass $m = 1.67 \times 10^{-27}$ kg and charge $e = 1.60 \times 10^{-19}$ C. The force due to the electric field has magnitude F_e

where

$$F_e = eE = (1.60 \times 10^{-19}E)\ \text{N},$$

whilst the gravitational force F_g is

$$F_g = mg = (1.67 \times 10^{-27}) \times 9.8\ \text{N}$$

$$= 1.63 \times 10^{-26}\ \text{N}.$$

Thus we see that $$F_e/F_g \approx 10^7 E\ ,$$

where E is measured in NC^{-1}. For instance, in an electric field of strength $10^{-4}\,\text{N C}^{-1}$, the force F_e is a factor 10^3 greater than the gravitational force F_g. This represents a very small electric field, so that for most practical applications the electric field dominates the gravitational field. The latter may therefore be neglected as in Ch. 5.2.2.

5.2.4 A Uniform Magnetic Field

Consider next a particle with mass m and charge q moving with velocity $\mathbf{v}$ in a constant uniform magnetic induction field $\mathbf{B}$. Choose axes $Oxyz$ such that Oz is parallel to the magnetic induction vector, and let v_x, v_y and v_z denote the components of the velocity vector. Thus we have

$$\mathbf{B} = (0, 0, B)\ ,\ \mathbf{v} = (v_x, v_y, v_z)\ .$$

From (5.17), the equation of motion is

$$\boxed{m\frac{d\mathbf{v}}{dt} = q(\mathbf{v} \times \mathbf{B})}\ . \tag{5.19}$$

Taking components we have

$$\dot{v}_x = \omega v_y \tag{5.20a}$$

$$\dot{v}_y = -\omega v_x \tag{5.20b}$$

$$\dot{v}_z = 0 \tag{5.20c}$$

where ω is the constant qB/m.

Integrating (5.20c) gives $$v_z = V \tag{5.21}$$

where V is a constant. Differentiating (5.20a) and (5.20b) with respect to time gives

$$\ddot{v}_x = \omega\dot{v}_y,\ \ddot{v}_y = -\omega\dot{v}_x\ .$$

Substituting for $\dot{v}_x$ and $\dot{v}_y$ we obtain

$$\ddot{v}_x = -\omega^2 v_x, \quad \ddot{v}_y = -\omega^2 v_y \ . \tag{5.22}$$

Integrating (5.22) for v_x gives

$$v_x = A\cos(\omega t) + A'\sin(\omega t) \ , \tag{5.23a}$$

where A and A' are arbitrary constants, and from (5.20a),

$$v_y = -A\sin(\omega t) + A'\cos(\omega t) \ . \tag{5.23b}$$

At this stage it is helpful to choose the Ox and Oy axes (not previously defined) such that the component of the initial velocity vector in the xy-plane is directed along the axis Oy; i.e.

$$v(0) = (0, U, V) \ ,$$

where U is a constant. The constants in (5.23) are readily determined and incorporating (5.21), we then have

$$v_x = U\sin(\omega t), \quad v_y = U\cos(\omega t), \quad v_z = V \ . \tag{5.24}$$

Finally, integrating (5.24) with the initial conditions

$$x(0) = x_0, y(0) = y_0, \ z(0) = z_0$$

gives

$$\boxed{\begin{aligned} x &= x_0 + (U/\omega)[1 - \cos(\omega t)] \\ y &= y_0 + (U/\omega)\sin(\omega t) \\ z &= z_0 + Vt \end{aligned}} \ . \tag{5.25}$$

Eliminating t from the first two of equations (5.25) gives the projection of the path on the xy-plane as

$$(x - x_0 - U/\omega)^2 + (y - y_0)^2 = U^2/\omega^2 \tag{5.26}$$

which is the equation of a circle. The particle therefore moves on the surface of the right circular cylinder whose cross-section is the circle represented by (5.26). The particle rotates uniformly about the axis of the cylinder ($x = x_0 + U/\omega$, $y = y_0$), at the same time moving parallel to this axis with uniform speed V. The complete trajectory is thus a right circular helix, as shown in Fig. 5.7.

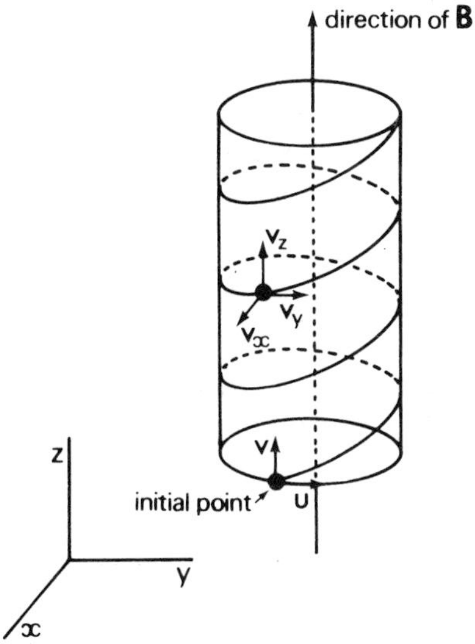

Fig. 5.7 The path of an electron moving in a constant magnetic induction field **B**

The particle *gyrates* in a spiralling motion about the direction of the magnetic induction vector. We call lines in this direction, i.e. the Oz direction, magnetic *field lines.* Thus the particle gyrates about the magnetic field lines. The radius of the cylinder on which the particle moves is called the *gyroradius,* r_ω. From (5.26) we have

$$r_\omega = U/\omega = Um/qB \; . \tag{5.27}$$

For a given particle, the gyroradius depends on the strength of the magnetic induction B and on the initial component of velocity perpendicular to **B**. The larger the magnetic field, the smaller the circles become. The frequency, $\omega = qB/m$, of the circular motion is called the *gyrofrequency.*

Before continuing with our main theory, we digress a little and consider the energy of the particle moving in a magnetic field. Since $v^2 = \mathbf{v}.\mathbf{v}$, we see that

$$\frac{d}{dt}(v^2) = 2\mathbf{v}.\frac{d\mathbf{v}}{dt} \; ,$$

and taking scalar product of (5.19) with $\mathbf{v}$ gives

$$\frac{d}{dt}(v^2) = \frac{2q}{m}\mathbf{v}.(\mathbf{v} \times \mathbf{B})$$

$$= 0 \, ,$$

since $\mathbf{v}$ is perpendicular to $\mathbf{v} \times \mathbf{B}$. Hence

$$\boxed{v^2 = \text{constant}} \,. \tag{5.28}$$

This is a conservation law which expresses the fact that, for motion in a magnetic field, kinetic energy ($\frac{1}{2} mv^2$) is conserved. As we shall see in the next chapter, this property is due to the fact that the force $\mathbf{v} \times \mathbf{B}$ is perpendicular to the direction of motion at any instant, and so does no 'work' in moving the particle. Equivalently we may say that the particle possesses no potential energy due to its position in a magnetic field.

We have seen, equation (5.21), that v_z is constant, and since $v^2 = v_x^2 + v_y^2 + v_z^2$ equation (5.28) implies that $(v_x^2 + v_y^2)$ is constant.

Thus

$$\boxed{v_x^2 + v_y^2 = U^2, \; v_z = V} \tag{5.29}$$

where U and V are constants representing the components of velocity perpendicular and parallel to the magnetic induction vector.

5.2.5 THE HALL DRIFT

We now consider the more general case in which both electric and magnetic fields are applied. Suppose a charged particle of mass m and charge q moves in a region pervaded by a constant electric field $\mathbf{E}$ and a constant magnetic induction field $\mathbf{B}$. The equation of motion is now

$$\frac{d\mathbf{v}}{dt} = \frac{q}{m}[\mathbf{E} + (\mathbf{v} \times \mathbf{B})] \,. \tag{5.30}$$

We again choose axes so that $\mathbf{B} = (0, 0, B)$ and the initial velocity vector is

$$\mathbf{v}(0) = (0, U, V) \,. \tag{5.31}$$

The electric field, referred to these axes is

$$\mathbf{E} = (E_x, E_y, E_z) \,.$$

Denoting the velocity vector at time t by $\mathbf{v} = (v_x, v_y, v_z)$, (5.30) can be written in component form

$$\dot{v}_x = \frac{q}{m}E_x + \omega v_y \tag{5.32a}$$

$$\dot{v}_y = \frac{q}{m}E_y - \omega v_x \tag{5.32b}$$

$$\dot{v}_z = \frac{q}{m}E_z \ , \tag{5.32c}$$

where $\omega = qB/m$.

Integrating (5.32c) and using (5.31), we have

$$v_z = V + qE_z t/m \ , \tag{5.33}$$

which represents motion in the z-direction with uniform acceleration qE_z/m. Differentiating (5.32a) and eliminating $\dot{v}_y$ from (5.32b) yields

$$\ddot{v}_x + \omega^2 v_x = \omega q E_y/m \ ,$$

which has solution

$$v_x = A\cos(\omega t) + A'\sin(\omega t) + qE_y/\omega m \ .$$

Equation (5.32a) now gives

$$v_y = -A\sin(\omega t) + A'\cos(\omega t) - qE_x/\omega m \ .$$

The constants, A and A', are determined from (5.31), giving

$$\left.\begin{aligned} v_x &= -\frac{qE_y}{\omega m}\cos(\omega t) + \left(U + \frac{qE_x}{\omega m}\right)\sin(\omega t) + \frac{qE_y}{\omega m} \\ v_y &= \left(U + \frac{qE_x}{\omega m}\right)\cos(\omega t) + \frac{qE_y}{\omega m}\sin(\omega t) - \frac{qE_x}{\omega m} \end{aligned}\right\} . \tag{5.34}$$

Substituting $q/\omega m = B$ and integrating we obtain

$$\left.\begin{aligned} x &= -\frac{E_y}{\omega B}\sin(\omega t) - \frac{1}{\omega}\left(U + \frac{E_x}{B}\right)\cos(\omega t) + \frac{E_y t}{B} + C \\ y &= \frac{1}{\omega}\left(U + \frac{E_x}{B}\right)\sin(\omega t) - \frac{E_y}{\omega B}\cos(\omega t) - \frac{E_x t}{B} + D \end{aligned}\right\} \tag{5.35}$$

where C and D are constants. From (5.35) we see that

$$\left(x - C - \frac{E_y t}{B}\right)^2 + \left(y - D + \frac{E_x t}{B}\right)^2 = R^2$$

where $$R^2 = [E_y{}^2 + (BU + E_x)^2]/\omega^2 B^2 \ .$$

The projection of the motion on the xy-plane is thus a 'circle' whose centre $(C + E_y t/B,\ D - E_x t/B)$ moves with velocity $\mathbf{v}_d$, where

$$\mathbf{v}_d = \frac{1}{B}(E_y,\ -E_x,\ 0)$$

i.e.
$$\boxed{\mathbf{v}_d = \frac{1}{B^2}(\mathbf{E} \times \mathbf{B})} \tag{5.36}$$

This velocity is called the *Hall drift* or the '**E** × **B**' drift. The addition of a constant electric field thus has two effects on the particle motion:

(1) The particle 'drifts' across the magnetic field with velocity $\mathbf{v}_d = (\mathbf{E} \times \mathbf{B})/B^2$ perpendicular to **B**, in addition to spiralling around the field lines. The direction of this drift is independent of the sign of the charge on the particle. The radius of gyration becomes

$$R = \frac{1}{\omega B}[E_y{}^2 + (BU + E_x)^2]^{\frac{1}{2}} \ ,$$

whereas the gyrofrequency ω is unaltered.

(2) A uniform acceleration qE_z/m in the direction of **B** is superimposed on the motion. This acceleration along the magnetic field (in a direction depending on the sign of the charge) is caused by the component of **E** parallel to **B**, and may be written as $(q/mB)(\mathbf{E}.\mathbf{B})\mathbf{B}$.

STEADY STATE

In many applications, it is assumed that there exists a state in which there is no particle acceleration along the magnetic field lines. In this context such a state is usually referred to as a 'steady-state'. From (5.30), taking the scalar product with **B**, we see that for a steady state, we require

$$\mathbf{E}.\mathbf{B} = 0 \ . \tag{5.37}$$

This means that **E** and **B** must be perpendicular vectors. The particle then moves in a helical path relative to axes which are drifting with the Hall velocity $(\mathbf{E} \times \mathbf{B})/B^2$. Thus a particle moving in crossed (i.e. perpendicular) electric and magnetic fields drifts in a direction perpendicular to *both* fields, at the same time spiralling around the magnetic field lines. If it has an initial velocity component V parallel to the magnetic field, it will also be moving with uniform speed V in this direction. This situation is depicted in Fig. 5.8.

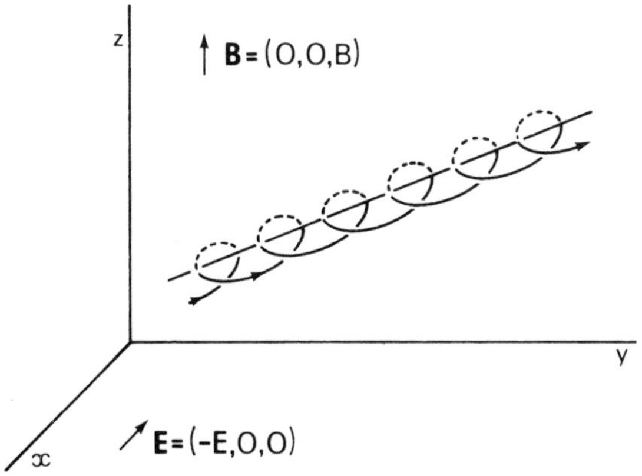

Fig. 5.8 Path of an electron in crossed electric and magnetic fields

5.2.6 STEADY STATE IN A UNIFORM FIELD OF FORCE: APPLICATION TO THE IONOSPHERE

Suppose that in the problem discussed in Ch. 5.2.5, we replace the electric field by a constant uniform field of force. The problem of a charged particle moving in a uniform magnetic induction field **B**, and acted upon by an additional constant force **F** is equivalent to the previous problem with the electric field $\mathbf{E} = \mathbf{F}/q$.

For 'steady state' motion we require

$$\mathbf{F}.\mathbf{B} = 0 \,, \tag{5.38}$$

and from (5.36) we see that the drift velocity, perpendicular to the magnetic field lines, is given by

$$(\mathbf{F} \times \mathbf{B})/qB^2 \,. \tag{5.39}$$

This is called a *force-induced drift.* In this case the direction of the drift depends on the sign of the charge, so that electrons and protons will drift in opposite directions. We will apply these results to the motion of charges in the ionosphere.

The *ionosphere* is a region in the upper atmosphere of the earth in which ionization processes are important. Ionization (the production of electrons and ions†) is caused by solar radiations incident in this region, which extends roughly from a height of about 50 km up to about 300 km. The ionosphere is subdivided into three main regions, according to the nature and extent of ionization, which depends on the constitution of the atmosphere and on the distribution of solar radiation. These regions are called the D, E and F regions, in ascending order of height. The F-region is sub-divided into two further regions, F_1 and F_2. These regions are illustrated in Fig. 5.9 which shows a sketch of electron density against height above the earth's surface. Note that the E, F_1 and F_2 regions all correspond to local maxima of the electron density.

† An ion is an atom or molecule which has lost or gained one or more of its outer electrons.

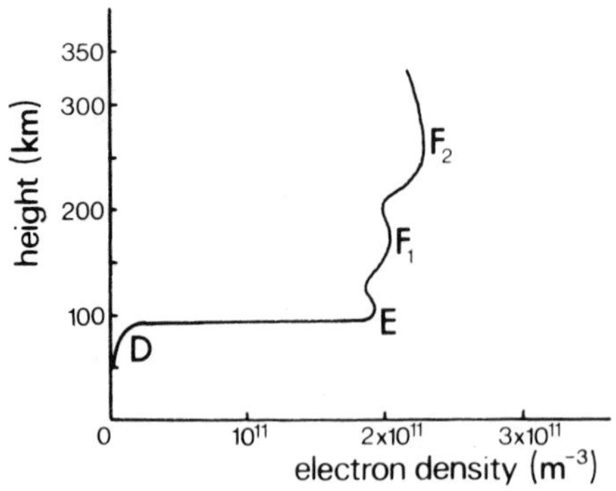

Fig. 5.9 Regions of the ionosphere

At a height of about 300 km, in the F_2 region, there are about 10^{12} ions and electrons per m^3 compared with about 10^{15} m^{-3} neutral atoms. Provided we neglect the effect of all collisions between ions and neutral atoms, the theory developed in Ch. 5.2.5, where we treated the case of an *isolated* charged particle, applies to the particles in the F_2 region.

Consider an ion or electron just formed in the F_2 region by absorption of solar radiation. The particle will experience forces due to the earth's magnetic induction field **B** and due to the gravitational attraction of the earth. In an *idealized model,* let us suppose that the direction of the magnetic field is uniform so that the magnetic field lines in the ionosphere are straight. The model is illustrated in Fig. 5.10, where **g** is the gravitational acceleration.

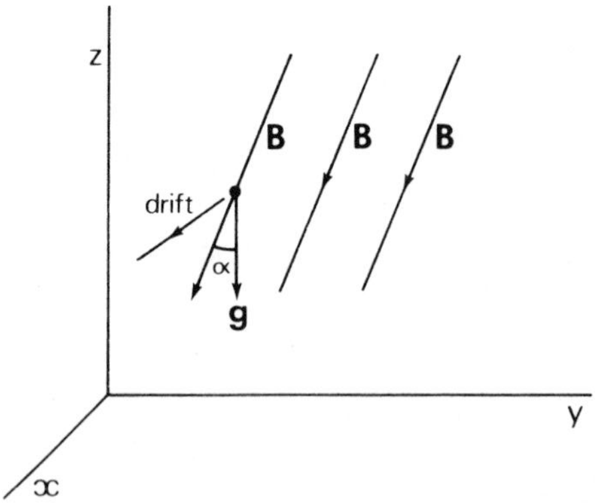

Fig. 5.10 Idealized model for the F_2 region of the ionosphere

Axes $Oxyz$ are defined so that

$$\mathbf{B} = (0, \ -B \sin \alpha, \ -B \cos \alpha) \ , \tag{5.40}$$

$$\mathbf{F} = m\mathbf{g} = (0, 0, \ -mg) \ . \tag{5.41}$$

In the absence of the magnetic field the particle would fall vertically. From Ch. 5.2.5 it follows that the combined effect of the magnetic field $\mathbf{B}$ and the constant gravitational field is:

(1) to cause spiralling around the field lines together with a drift with velocity

$$\begin{aligned} \mathbf{v}_d &= (\mathbf{F} \times \mathbf{B})/qB^2 \\ &= [\ -(mg \sin \alpha)/qB, \ 0, 0] \ . \end{aligned} \tag{5.42}$$

This is a horizontal force induced drift, the direction of which depends on whether the particle is an ion or an electron.

(2) to produce an acceleration along the field lines of magnitude

$$(\mathbf{F}.\mathbf{B})/mB = g \cos \alpha \ . \tag{5.43}$$

A steady state is achieved if the field lines are horizontal (i.e. $\mathbf{F}.\mathbf{B} = 0$). In this case we have the rather surprising result that the particle does not fall to the ground, but spirals around the magnetic field lines, drifting horizontally with speed mg/qB. Observations in fact show that the horizontal drift is small, indicating that the predominant factor in most ionospheric motions is the acceleration of particles along the field lines.

5.2.7 THE MAGNETOSPHERE

Although in this book we cannot pursue further the fascinating study of charged particle motion, we conclude this section by mentioning the scope of application of this theory. We have already met one region of interest above the earth's surface, namely the ionosphere. In Fig. 5.11 we illustrate all the important regions in the vicinity of the earth. The dashed lines in the diagram are magnetic field lines. The magnetic field of the earth is approximately a *dipole,* but due to solar particles emitted continuously from the sun (called the solar wind) the earth's magnetic field is confined inside a region called the magnetosphere. There is still much doubt as to whether the magnetic field is completely confined, and observations from satellite experiments are continuing to provide plenty of information. There is no doubt though that the earth's magnetic field is compressed on the day-time side and inflated on the night-time side of the magnetosphere, as illustrated in Fig. 5.11.

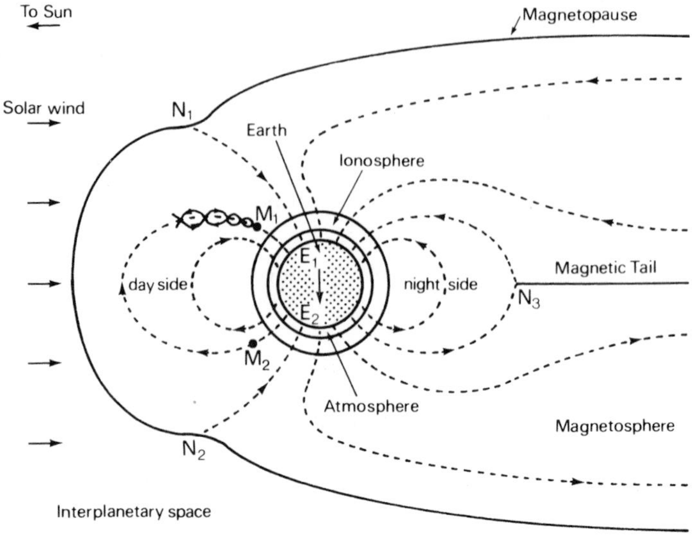

Fig. 5.11 Topology of the magnetosphere: Meridian cross-section

The points N_1, N_2 and N_3 are called *neutral points,* for at these points $\mathbf{B} = \mathbf{0}$. The field lines through these points divide the magnetosphere into distinct regions. In the shaded region the field lines start and end at the earth's dipole, and are called *closed* field lines. In these regions large numbers of charged particles are observed. As illustrated in Ch. 5.2.4 the particles spiral around the magnetic field lines. Particles are also reflected at *mirror* points, M_1 and M_2 in Fig. 5.11, and consequently are 'trapped' in these regions. The regions of trapped particles are called the Van Allen radiation belts. In the unshaded region the field lines are *open* and form the earth's *magnetic tail* which stretches back into interplanetary space possibly for many thousands of earth radii.

The field lines separating the open and closed regions intersect the earth in two ovals (remember the configuration in Fig. 5.11 is really three dimensional), one around the North pole and the other around the South pole. These ovals are closely related to the observed positions of the *aurora*. It is conjectured that high speed electrons and protons enter the magnetosphere from the solar wind at N_1 and N_2, where the magnetic field is zero, and spiral around the field lines $N_1 E_1$ and $N_2 E_2$, causing the aurora as they penetrate the upper atmosphere.

In this section we have presented some of the observed phenomena and mentioned briefly some of the theories associated with the interaction of the solar wind with the geomagnetic field. In order to investigate the theory we must set up a mathematical model, solve the equations and compare the theoretical results with the many observations that have now been made by satellites, rockets and balloons. This is an area of much current research.

Worked Examples

EXAMPLE 5.1:

A ball is thrown vertically downwards from the top of a tall building. Assuming a model with constant gravity and air resistance proportional to its speed, show that if the building is sufficiently tall, the ball's velocity on hitting the ground is approximately independent of its initial speed.

SOLUTION:

The equation of motion is $m\ddot{\mathbf{r}} = m\mathbf{g} - mk\mathbf{v}$

where $\mathbf{r} = y\mathbf{j}$, $\mathbf{v} = \dot{y}\mathbf{j}$, $\mathbf{g} = -g\mathbf{j}$, $\mathbf{j}$ being a unit vector in the upward vertical direction. Thus

$$\ddot{y} = -g - k\dot{y} ,$$

and writing $v = \dot{y}$, $\dot{v} + kv = -g .$

This has solution, $v = Ae^{-kt} - g/k$

where the constant of integration, A, is determined from the initial conditions.

Suppose the ball is given a downwards velocity u, so that at $t = 0$, $v = -u$. Then

$$-u = A - g/k$$

and $v = (g/k - u)e^{-kt} - g/k .$

If the building is sufficiently tall, the time of flight will be large, and we see that v will approach $(-g/k)$ for large t. Thus the particle will tend to move downwards with speed approaching g/k, which is independent of the initial speed u.

A second approach would be to write the equation of motion in the form

$$\frac{dv}{dt} = \frac{dv}{dy}\frac{dy}{dt} = v\frac{dv}{dy} = -(g + kv)$$

where we are now solving for v as a function of y, and not t. Thus

$$\int \frac{vdv}{(v + g/k)} = -\int kdy$$

i.e. $$\int \frac{(v + g/k - g/k)}{(v + g/k)}dv = -ky + A$$

giving $$v - (g/k)\log(v + g/k) = -ky + A\ .$$

Initially, $v = -u$, and $y = h$, say, so that

$$A = -u - (g/k)\log(-u + g/k) + kh\ .$$

The velocity when $y = 0$ is then determined from

$$v - (g/k)\log(v + g/k) = -u - (g/k)\log(-u + g/k) + kh$$

where h is a large constant. Thus the right hand side is large provided kh is large, and to make the left hand side large and positive, it is clear that $v + g/k \approx 0$ i.e. $v \to (-g/k)$, as found above.

EXAMPLE 5.2:

A gun is situated on the edge of a vertical cliff of height h. A ship is at a horizontal distance R from the foot of the cliff. Show that the ship is in range of the gun if

$$R \leqslant (V^4/g^2 + 2V^2h/g)^{1/2}$$

where V is the initial speed of the gun's shell.

If shells are fired from the ship with initial speed U, show that the cliff gun is in range of the ship if

$$R \leqslant (U^4/g^2 - 2U^2h/g)^{1/2}\ .$$

SOLUTION:

The situation is illustrated in Fig. 5.12, where Cartesian axes Oxy are chosen with origin at the foot of the cliff.

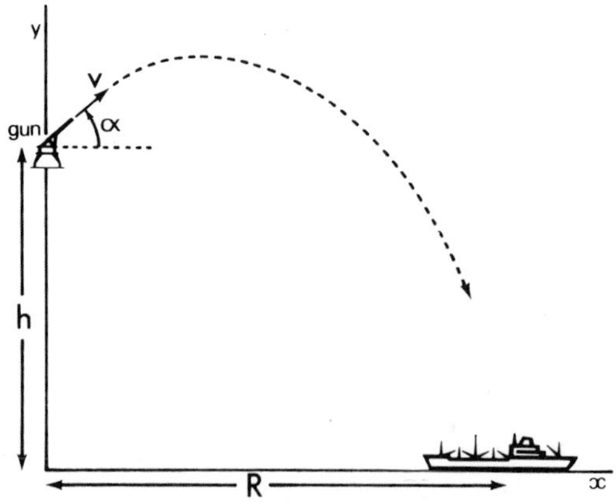

Fig. 5.12 Illustration of example 5.2

The shell's equation of motion is $\ddot{\mathbf{r}} = -g\mathbf{j}$

where $r = x\mathbf{i} + y\mathbf{j}$. Thus $\dot{\mathbf{r}} = -gt\mathbf{j} + \mathbf{a}$

where the arbitrary constant $\mathbf{a}$ is determined from the initial condition $\dot{\mathbf{r}}(0) = \mathbf{V} = \mathbf{a}$. Thus, integrating again

$$\mathbf{r} = -\tfrac{1}{2}gt^2\mathbf{j} + t\mathbf{V} + \mathbf{b} ,$$

where the arbitrary constant $\mathbf{b}$ is determined from the initial condition

$$\mathbf{r}(0) = h\mathbf{j} = \mathbf{b} .$$

Hence

$$\mathbf{r} = -\tfrac{1}{2}gt^2\mathbf{j} + t\mathbf{V} + h\mathbf{j} ,$$

and taking components

$$x = (V\cos\alpha)t, \; y = -\tfrac{1}{2}gt^2 + (V\sin\alpha)t + h .$$

The trajectory equation is then given by

$$y = x\tan\alpha - \frac{gx^2}{(2V^2\cos^2\alpha)} + h .$$

The ship is in range of the gun if the point given by $x = R$, $y = 0$ lies on the trajectory for some real value of α. Thus we require a real solution for $\tan\alpha$, where

$$0 = R\tan\alpha - \frac{gR^2}{2V^2}(1 + \tan^2\alpha) + h$$

or

$$\frac{gR^2}{2V^2}\tan^2\alpha - R\tan\alpha + \frac{gR^2}{2V^2} - h = 0 .$$

Solving this quadratic for $\tan\alpha$, real values will be obtained if

$$R^2 \geqslant \frac{2gR^2}{V^2}\left(\frac{gR^2}{2V^2} - h\right)$$

i.e.

$$\frac{V^2}{2g} \geqslant \frac{gR^2}{2V^2} - h$$

or

$$R^2 \leqslant (2V^2/g)(h + V^2/2g) .$$

Thus the ship is in range if

$$R \leqslant (V^4/g^2 + 2V^2h/g)^{1/2} .$$

The trajectory equation for shells fired from the ship with velocity **U** will be given by

$$\mathbf{r} = -\tfrac{1}{2}gt^2\mathbf{j} + t\mathbf{U} + \mathbf{b} ,$$

where **b** is now determined from the initial condition

$$\mathbf{r}(0) = R\mathbf{i} = \mathbf{b} .$$

Thus
$$\mathbf{r} = -\tfrac{1}{2}gt^2\mathbf{j} + t\mathbf{U} + R\mathbf{i} ,$$

and taking components

$$x = (U\cos\beta)t + R, \; y = -\tfrac{1}{2}gt^2 + (U\sin\beta)t .$$

The trajectory equation is then given by

$$y = -\frac{g(x - R)^2}{2U^2\cos^2\beta} + (x - R)\tan\beta .$$

The cliff gun is in range of the ship's gun if this equation has a real solution for the initial projection angle, β, when $y = h$ and $x = 0$. Thus

$$h = -R\tan\beta - \frac{gR^2}{2U^2}(1 + \tan^2\beta)$$

or
$$\frac{gR^2}{2U^2}\tan^2\beta + R\tan\beta + h + \frac{gR^2}{2U^2} = 0 .$$

This has real solutions for $\tan\beta$ if

$$R^2 \geqslant \frac{2gR^2}{U^2}\left(\frac{gR^2}{2U^2} + h\right)$$

which simplifies to

$$R \leqslant (U^4/g^2 - 2U^2h/g)^{1/2} .$$

Note that if the guns are both firing with the same speed, the ship will be in range of the cliff gun before the cliff gun is in range of the ship's gun (unless $h = 0$).

EXAMPLE 5.3:

A charged particle of mass m and charge q moves under the influence of a uniform constant electric field **E**. If its initial position and velocity are **a** and **u** respectively, show that at any subsequent time t, its position vector is given by

$$\mathbf{r} = \frac{qt^2}{2m}\mathbf{E} + t\mathbf{u} + \mathbf{a} .$$

SOLUTION:

From (5.18) the particle's equation of motion is

$$\ddot{\mathbf{r}} = \frac{q}{m}\mathbf{E}$$

whence, integrating,

$$\dot{\mathbf{r}} = \frac{qt}{m}\mathbf{E} + \mathbf{c} .$$

The constant vector **c** is determined from the initial condition, $\dot{\mathbf{r}}(0) = \mathbf{u}$, so that $\mathbf{u} = \mathbf{c}$. Hence

$$\dot{\mathbf{r}} = \frac{qt}{m}\mathbf{E} + \mathbf{u} ,$$

and, integrating again,

$$\mathbf{r} = \frac{qt^2}{2m}\mathbf{E} + t\mathbf{u} + \mathbf{d} .$$

The constant vector **d** is determined from the initial condition $\mathbf{r}(0) = \mathbf{a}$, so that $\mathbf{a} = \mathbf{d}$, and

$$\mathbf{r} = \frac{qt^2}{2m}\mathbf{E} + t\mathbf{u} + \mathbf{a} .$$

EXAMPLE 5.4

A beam of electrons passes through a small aperture into a uniform magnetic field of magnitude B. The electrons have speed u at the aperture and are directed almost parallel to the magnetic field. Neglecting interactions between the electrons, show that the beam will focus at a distance approximately $2\pi mu/eB$ from the aperture, where m and $(-e)$ denote electron mass and charge respectively.

SOLUTION:

Choosing axes $Oxyz$ with origin at the aperture and such that the magnetic field is directed parallel to Oz we have

$$\mathbf{B} = (0, 0, B) , \quad \mathbf{v} = (v_x, v_y, v_z) .$$

The equation of motion for a single electron (5.19) is

$$m\frac{d\mathbf{v}}{dt} = -e(\mathbf{v} \times \mathbf{B}) .$$

Thus, taking components we obtain

$$\dot{v}_x = -\omega v_y, \quad \dot{v}_y = \omega v_x, \quad \dot{v}_z = 0 ,$$

where $\omega = eB/m$.

Integrating twice, the third component gives

$$z = Wt$$

where we have applied the initial conditions

$$\mathbf{r}(0) = \mathbf{0}, \quad \mathbf{v}(0) = (U, V, W) .$$

Note that $U^2 + V^2 + W^2 = u^2$, the given initial speed. Differentiating the first component of the equation of motion gives

$$\ddot{v}_x = -\omega \dot{v}_y = -\omega^2 v_x ,$$

which has solution

$$v_x = \alpha \cos \omega t + \beta \sin \omega t$$

where α and β are arbitrary constants.

From the equation $\dot{v}_x = -\omega v_y$, we obtain

$$v_y = \alpha \sin \omega t - \beta \cos \omega t ,$$

and applying the initial condition $\mathbf{v}(0) = (U, V, W)$ we have $\alpha = U$ and $\beta = -V$. This gives

$$v_x = U \cos \omega t - V \sin \omega t$$

$$v_y = U \sin \omega t + V \cos \omega t .$$

Integrating again, and using the condition $\mathbf{r}(0) = \mathbf{0}$, finally gives

$$x = \frac{1}{\omega}\left\{U \sin \omega t - V(1 - \cos \omega t)\right\}$$

$$y = \frac{1}{\omega}\left\{V \sin \omega t + U(1 - \cos \omega t)\right\} .$$

The distance of the electron from the z-axis after time t is thus given by $(x^2 + y^2)^{1/2}$, where

$$x^2 + y^2 = \frac{(U^2 + V^2)}{\omega^2}\left\{\sin^2 \omega t + (1 - \cos \omega t)^2\right\}$$

$$= 2\frac{(U^2 + V^2)}{\omega^2}(1 - \cos \omega t) \; .$$

The beam will focus when $(x^2 + y^2)$ is least; i.e. when $\cos \omega t = 1$. This first occurs when $t = 2\pi/\omega$. At this instant we have

$$z = Wt = 2\pi W/\omega \; ,$$

where $U^2 + V^2 + W^2 = u^2$. Since U and V are small in comparison with W (the electron's initial velocity is almost parallel to the z-axis), we have

$$u = (U^2 + V^2 + W^2)^{1/2} \approx W \; .$$

Thus the beam focusses at a distance approximately $2\pi u/\omega = 2\pi mu/eB$ from the aperture.

Problems

5.1 A ball is projected with speed u from a point O on level ground towards a vertical wall of height h, which is distance c from O. Prove that if

$$u^2/g < h + (h^2 + c^2)^{1/2}$$

the ball cannot go over the wall. [Neglect all forces acting on the particle except gravity.]

5.2 A projectile is fired from a point O with velocity u at an angle of elevation α above the horizontal. Neglecting all forces except gravity, find the equation of the particle's trajectory, taking O as origin and using xy-coordinates with the y-axis vertically upwards. Deduce that the parabola of safety is

$$x^2 = 4d(d - y)$$

where $u^2 = 2gd$.

If the path of the projectile touches the parabola of safety at a point P(a,b), show that the angle of projection is given by

$$\tan \alpha = 2d/a \; .$$

Show also that in order to reach the maximum range on a plane through O inclined at an angle β above the horizontal, the angle α of projection measured from the horizontal, must satisfy the equation

$$\tan\beta + \cos 2\alpha = 0 .$$

Find the maximum range up this plane.

5.3 A projectile of mass m is fired from a point O with a speed u at an angle α to the horizontal. The particle moves under gravity in a medium which produces a retarding force $k\mathbf{v}$ per unit mass, where $\mathbf{v}$ is the particle's velocity. Choosing axes Oxy so that the y-axis is vertically upwards and the motion takes place in the Oxy plane, show that

(i) the equation of the trajectory of the projectile is given by

$$y = (u\sin\alpha + g/k)\,[x/(u\cos\alpha)] + g\log[1 - xk/(u\cos\alpha)]/k^2 ;$$

(ii) if the particle returns to the level of projection after a time τ, then τ is a solution of

$$k\tau = [(uk/g)\sin\alpha + 1]\,(1 - e^{-k\tau}) ;$$

(iii) if k is very small, the time taken for the projectile to reach its greatest height is given by

$$[(u\sin\alpha)/g]\,[1 - (ku\sin\alpha)/2g] + O(k^2) .$$

Show that, in all three parts, letting $k \to 0$ results in the expected formulae for unresisted projectile motion.

5.4 A particle is projected under gravity in a medium whose resistance per unit mass is kv^n, where v is the particle's speed and k and n are constants. If ψ is the angle between the particle's direction of motion and the horizontal show that

(i) the horizontal velocity, u, of the particle satisfies the differential equation

$$\frac{du}{d\psi} = (ku^{n+1}\sec^{n+1}\psi)/g ;$$

(ii) if x and y are the horizontal and vertical coordinates of the particle then

$$\frac{d^2y}{dx^2} = -\frac{g}{u^2} .$$

5.5 For the problem described in problem 5.4, show that, if $n = 4$

$$\frac{d^3y}{dx^3} = -2gk\left\{1 + \left(\frac{dy}{dx}\right)^2\right\}^{3/2} .$$

If the particle is projected from the origin of coordinates with speed u_0 at an angle α to the x-axis, find the *three* initial conditions appropriate for the solution of the above third order differential equation.

5.6* A charged particle of mass m and charge q moves with velocity $\mathbf{v}$ in a constant uniform magnetic field $\mathbf{B}$. Rectangular Cartesian coordinates $Oxyz$ are chosen such that $\mathbf{B}$ is parallel to Oz. If initially the position vector $\mathbf{r}$ of the particle is given by $\mathbf{r}_0 = (a, b, c)$, and if initially the velocity is given by $\mathbf{v}_0 = (0, U, V)$, solve the equation of motion and determine $\mathbf{r}$.

If $\mathbf{v}$ is initially perpendicular to $\mathbf{B}$, show that the particle describes a circle of radius mU/qB.

5.7* A charged particle of mass m and charge q moves with velocity $\mathbf{v}$ in a magnetic field $\mathbf{B}$ and electric field $\mathbf{E}$. If, referred to rectangular axes $Oxyz$,

$$\mathbf{B} = (0, 0, B), \quad \mathbf{E} = (E_x, 0, E_z)$$

where B, E_x and E_z are constants, show that

$$\mathbf{v} = [(U + E_x/B)\sin(\omega t),\ (U + E_x/B)\cos(\omega t) - E_x/B,\ V + qE_z t/m]$$

where $\mathbf{v}(0) = (0, U, V)$ and $\omega = qB/m$. If $\mathbf{r} = (x, y, z)$ is the position vector of the charged particle, and if initially $\mathbf{r} = (a, b, 0)$ show that

$$\left[x - a + \frac{1}{\omega}(U + E_x/B)^2\right] + [y - b + E_x t/B]^2 = \frac{1}{\omega^2}(U + E_x/B)^2 .$$

Find z and describe the nature of the motion of the particle.

5.8* A charged particle of mass m and charge q is moving with velocity $\mathbf{v}$ in a magnetic field which has constant direction and whose magnitude B is a function of time t. Rectangular Cartesian axes $Oxyz$ are chosen so that the direction of $\mathbf{B}$ is parallel to Oz. If $\mathbf{v} = (v_x, v_y, v_z)$, show that v_x satisfies

$$\frac{d^2 v_x}{dt^2} - \frac{1}{B}\frac{dB}{dt}\frac{dv_x}{dt} + \omega^2 v_x = 0 ,$$

where $\omega = qB/m$.

If $B = B_0\tau/(t + \tau)$, where B_0 and τ are positive constants, show that v_x satisfies

$$(t + \tau)^2\frac{d^2 v_x}{dt^2} + (t + \tau)\frac{dv_x}{dt} + \omega^2\tau^2 v_x = 0 ,$$

and verify that $\quad \mathbf{v} = (u\cos p,\ - u\sin p,\ 0)$,

where $p = \omega\tau\log[(t + \tau)/\tau]$ and initially $\mathbf{v} = (u, 0, 0)$.

Chapter 6

Conservation Theorems

We have already met a simple example of the use of a conservation law in Ch. 3.5.3. Here we were able to determine the height reached by a ball thrown vertically upwards by discovering a quantity which remains constant, i.e. is *conserved*, throughout the motion. The resulting *conservation law* (3.14) enabled us to simplify the working of the problem by by-passing unimportant intermediate results such as the time at which the greatest height is reached. A more general form of this type of conservation law (conservation of energy in this instance) was derived in Worked Example 3.4.as a first integral of the equation of motion.

It seems natural to ask whether we can formulate any further conservation laws which are applicable to a wide range of situations. The motivation for this is indicated by the simple examples above. In addition to simplifying the working, a conservation law can often give valuable insight into a more complicated system, enabling general inferences to be made even when the detailed mathematics is highly complex.

In constructing a conservation law we are looking for a fundamental quantity whose value remains constant under certain conditions. The law can then be used to facilitate the solution of a problem, often shortening the solution considerably, whenever the specified conditions are satisfied. If the conditions are violated, then the conservation law is of course invalidated and a more basic approach must be employed.

We consider three conservation laws relating to the 'Newtonian'† motion of a single particle, namely the conservation of linear momentum, angular momentum and energy. There are many other situations in which conservation laws are helpful; for example the equation of continuity in fluid flow (based on the conservation of mass) or the relationship between quantities such as velocity and pressure across a shock wave. These however are beyond the scope of this book.

6.1 Linear Momentum

Our first conservation law arises as a direct consequence of Newton's second law in the case of zero force.

† Assuming Newton's laws are valid.

THEOREM 6.1: CONSERVATION OF LINEAR MOMENTUM

If the total force acting on a particle is zero, the linear momentum of the particle is conserved.

PROOF:

Suppose the particle, mass m, moves on a curve, $\mathscr{C}$, which referred to an inertial frame of reference has parametric equation

$$\mathbf{r} = \mathbf{r}(t), \quad 0 \leqslant t < \infty, \tag{6.1}$$

t being the time. If $\mathbf{p} = m\dot{\mathbf{r}}$ is the linear momentum of the particle, Newton's second law gives

$$\frac{d\mathbf{p}}{dt} = \mathbf{F},$$

$\mathbf{F}$ being the total force acting. But $\mathbf{F} = \mathbf{0}$, so that

$$\boxed{\frac{d\mathbf{p}}{dt} = \mathbf{0}.} \tag{6.2}$$

Integrating, we see that $\mathbf{p}$ is a constant vector, and this completes the proof.

COROLLARY:

If the total force acting *in any given direction* is zero, the linear momentum in that direction is conserved.

Before introducing our second conservation theorem we make one simple definition, applicable to all vectors which have a line of action.

DEFINITION:

The *moment of a vector*, $\mathbf{F}$ say, with the line of action $\mathcal{L}$, *about a point* A, which has position vector $\mathbf{a}$, is defined as

$$\mathbf{M}_\mathrm{A} = (\mathbf{r} - \mathbf{a}) \times \mathbf{F}, \tag{6.3}$$

where $\mathbf{r}$ is the position vector of any point, P, on $\mathcal{L}$.

Is this a meaningful definition? It will only make sense if $\mathbf{M}_\mathrm{A}$ does *not* depend on the position vector $\mathbf{r}$ of the point P on $\mathcal{L}$ (see Fig. 6.1).

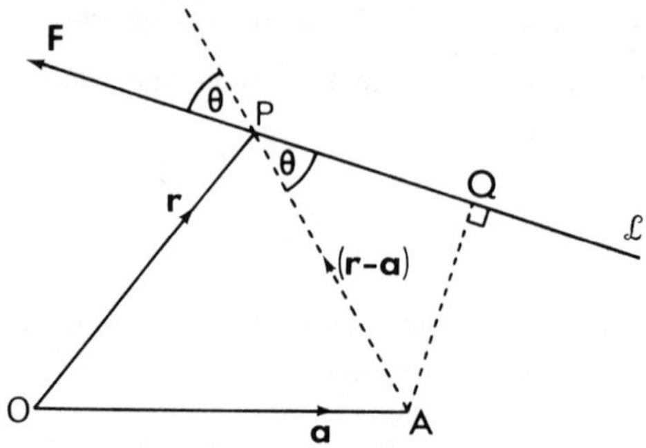

Fig. 6.1 Moment of vector **F** *about* A

If θ is the angle between **F** and $(\mathbf{r}-\mathbf{a})$, we see that

$$|\mathbf{M}_A| = |\mathbf{r}-\mathbf{a}|\ |\mathbf{F}|\sin\theta$$

$$= AQ \times |\mathbf{F}|,$$

where Q is the foot of the perpendicular from A to the line of action of **F**. Hence $|\mathbf{M}_A|$ is independent of P, and clearly the direction of $\mathbf{M}_A$ is also independent of P. Thus the definition is meaningful and we note that $|\mathbf{M}_A|$ is the product of the magnitude of **F** and its shortest distance from A.

6.2 Angular Momentum

Angular momentum is an important vector quantity which we will find very useful in dealing with space dynamics (Ch. 7). Suppose a particle, P, of mass m is moving on a curve, as defined by (6.1).

DEFINITION:

Let A be a point with position vector **a**, relative to the fixed origin O. The *angular momentum about* A of the particle is defined by

$$\boxed{\mathbf{H}_A = m(\mathbf{r}-\mathbf{a}) \times \dot{\mathbf{r}}} \quad . \tag{6.4}$$

Fig. 6.2 illustrates the configuration. Note that $\mathbf{H}_A$ is a vector, perpendicular to both $(\mathbf{r}-\mathbf{a})$ and $\dot{\mathbf{r}}$. In terms of linear momentum, $\mathbf{p} = m\dot{\mathbf{r}}$, we see that

$$\mathbf{H}_A = (\mathbf{r}-\mathbf{a}) \times \mathbf{p}.$$

From (6.3) we see that $\mathbf{H}_A$ is the moment of linear momentum about A.

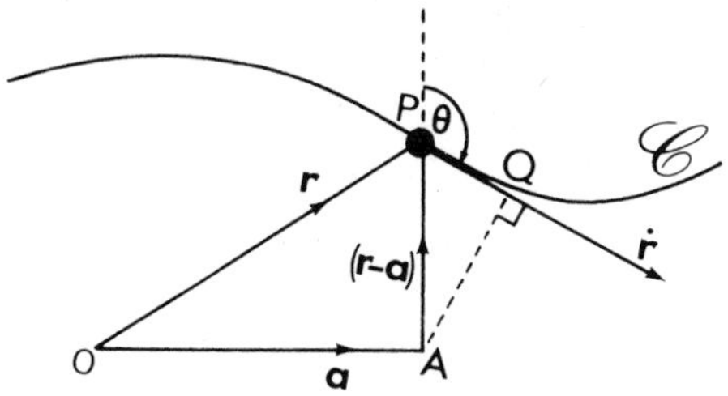

Fig. 6.2 Angular momentum about A

Angular momentum is defined about a specific point, and the expression 'angular momentum' strictly has no unique meaning. However, if A is coincident with the origin O,

$$\boxed{\mathbf{H}_o = m\mathbf{r} \times \dot{\mathbf{r}}} \tag{6.5}$$

and we often refer to angular momentum about O as simply 'angular momentum' (this being assumed to be about O).

In order to relate angular momentum to the equation of motion, we must introduce the force acting on the particle.

DEFINITION:

If **F** is the total force acting on P, we define the *torque* (or *moment* of **F**) about the point A as

$$\boxed{\mathbf{G}_A = (\mathbf{r} - \mathbf{a}) \times \mathbf{F}.} \tag{6.6}$$

Fig. 6.3 illustrates the configuration. If α is the angle between

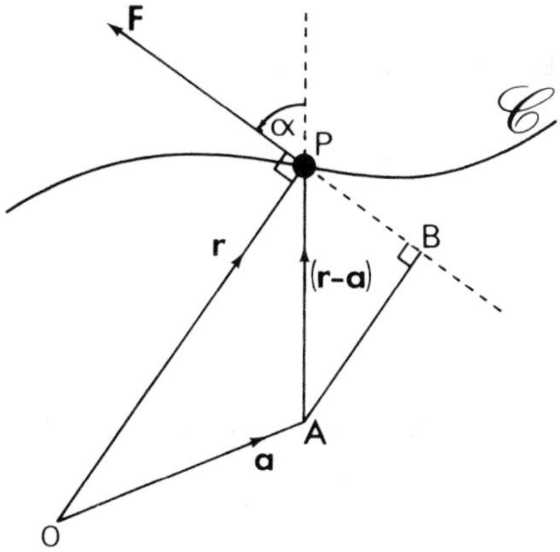

Fig. 6.3 Torque about A

the vectors $(\mathbf{r}-\mathbf{a})$ and $\mathbf{F}$, we note that $\mathbf{G}_A$ is a vector perpendicular to both $(\mathbf{r}-\mathbf{a})$ and $\mathbf{F}$ and that the magnitude of $\mathbf{G}_A$ is given by

$$G_A = |\mathbf{r}-\mathbf{a}|\ |\mathbf{F}|\ |\sin\alpha|$$

$$= |\mathbf{F}| \times AB\,.$$

If A is coincident with O,

$$\boxed{\mathbf{G}_o = \mathbf{r} \times \mathbf{F}.} \tag{6.7}$$

LEMMA 6.1:

The rate of change of angular momentum about any *fixed* point A is equal to the torque about A, i.e.

$$\boxed{\frac{d\mathbf{H}_A}{dt} = \mathbf{G}_A\,.} \tag{6.8}$$

PROOF:

Differentiating (6.4),

$$\frac{d\mathbf{H}_A}{dt} = m\dot{\mathbf{r}} \times \dot{\mathbf{r}} + (\mathbf{r}-\mathbf{a}) \times \frac{d}{dt}(m\dot{\mathbf{r}})$$

$$= (\mathbf{r}-\mathbf{a}) \times \frac{d\mathbf{p}}{dt}\,.$$

But the equation of motion of the particle is

$$\frac{d\mathbf{p}}{dt} = \mathbf{F},$$

so that

$$\frac{d\mathbf{H}_A}{dt} = (\mathbf{r}-\mathbf{a}) \times \mathbf{F}$$

$$= \mathbf{G}_A\,,$$

using (6.6).

We note that A is any fixed point, and in particular if A is the origin,

$$\boxed{\frac{d\mathbf{H}_o}{dt} = \mathbf{G}_o\,.} \tag{6.9}$$

The above lemma now leads to our second conservation theorem:

THEOREM 6.2: CONSERVATION OF ANGULAR MOMENTUM

If the torque about A of the forces acting on a particle is zero, the particle's angular momentum about A is conserved.

PROOF:
From (6.8), if $\mathbf{G}_A = \mathbf{0}$, $d\mathbf{H}_A/dt = \mathbf{0}$, and the result follows.

The torque of $\mathbf{F}$ about A is zero when either

(i) $\mathbf{F} = \mathbf{0}$, or

(ii) the line of action of $\mathbf{F}$ passes through A

The actual expression for the angular momentum vector will depend on the co-ordinate system used. We will evaluate angular momentum about O for *motion in a plane* using Cartesian coordinates (x,y) and plane polar coordinates (R, ϕ), as illustrated in Fig. 6.4.

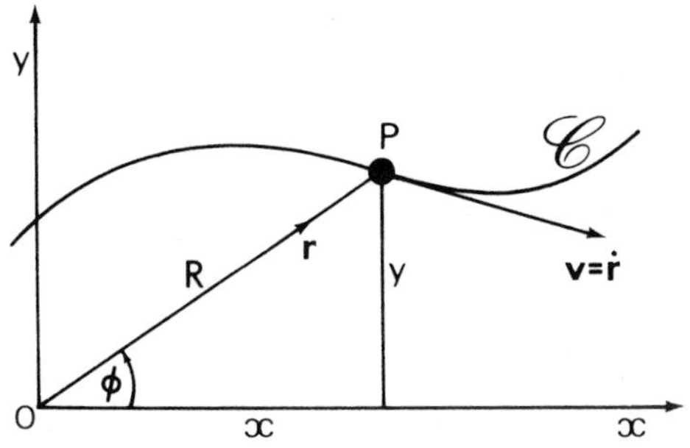

Fig. 6.4 Coordinate systems

Since $\mathbf{H}_o$ is perpendicular to both $\dot{\mathbf{r}}$ and $\mathbf{r}$, for motion in the plane Oxy, $\mathbf{H}_o$ is either parallel or anti-parallel to Oz.

(a) Rectangular Cartesian coordinates (x, y, z)

The path of the particle is described by

$$\mathbf{r} = (x, y, 0)$$

where x and y are functions of time t. Hence

$$\dot{\mathbf{r}} = (\dot{x}, \dot{y}, 0)$$

and

$$\mathbf{H}_o = m\mathbf{r} \times \dot{\mathbf{r}}$$

i.e.

$$\boxed{\mathbf{H}_o = m(0, 0, x\dot{y} - \dot{x}y)} \quad . \tag{6.10}$$

(b) Cylindrical Polar Coordinates (R, ϕ, z)

The path of the particle is described by

$$\mathbf{r} = R\mathbf{e}_R,$$

where R and ϕ are functions of time t. In Ch. 2.3 it was shown that

$$\dot{\mathbf{r}} = \dot{R}\mathbf{e}_R + R\dot{\phi}\mathbf{e}_\phi ,$$

giving

$$\mathbf{H}_o = m(R\mathbf{e}_R) \times (\dot{R}\mathbf{e}_R + R\dot{\phi}\mathbf{e}_\phi)$$

i.e.

$$\boxed{\mathbf{H}_o = mR^2\dot{\phi}\,\mathbf{e}_z} \quad . \tag{6.11}$$

EXAMPLE:

A particle, P, of constant mass m is moving in a plane subject to a force directed towards a fixed point, O, in the plane. Choosing plane polar coordinates (R, ϕ) such that $R = \text{OP}$ and ϕ is measured from some fixed line through the origin, show that $R^2\dot{\phi}$ is constant.

SOLUTION:

Since the line of action of the force is through O, the torque about O is zero, i.e. $\mathbf{G}_o = \mathbf{0}$. From Theorem 6.2, $d\mathbf{H}_o/dt = \mathbf{0}$, and integrating, $\mathbf{H}_o$ is a constant vector. Hence $|\mathbf{H}_o|$ is constant and, from (6.11),

$$R^2\dot{\phi} = \text{constant} .$$

6.3 Work Done

In Ch. 3.5 and Ch. 5.2 we met the idea of energy conservation. This will provide our third conservation theorem, but before we can state and prove this result we have several definitions to make. If a particle of mass m is moving with velocity $\mathbf{v}$, its kinetic energy, T, is defined by

$$\boxed{T = \tfrac{1}{2}\, m\mathbf{v}.\mathbf{v}} \quad . \tag{6.12}$$

This is a scalar quantity and is the energy associated with the actual motion of the particle. We now consider the forces acting on the particle.

If a man is climbing a ladder carrying a load of bricks, he will do work in order to reach the top of the ladder. He has to lift both the load and himself against the pull of gravity. This gives an intuitive idea of 'work'. We will make a precise definition and then analyse its meaning.

Suppose a particle is moving on a curve, $\mathscr{C}$, with parametric equation

$$\mathbf{r} = \mathbf{r}(t), \quad t_o \leqslant t \leqslant t_1 ,$$

and is subject to a force $\mathbf{F}$.

DEFINITION:

The work done, W, on the particle by the force $\mathbf{F}$ as the particle moves from A to B is defined as

$$\boxed{W = \int_{\mathscr{C}_{AB}} \mathbf{F}.d\mathbf{r}} \quad . \tag{6.13}$$

where $\int_{\mathscr{C}_{AB}}$ means the line integral of $\mathbf{F}.d\mathbf{r}$ along the curve $\mathscr{C}$ from A to B.

For readers unfamiliar with line integrals we have included a brief discussion in Appendix 1.

To interpret this definition of work done, let us look at a simple example of a particle, P, moving from O(0,0,0) to A(a,a,0) along the straight line $\mathscr{C}$, which has parametric equation $\mathbf{r} = (t,t,0)$, $0 \leqslant t \leqslant a$. Suppose the force acting on P is given by

$$\mathbf{F} = (F, 0, 0)$$

where F is a constant. The situation is illustrated in Fig. 6.5.

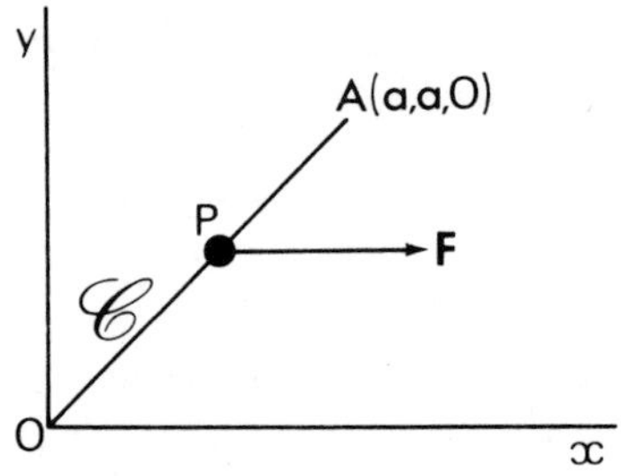

Fig. 6.5 Particle moving from O *to* A

By definition of $\mathscr{C}$, the line element of $\mathscr{C}$ is

$$d\mathbf{r} = (1, 1, 0)dt$$

and using (6.13)

$$W = \int_{\mathscr{C}_{OA}} (F, 0, 0).(1, 1, 0)dt$$
$$= \int_0^a F\,dt$$
$$= Fa.$$

We see that the work done is the product of the magnitude of the force and the distance moved by the particle *in the direction of the force.* The S.I. unit of work done is Nm or kg $m^2 s^{-2}$. In Ch. 3.3, we defined this unit as the joule, so that

$$1 \text{ joule} = 1 \text{ Newton metre.} \qquad (6.14)$$

EXAMPLE:

Find the work done by the force $\mathbf{F} = (0, 0, -mg)$ in moving a particle of mass m from O(0, 0, 0) to A(1, 1, 1) along the curve $x = t$, $y = t^2$, $z = t^3$, t being a parameter.

SOLUTION:

The parametric equation of the curve is

$$\mathbf{r} = (t, t^2, t^3), \quad 0 \leqslant t \leqslant 1$$

so that

$$d\mathbf{r} = (1, 2t, 3t^2)dt.$$

Hence
$$\begin{aligned} W &= \int_0^1 (0, 0, -mg).(1, 2t, 3t^2)\,dt \\ &= \int_0^1 (-3mgt^2)\,dt \\ &= -3mg \int_0^1 t^2\,dt \\ &= -mg = mg(-1). \end{aligned}$$

We again note that the work done is the product of the magnitude of the force and the distance moved by the particle in the direction of the force.

6.4 Conservative Forces

DEFINITION:

If the work done by a force **F** against a particle moving from A to B is *independent* of the path taken by the particle between A and B, the force is said to be *conservative.*

Let us look at the force described in the problem above, where $\mathbf{F} = (0, 0, -mg)$ and the particle moves from O(0, 0, 0) to A(1, 1, 1). Using the curve described above, we showed that

$$W_1 = -mg.$$

Consider a second path from O to A described by the parametric equation

$$\mathbf{r} = (t, t, t), \quad 0 \leqslant t \leqslant 1.$$

Then the work done is given by

$$\begin{aligned} W_2 &= \int_0^1 (0, 0, -mg).(1, 1, 1)\,dt \\ &= \int_0^1 (-mg)\,dt = -mg. \end{aligned}$$

Consider yet a third path from O to A described by the parametric equation

$$\mathbf{r} = (t^2, t, 2t^2 - t), \quad 0 \leqslant t \leqslant 1.$$

In this case the work done is

$$\begin{aligned} W_3 &= \int_0^1 (0, 0, -mg).(2t, 1, 4t-1)\,dt \\ &= -(mg)\int_0^1 (4t-1)\,dt \\ &= -mg\,[2t^2 - t]_0^1 = -mg. \end{aligned}$$

For each path $\mathscr{C}$ considered above between O and A, the work done was always given by $W = -mg$. Analysing the situation further, let

$$\mathbf{r} = (x, y, z)$$

be any curve $\mathscr{C}$, between O and A, where x, y and z are functions of some parameter t (not necessarily time), $t_0 \leqslant t \leqslant t_1$, so that $\mathbf{r}(t_0)$, $\mathbf{r}(t_1)$ correspond to O and A respectively. Then the work done is given by

$$\begin{aligned} W &= \int_{\mathscr{C}_{OA}} \mathbf{F}.d\mathbf{r} \\ &= \int_{t_0}^{t_1} (0, 0, -mg).\left(\frac{dx}{dt}, \frac{dy}{dt}, \frac{dz}{dt}\right) dt \\ &= \int_{t_0}^{t_1} (-mg)\frac{dz}{dt}\, dt \\ &= (-mg)\int_0^1 dz \quad (\text{since } z(t_0) = 0, \;\; z(t_1) = 1) \\ &= -mg. \end{aligned}$$

This shows that the work done is *independent of the path* taken between O and A. Hence the constant gravitational force is a conservative force.

In most cases, it is not so obvious or straightforward to prove that the force is conservative. But vector field theory tells us that if $\mathbf{F}$ satisfies

$$\text{curl } \mathbf{F} = \mathbf{0} \tag{6.15}$$

(see Appendix 2 for definition of curl) then provided $\mathbf{F}$ is well behaved the line integral

$$\int_{\mathscr{C}_{AB}} \mathbf{F}.d\mathbf{r}$$

is independent of the path $\mathscr{C}$ between A and B. This result can soon be deduced. For if (6.15) holds it can be proved that if $\mathbf{F}$ is well-behaved there exists a scalar function Ω such that

$$\boxed{\mathbf{F} = \text{grad } \Omega\dagger} \tag{6.16}$$

(The proof of this result is outside the scope of this book.)

† See Appendix 2 for definition of grad.

Then

$$W = \int_{\mathscr{C}_{AB}} \mathbf{F}.d\mathbf{r} = \int_{\mathscr{C}_{AB}} \left(\frac{\partial\Omega}{\partial x}, \frac{\partial\Omega}{\partial y}, \frac{\partial\Omega}{\partial z}\right).\left(\frac{dx}{dt}; \frac{dy}{dt}, \frac{dz}{dt}\right)dt$$

$$= \int_{t_0}^{t_1} \left(\frac{\partial\Omega}{\partial x}\frac{dx}{dt} + \frac{\partial\Omega}{\partial y}\frac{dy}{dt} + \frac{\partial\Omega}{\partial z}\frac{dz}{dt}\right)dt$$

$$= \int_{t_0}^{t_1} \frac{d\Omega}{dt}\,dt \text{ (chain rule for partial differentiation)}$$

$$= \int_{\mathscr{C}_{AB}} d\Omega$$

$$= \Omega_B - \Omega_A \tag{6.17}$$

where Ω_B, Ω_A denote Ω evaluated at the points B and A respectively. Hence W is independent of the curve $\mathscr{C}$ between A and B. So we now have a simple test to determine whether or not **F** is conservative; i.e. the vector field **F** is conservative if curl $\mathbf{F} = \mathbf{0}$. The converse result is also true; i.e. if **F** is conservative then curl $\mathbf{F} = \mathbf{0}$.

EXAMPLE:

A particle of mass m moving in a plane is subject to a force **F** given by

$$\mathbf{F} = mf(R)\mathbf{e}_R$$

where (R,ϕ) are plane polar coordinates, and f is a function of R. Show that **F** is conservative.

SOLUTION:

F is conservative if curl $\mathbf{F} = \mathbf{0}$. But

$$\text{curl}\,\mathbf{F} = \text{curl}\{mf(R)\mathbf{e}_R\}$$

$$= m\,\text{curl}\{f(R)\mathbf{e}_R\}$$

$$= \frac{m}{R}\begin{vmatrix} \mathbf{e}_R & R\mathbf{e}_\phi & \mathbf{e}_z \\ \dfrac{\partial}{\partial R} & \dfrac{\partial}{\partial \phi} & \dfrac{\partial}{\partial z} \\ f(R) & 0 & 0 \end{vmatrix}$$

$$= \frac{m}{R}(0, 0, 0)$$
$$= \mathbf{0}.$$

[The expression for curl **F** in cylindrical polar coordinates is given in Appendix 3.]

6.5 Potential Energy

We are now in a position to define potential energy. If the force is conservative, curl $\mathbf{F} = \mathbf{0}$ and there exists a scalar function Ω such that

$$\mathbf{F} = \text{grad}\ \Omega\,. \tag{6.18}$$

The function Ω is known as a scalar potential for **F** and is unique apart from an arbitrary additive constant.

DEFINITION:

In a conservative field of force **F**, the potential energy V of a particle situated at P is defined as the work done by **F** in moving the particle from P to some fixed position O.

Thus

$$V_{\mathrm{P}} = \int_{\mathscr{C}_{\mathrm{PO}}} \mathbf{F}.d\mathbf{r}$$
$$= \int_{\mathscr{C}_{\mathrm{PO}}} \text{grad}\ \Omega.d\mathbf{r}$$
$$= \Omega_{\mathrm{O}} - \Omega_{\mathrm{P}}\,.$$

using (6.17). Since **F** is conservative, the value of V_{P} is independent of the curve chosen between P and O. The potential energy is closely related to the scalar potential Ω, for, dropping the suffix P, we have

$$\boxed{V = -\Omega + \Omega_{\mathrm{O}}}\,. \tag{6.19}$$

Since O can be any fixed point, the potential energy at any point P is unique to within an additive constant. In any specific problem all potential energies must be referred to the same reference point or level. We also note that the potential energy is a function of position only (relative to some fixed point) so that we may write

$$V = V(\mathbf{r}),$$

which indicates that V is a function of x, y and z only. Potential energy is a measure of the energy possessed by a particle due to its position in a conservative field of force.

EXAMPLE:

A particle is projected upwards under gravity with initial speed u. Find the potential energy of the particle when it has reached a height h.

SOLUTION:

Choosing the x-axis vertically upwards,

$$\mathbf{F} = m(-g, 0, 0) .$$

We have already seen that $\mathbf{F}$ is conservative, and clearly

$$\mathbf{F} = \text{grad}\,(-mgx).$$

Hence we can take $\Omega = -mgx + k$, where k is a constant. Taking the fixed point O as the origin, (6.19) gives

$$V = -(-mgx + k) + (-0 + k)$$

i.e.

$$V = mgx.$$

Thus if the particle is at a height h above the origin, its potential energy is given by

$$\boxed{V = mgh} \quad . \tag{6.20}$$

The position O is called the zero potential energy level, for at O, (6.19) shows that $V = 0$. In the example above we have chosen zero potential energy level to be the initial point of projection. If we choose O to be a point on the particle's path, distance d above the initial point, we see that

$$V = -(-mgx + k) + (-mgd + k)$$

$$= mg\,(x - d);$$

so that, at a point height h above the initial point

$$V = mgh - mgd.$$

Comparing with (6.20) we see as expected that V is unique to within an additive constant.

EXAMPLE:

If the force acting on a particle of mass m moving in a plane is of the form

$$\mathbf{F} = mf(R)\mathbf{e}_R , \tag{6.21}$$

in the usual notation, find the potential energy of the particle.

SOLUTION:

We have already shown that curl $\mathbf{F} = \mathbf{0}$, so that $\mathbf{F}$ is conservative, and there exists a scalar function Ω which satisfies

$$\mathbf{F} = \text{grad } \Omega.$$

Evaluating grad Ω in terms of cylindrical polar coordinates (R, ϕ, z) (see Appendix 3),

$$\mathbf{F} = \frac{\partial \Omega}{\partial R}\mathbf{e}_R + \frac{1}{R}\frac{\partial \Omega}{\partial \phi}\mathbf{e}_\phi + \frac{\partial \Omega}{\partial z}\mathbf{e}_z$$

and equating to the given value of $\mathbf{F}$,

$$mf(R)\mathbf{e}_R = \frac{\partial \Omega}{\partial R}\mathbf{e}_R + \frac{1}{R}\frac{\partial \Omega}{\partial \phi}\mathbf{e}_\phi + \frac{\partial \Omega}{\partial z}\mathbf{e}_z .$$

This is clearly satisfied by taking

$$mf(R) = \frac{\partial \Omega}{\partial R}$$

where Ω is a function of R alone. Hence, on integrating from $R = R_0$ to R,

$$\Omega = \int_{R_0}^{R} mf(R)dR ,$$

and the potential energy, unique to within an additive constant, is given by

$$\boxed{V = -m\int f(R)dR} \tag{6.22}$$

We have defined potential energy only when the force $\mathbf{F}$ is conservative. When $\mathbf{F}$ is not conservative, we can find the work done by the force along any specified path but it is not possible to introduce a scalar potential function as defined in (6.18). In the next section we will see more clearly the meaning of 'conservative' forces, but we will finish this section by pointing out an important class of conservative forces.

EXAMPLE:

A particle is moving along a straight line subject to a force $f(x)$, directed along the line, where x is the distance of the particle from a fixed point O on the line. Show that this force is conservative and find the particle's potential energy.

SOLUTION:

Introducing rectangular Cartesian axes $Oxyz$, the force can be written as

$$\mathbf{F} = (f, 0, 0)$$

where f is a function of x alone. Clearly

$$\text{curl}\,\mathbf{F} = \begin{vmatrix} \mathbf{i} & \mathbf{j} & \mathbf{k} \\ \dfrac{\partial}{\partial x} & \dfrac{\partial}{\partial y} & \dfrac{\partial}{\partial z} \\ f(x) & 0 & 0 \end{vmatrix}$$

$$= \mathbf{0},$$

and so $\mathbf{F}$ is conservative. To find the potential energy, we must find a scalar function Ω such that

$$\mathbf{F} = \text{grad}\,\Omega = \left(\frac{\partial\Omega}{\partial x}, \frac{\partial\Omega}{\partial y}, \frac{\partial\Omega}{\partial z}\right).$$

Clearly, we can choose Ω so that

$$\frac{d\Omega}{dx} = f(x),$$

Ω being a function of x alone. Thus integrating from $x = x_0$ to x,

$$\Omega = \int_{x_0}^{x} f(x)\,dx$$

and we can take $V = -\Omega + \Omega_A$, where A is the zero potential energy level. If A is the point $x = a$, we see that

$$\boxed{V = -\int_a^x f(x)\,dx} \quad . \tag{6.23}$$

6.6 Conservation of Energy

Suppose a particle P is moving on curve $\mathscr{C}$ with parametric equation

$$\mathbf{r} = \mathbf{r}(t), \quad t_0 \leqslant t \leqslant t_1,$$

subject to a conservative force $\mathbf{F}$. The equation of motion of the particle, assuming constant mass, is

$$m\ddot{\mathbf{r}} = \mathbf{F}. \tag{6.24}$$

Taking the scalar product with $\dot{\mathbf{r}}$ throughout (6.24) gives

$$m\ddot{\mathbf{r}}.\dot{\mathbf{r}} = \mathbf{F}.\dot{\mathbf{r}}\,.$$

The left hand side can be written as

$$\frac{dT}{dt} = \frac{d}{dt}\left(\tfrac{1}{2}m\,\dot{\mathbf{r}}.\dot{\mathbf{r}}\right),$$

where T is the kinetic energy of the particle, giving

$$\frac{dT}{dt} = \mathbf{F}.\dot{\mathbf{r}}\,.$$

Since $\mathbf{F}$ is conservative, there exists a scalar function Ω such that

$$\mathbf{F} = \text{grad}\,\Omega\,.$$

Thus

$$\begin{aligned}\frac{dT}{dt} &= \text{grad}\,\Omega\,.\,\dot{\mathbf{r}}\\ &= \left(\frac{\partial\Omega}{\partial x},\frac{\partial\Omega}{\partial y},\frac{\partial\Omega}{\partial z}\right).\left(\frac{dx}{dt},\frac{dy}{dt},\frac{dz}{dt}\right)\\ &= \frac{\partial\Omega}{\partial x}\frac{dx}{dt}+\frac{\partial\Omega}{\partial y}\frac{dy}{dt}+\frac{\partial\Omega}{\partial z}\frac{dz}{dt}\\ &= \frac{d\Omega}{dt}\ \text{(chain rule)}.\end{aligned}$$

Now the potential energy is defined as

$$V = -\,\Omega + \text{constant},$$

so that

$$\frac{dT}{dt} = -\frac{dV}{dt}$$

i.e.

$$\frac{d}{dt}(T+V) = 0$$

This leads to our third conservation theorem.

THEOREM 6.3: CONSERVATION OF ENERGY

If the force $\mathbf{F}$ acting on a particle is conservative, the sum of the kinetic and potential energies remains constant throughout the motion,

i.e.

$$\boxed{T + V = \text{constant}} \quad . \tag{6.25}$$

This means that *mechanical energy is conserved provided* **F** *is a conservative force.* Thus, whenever we are dealing with conservative forces the conservation of energy provides us with an immediate first integral of the equations of motion, removing the need to consider the latter equations themselves. The term 'conservative force' expresses the property of the force which ensures that a potential energy exists and hence the total energy is 'conserved'. If **F** is a non-conservative force, a mechanical energy conservation theorem does not exist and we must investigate the equation of motion directly. Frictional forces dependent on velocity are an important class of non-conservative forces.

EXAMPLE:

A satellite of mass m is moving in a plane under a force per unit mass

$$\mathbf{F} = -\frac{k}{R^2}\mathbf{e}_R$$

where (R, ϕ) are plane polar coordinates. The satellite, with initial speed u, moves along the straight line $\phi = \phi_0$, a constant, from its initial point $R = a,\ \phi = \phi_0$ in the direction away from the origin. Show that if $u \geqslant (2k/a)^{1/2}$, the satellite will escape to infinity.

SOLUTION:

The force **F** is of the form $f(R)\mathbf{e}_R$, which we have already shown to be conservative. Thus theorem 6.3 holds, and if v is the particle's speed along the straight line

$$\tfrac{1}{2}mv^2 + V = \text{constant}$$

for the motion, V being the particle's potential energy. But from (6.22),

$$\begin{aligned} V &= -m\int(-k/R^2)\,dR \\ &= -mk/R\,. \end{aligned}$$

Hence

$$\tfrac{1}{2}mv^2 - mk/R = \text{constant}.$$

Evaluating the constant at the initial point, we see that

$$\tfrac{1}{2}mv^2 - mk/R = \tfrac{1}{2}mu^2 - mk/a$$

i.e.
$$v^2 = 2k\left(\frac{1}{R} - \frac{1}{a}\right) + u^2 .$$

If the satellite just reaches infinity, i.e. if $R \to \infty$ as $v \to 0$, we see that

$$u^2 = 2k/a$$

Thus if $u \geqslant (2k/a)^{1/2}$, the satellite will escape to infinity.

[The critical speed $u_c = (2k/a)^{1/2}$ is called the *escape velocity.* If the satellite's initial speed is greater than u_c, it will be able to escape from this field of force.]

Worked Examples

EXAMPLE 6.1:

A particle of unit mass moves under a force $\mathbf{F}$ such that its position vector relative to a fixed origin O at time t is

$$\mathbf{r} = (2t^3, t^4, 3t^2).$$

Calculate the angular momentum about O and about the point A which has position vector $(t, 1, 0)$. Determine also the moment of the force $\mathbf{F}$ about O and about A and discuss the results in relation to lemma 6.1.

SOLUTION:

Differentiating the position vector

$$\mathbf{r} = (2t^3, t^4, 3t^2)$$

we obtain
$$\mathbf{v} = \dot{\mathbf{r}} = (6t^2, 4t^3, 6t)$$

and
$$\ddot{\mathbf{r}} = (12t, 12t^2, 6) .$$

Also, $\mathbf{F} = \ddot{\mathbf{r}} = 6(2t, 2t^2, 1)$, since the particle is of unit mass. The angular momentum about O is

$$\mathbf{H}_o = \mathbf{r} \times (m\mathbf{v}) = \mathbf{r} \times \mathbf{v} = (-6t^5, 6t^4, 2t^6) .$$

The angular momentum about A is

$$\mathbf{H}_A = (\mathbf{r} - \mathbf{a}) \times \mathbf{v}$$

where
$$\mathbf{r} - \mathbf{a} = (2t^3 - t, t^4 - 1, 3t^2) .$$

Thus $$\mathbf{H}_A = (-6t^5 - 6t,\ 6t^4 + 6t^2,\ 2t^6 - 4t^4 - 6t^2).$$

The moment of force about O is

$$\mathbf{G}_o = \mathbf{r} \times \mathbf{F} = 6(-5t^4,\ 4t^3,\ 2t^5).$$

Also, we have

$$\begin{aligned}\frac{d\mathbf{H}_o}{dt} &= \frac{d}{dt}(-6t^5,\ 6t^4,\ 2t^6)\\ &= (-30t^4, 24t^3,\ 12t^5)\\ &= 6(-5t^4, 4t^3, 2t^5).\end{aligned}$$

Thus, as expected, we see that

$$\frac{d\mathbf{H}_o}{dt} = \mathbf{G}_o$$

in accordance with lemma 6.1.

The moment of force about A is

$$\begin{aligned}\mathbf{G}_A &= (\mathbf{r} - \mathbf{a}) \times \mathbf{F}\\ &= (2t^3 - t,\ t^4 - 1,\ 3t^2) \times 6(2t, 2t^2, 1)\\ &= 6(-5t^4 - 1, 4t^3 + t,\ 2t^5 - 2t^3 + 2t).\end{aligned}$$

Differentiating $\mathbf{H}_A$ we obtain

$$\frac{d\mathbf{H}_A}{dt} = (-30t^4 - 6,\ 24t^3 + 12t,\ 12t^5 - 16t^3 - 12t)$$

from which it is clear that

$$\frac{d\mathbf{H}_A}{dt} \neq \mathbf{G}_A.$$

In this case, A is a *moving* point and lemma 6.1 does not necessarily hold.

EXAMPLE 6.2:

A particle of mass m is moving from the origin O to a point A which has coordinates (1, 1, 1) subject to a force $\mathbf{F}$ which is given by

$$\mathbf{F} = (y + z,\ z + x,\ x + y).$$

Determine the work done when the particle moves on the following curves from O to A.

(i) the straight line from O to A ;

(ii) the curve $\mathbf{r} = (t, t^2, t^3) \quad 0 \leqslant t \leqslant 1$;

(iii) the straight line from O to (1, 0, 0) followed by the straight line from (1, 0, 0) to (1, 1, 1).

SOLUTION:

(i) A parametric equation for the straight line from O to (1, 1, 1) is

$$\mathbf{r} = (t, t, t), \qquad 0 \leqslant t \leqslant 1.$$

On this straight line, we thus have $\mathbf{F} = (2t, 2t, 2t)$, $d\mathbf{r}/dt = (1, 1, 1)$, so that the work done is given by

$$\begin{aligned} W = \int_{\mathrm{OA}} \mathbf{F}.d\mathbf{r} &= \int_0^1 (2t, 2t, 2t) \,.\, \frac{d\mathbf{r}}{dt}\, dt \\ &= \int_0^1 (2t, 2t, 2t) \,.\, (1, 1, 1)\, dt \\ &= \int_0^1 6t\, dt = [3t^2]_0^1 \\ &= 3\,. \end{aligned}$$

(ii) Evaluating on the curve $\mathbf{r} = (t, t^2, t^3), \quad 0 \leqslant t \leqslant 1,$

$$\begin{aligned} W = \int_{\mathrm{OA}} \mathbf{F}.d\mathbf{r} &= \int_0^1 (t^2 + t^3,\ t^3 + t,\ t + t^2) \,.\, (1,\ 2t,\ 3t^2)\, dt \\ &= \int_0^1 [t^2 + t^3 + 2t(t^3 + t) + 3t^2(t + t^2)]\, dt \\ &= \int_0^1 (5t^4 + 4t^3 + 3t^2)\, dt \\ &= [t^5 + t^4 + t^3]_0^1 = 3. \end{aligned}$$

(iii) Here $$W = \int_{\mathscr{C}_1} \mathbf{F}.d\mathbf{r} + \int_{\mathscr{C}_2} \mathbf{F}.d\mathbf{r}\,,$$

where $\mathscr{C}_1$ has parametric equation

$$\mathbf{r} = (t, 0, 0), \qquad 0 \leqslant t \leqslant 1$$

and $\mathscr{C}_2$ has parametric equation

$$\mathbf{r} = (1, \theta, \theta), \qquad 0 \leqslant \theta \leqslant 1.$$

Thus

$$\begin{aligned} W &= \int_0^1 (0,1,1).(1,0,0)dt + \int_0^1 (2\theta, 1+\theta, 1+\theta).(0,1,1)d\theta \\ &= 0 + \int_0^1 2(1+\theta)d\theta \\ &= [2\theta + \theta^2]_0^1 \\ &= 3. \end{aligned}$$

[Note that the value of W is the same on all three paths, which tempts one to conjecture that **F** is conservative.]

EXAMPLE 6.3:

By evaluating curl **F**, show that the force in Worked Example 6.2 is conservative. Find a scalar function Ω such that

$$\mathbf{F} = \text{grad } \Omega$$

and hence deduce the potential energy associated with this force.

SOLUTION:

If $\mathbf{F} = (y+z,\ z+x,\ x+y)$, then

$$\begin{aligned} \text{curl } \mathbf{F} &= \begin{vmatrix} \mathbf{i} & \mathbf{j} & \mathbf{k} \\ \dfrac{\partial}{\partial x} & \dfrac{\partial}{\partial y} & \dfrac{\partial}{\partial z} \\ y+z & z+x & x+y \end{vmatrix} \\ &= \left\{ \frac{\partial}{\partial y}(x+y) - \frac{\partial}{\partial z}(z+x),\ 1-1,\ 1-1 \right\} \\ &= \mathbf{0}. \end{aligned}$$

Thus **F** is conservative, and there exists a scalar function Ω such that

$$\mathbf{F} = \text{grad } \Omega = \left(\frac{\partial \Omega}{\partial x}, \frac{\partial \Omega}{\partial y}, \frac{\partial \Omega}{\partial z} \right).$$

Hence
$$\frac{\partial \Omega}{\partial x} = y + z\,,$$
$$\frac{\partial \Omega}{\partial y} = z + x\,,$$
$$\frac{\partial \Omega}{\partial z} = x + y\,,$$

We can integrate the first equation partially with respect to x to give

$$\Omega = xy + zx + G(y, z)\,,$$

where G is an arbitrary function of y and z (which is the equivalent of the constant of integration in ordinary integration). But Ω must satisfy the second equation; i.e.

$$z + x = \frac{\partial \Omega}{\partial y} = x + \frac{\partial G}{\partial y}\,.$$

Thus
$$\frac{\partial G}{\partial y} = z\,,$$

and integrating this equation partially with respect to y gives

$$G = yz + H(z)$$

where H is an arbitrary function of z.

Thus
$$\Omega = xy + zx + yz + H(z)\,,$$

and to satisfy the third equation,

$$x + y = \frac{\partial \Omega}{\partial z} = x + y + \frac{dH}{dz}\,.$$

Hence $\dfrac{dH}{dz} = 0$ giving $H =$ constant and

$$\Omega = xy + yz + zx + \text{constant.}$$

From (6.19), the potential energy is given by

$$V = -(xy + yz + zx) + \text{constant.}$$

EXAMPLE 6.4

Find the potential energy of the spring mass system as illustrated in Fig. 3.8 when the mass is displaced a distance x from the equilibrium position.

SOLUTION:

The force acting on the particle in the horizontal direction is $T = \lambda x/a$ directed towards the origin, so that

$$\mathbf{F} = \left(-\frac{\lambda x}{a}, 0, 0\right) = \text{grad}\left(-\frac{\lambda x^2}{2a}\right).$$

Thus the potential energy is given by $V = \lambda x^2/2a$ (choosing zero potential energy when there is no extension).

EXAMPLE 6.5:

A pendulum consists of a mass m which is attached to a spring of modulus λ and natural length l, the other end of the spring being fixed at a point O. The point O is vertically above a frictionless horizontal channel along which the mass is constrained to move. The configuration is illustrated in Fig. 6.6.

The mass is released from rest at A, where $AB = a$, and B is vertically below O. Determine the speed of the mass at B, given that $OB = b$, and assuming the spring remains straight during the motion.

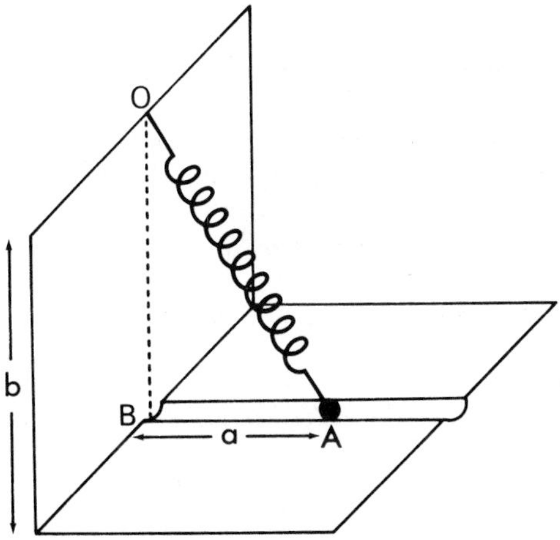

Fig. 6.6 Pendulum in example 6.5

SOLUTION:

Initially $OA = (a^2 + b^2)^{½}$, so that the initial extension of the spring is $(a^2 + b^2)^{½} - l$. Taking AB as the zero gravitational potential energy level, the initial energy of the system is the potential energy of the spring, which, using worked Example 6.4, is given by

$$E = \lambda\,[(a^2 + b^2)^{½} - l]^2/(2l).$$

In the subsequent motion, energy is conserved, so that

$$\frac{1}{2}\,mv^2 + P.E. \;=\; \lambda\,[(a^2+b^2)^{1/2} - l]^2/(2l)\,,$$

where v is the speed of the mass. When the mass reaches B, the extension of the spring is $b - l$, so that the potential energy is $\lambda(b-l)^2/(2l)$.

Hence $$v^2 \;=\; \left(\frac{\lambda}{ml}\right)\left\{[(a^2+b^2)^{1/2} - l]^2 - (b-l)^2\right\}$$

and v is determined.

EXAMPLE 6.6:

A particle of mass m is suspended by an elastic *string* of length l. In equilibrium the string is extended a distance d.

The mass is pulled down a further distance b and released. Find the maximum height reached by the mass in its subsequent motion.

SOLUTION:

The system is illustrated in Fig. 6.7, with the mass displaced a distance x below its equilibrium position. In equilibrium, $\lambda d/l \;=\;$ tension $\;=\; mg$, so that $(\lambda/l) \;=\; (mg/d)$. Energy will be conserved during the motion, and the initial energy will be due to the string's extension and the gravitational potential energy;

i.e. $$E \;=\; \lambda\,(d+b)^2/(2l) \;-\; mgb.$$

Here the equilibrium position is chosen as the zero level of gravitational potential energy.

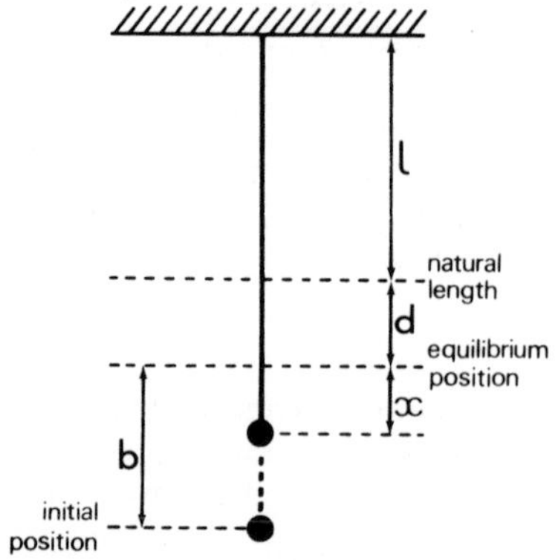

Fig. 6.7 Vertical elastic string in example 6.6

In the subsequent motion, the particle will move vertically upwards until either it comes to rest with the string still extended, or the string becomes slack, in which case the particle will move freely upwards under gravity until it comes to rest. Let us consider the limiting case in which the particle comes to rest just as the string is slack (i.e. when $x = -d$). The final energy is then mgd (since the particle is instantaneously at rest at height d above the equilibrium position) and, since energy is conserved,

$$mgd = \frac{\lambda}{2l}(d+b)^2 - mgb$$

or

$$mgd = \left(\frac{mg}{2d}\right)(d+b)^2 - mgb$$

i.e.

$$2d^2 = (d+b)^2 - 2bd\,,$$

which gives $d^2 = b^2$ i.e. $b = d$. Thus the limiting case occurs when the mass is initially displaced a distance d below its equilibrium position.

Now if $b < d$, the particle will come to rest before the string is slack. Suppose this occurs when the extension is a(i.e. $x = a - d$), so that, equating initial and final energies,

$$\left(\frac{\lambda}{2l}\right)(d+b)^2 - mgb = \left(\frac{\lambda}{2l}\right)a^2 + (d-a)\,,$$

since the particle is at height $(d-a)$ above the equilibrium position.

Thus

$$(d+b)^2 - 2bd = a^2 + 2d(d-a)$$

i.e.

$$a^2 - 2da + d^2 - b^2 = 0.$$

Solving for a,

$$a = d \pm [d^2 - (d^2 - b^2)]^{1/2} = d \pm b.$$

The value $a = d + b$ is the initial extension, and the required extension is $a = d - b\ (>0)$.

If $b > d$, the string becomes slack and the particle moves under gravity. If it instantaneously comes to rest at a height $(h + d)$ above the equilibrium level, we have

$$\left(\frac{\lambda}{2l}\right)(d+b)^2 - mgb = mg(h+d)$$

$$(d+b)^2 - 2bd = 2d(h+d)$$

i.e.

$$2hd = b^2 - d^2$$

or

$$h = (b^2 - d^2)/(2d) \quad (> 0) .$$

Problems

6.1 A particle of mass m is moving in three dimensional space. Using rectangular Cartesian coordinates Oxyz, show that the particle's angular momentum about O is given by

$$\mathbf{H}_o = m(y\dot{z} - \dot{y}z,\ z\dot{x} - \dot{z}x,\ x\dot{y} - \dot{x}y).$$

6.2 If a particle is moving on a sphere of radius a, centre O, show that its angular momentum about O is given by

$$\mathbf{H}_o = ma^2(\dot{\theta}\,\mathbf{e}_\phi - \sin\theta\,\dot{\phi}\mathbf{e}_\theta) ,$$

where (r, θ, ϕ) are spherical polar coordinates, origin O.
[See Worked Example 2.9.]

6.3 Referred to rectangular Cartesian coordinates, a particle of mass m moves from the origin O to a point A(2, 1, 2), along the curve

$$\mathbf{r} = (2t, t^3, 2t), \quad 0 \leqslant t \leqslant 1.$$

If the force acting on the particle is **F**, where

$$\text{(i)} \quad \mathbf{F} = (x^2y,\ yz,\ x);$$

$$\text{(ii)} \quad \mathbf{F} = (y, x, 1),$$

evaluate the work done by **F** in each case.

6.4 Repeat Problem 6.3, using the curves:

(a) $$\mathbf{r} = (2t, t, 2t), \quad 0 \leqslant t \leqslant 1;$$

(b) $$\mathbf{r} = \{2(t-1)^2,\ t-1,\ 2(t-1)\}, \quad 1 \leqslant t \leqslant 2.$$

6.5 Would you be tempted to conjecture that one of the forces considered in Problem 6.3 is conservative? Prove or disprove your conjecture by evaluating curl **F**, and if **F** is conservative, find Ω such that $\mathbf{F} = \text{grad } \Omega$.

6.6 If the force acting on a particle of mass m is given by

$$\mathbf{F} = (yz, zx, xy),$$

find the potential energy of the particle.

6.7* An electron of charge q and mass m is moving under the influence of a magnetic induction field **B**. Show that the kinetic energy of the charge is conserved.

If a constant uniform electric field **E** is also present, show that the kinetic energy of the particle is conserved provided the motion of the particle lies in the plane perpendicular to the direction of **E**.

6.8 A linear spring, of modulus λ and natural length l, is fixed at one end and has a mass m attached to the other end. The spring and mass lie on a horizontal table. If the mass is extended a distance a along the line of the spring, find the particle's potential energy, assuming that the zero potential energy level corresponds to the spring being unextended. Write down the equation of energy conservation, and *hence* find the particle's displacement at any subsequent time t after the particle has been released.

6.9 A satellite of mass m is moving along a straight line under the influence of a force $k/x^n (n > 1)$ per unit mass, directed along the straight line and towards a fixed point O, where x is the distance of the particle from O. If initially the satellite is projected from a point distance a from O, show that its escape velocity is $\{2k/(n-1)a^{n-1}\}^{1/2}$.

Chapter 7

Space Dynamics

The expression 'space dynamics' normally refers to the nature and motion of stellar phenomena and related topics, including celestial mechanics, orbit theory and space navigation. In this chapter we consider only two main branches of space dynamics; namely, the dynamics of the solar system and the application of orbit theory to satellite dynamics.

7.1 A Short History of Man's Interest in the Universe

Man has an inner urge to explore and conquer, described simply by the famous mountaineer George Leigh-Mallory, who, when asked why he wanted to climb Everest, replied 'Because it is there!' One of man's earliest reasons for wanting to explore and understand the heavens was his need for a supernatural being to control his destiny. Not surprisingly he assigned his gods to the winds, storms and other atmospheric events, and to the motions of the sun, moon and planets. These phenomena affected him, but he had no control over them. The names assigned to the days of the week are closely connected with the planets. For if we write down the planets together with the sun and moon in order of decreasing distance from the Earth, as supposed in 3000 B.C., we obtain:

Saturn, Jupiter, Mars, Sun, Venus, Mercury, Moon.

Taking the first name, and then missing two and continuing in this way gives a derivation (in French or English) of the days of the week.

In 300 B.C., Aristarchus developed a theory in which the sun and stars were fixed and the Earth revolved in a circular orbit about the sun. Unfortunately he was way ahead of his time and the leading philosophers treated the theory with contempt. The most popular theory was that the Earth was the central element of the universe and the planets moved round the Earth. In fact, in 130 B.C. Hipparchus explained the motion of the sun, moon and planets by supposing that each moved in an orbit or epicycle which was carried round the Earth in a larger circular orbit. This is illustrated in Fig. 7.1. This idea was further advanced by the publication in A.D. 150 by Ptolemy of an encyclopedia of astronomy which used the epicycle theory to predict the motions of the planets. Surprisingly, some of these results were in fact very accurate and his predicted distance to the moon was in error by only 2%.

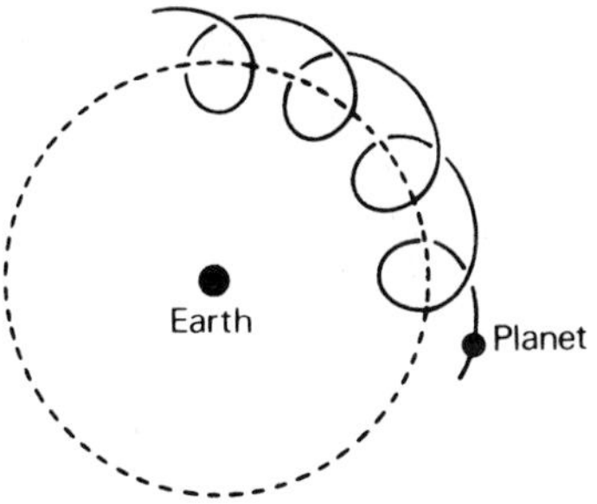

Fig. 7.1 Epicycle motion

But there were still no physical principles upon which the motions were based, and this picture remained virtually unchanged right through the middle ages up to the Renaissance.

Early in the sixteenth century, Copernicus put forward a scheme with the sun as centre, the planets moving in circular orbits round the sun, and the moon moving round the Earth. He also supposed that the stars lay on a sphere of very large radius. As expected the theory was not well received especially from the religious point of view.

In 1601 the German astronomer and science-fiction writer JOHANN KEPLER (1571-1630) became the director of the Prague Observatory on the unexpected death of Tycho Brahe. Kepler had previously been assistant to Brahe and had helped in collecting thirteen years of observations on the relative motion of the planet Mars. By 1609, Kepler had formulated his first two laws:

(1) Each planet moves in an ellipse with the sun at one focus;

(2) For each planet, the line from the sun to the planet sweeps out equal areas in equal times.

Kepler spent the next decade in verifying the above two laws for other planets and in formulating a third observational law, published in 1619 and dedicated to James I of England:

(3) The squares of the orbital periods of the planets vary as the cubes of their mean distances from the sun.

As a result of Galileo's observations of the four moons of Jupiter, the Copernican heliocentric theory was finally accepted and eventually Newton's theory of universal gravitation produced a theoretical principle on which to base Kepler's three observational laws. From these laws, Newton deduced that the acceleration of any planet in orbit is proportional to the inverse square of its distance from the sun. Later he published his *law of gravitation* between particles, which states that any two particles attract one another with a force of magnitude

$$\gamma m_1 m_2 / r^2 \ , \tag{7.1}$$

where m_1, m_2 are the masses of the particles, r is the distance between them and γ is a universal gravitational constant.

Note that with the assumption of the law of gravitation we can now justify our use of a constant gravitational field for near Earth motions. For if m is the mass of the particle concerned and M_E the mass of the Earth, (7.1) tells us that the force on the particle is

$$\gamma M_E m/r^2$$

where r is the distance between the particle and the centre of the Earth. Hence the acceleration experienced by the particle is

$$g(r) = \frac{\gamma M_E}{r^2} . \tag{7.2}$$

A particle situated on the surface of the Earth will experience an acceleration

$$g(R_E) = \frac{\gamma M_E}{R_E^{\ 2}}$$

where R_E is the radius of the Earth (6.4×10^3 km) and $g(R_E) \approx 9.81 \text{ m s}^{-2}$.

At a height of 100 km above the Earth's surface the acceleration is given by

$$g(R_E + 100) = \frac{\gamma M_E}{(R_E + 100)^2} .$$

Hence we have

$$\frac{g(R_E + 100)}{g(R_E)} = \frac{R_E^{\ 2}}{(R_E + 100)^2} = \left(1 + \frac{100}{R_E}\right)^{-2}$$

where

$$\frac{100}{R_E} = \frac{10^{-2}}{6{\cdot}4} \approx 0{\cdot}016 .$$

Using the first two terms of the binomial expansion we obtain

$$\frac{g(R_E + 100)}{g(R_E)} \approx 1 - 2(0{\cdot}016) = 0{\cdot}97 , \tag{7.3}$$

whence

$$g(R_E + 100) = 0{\cdot}97 \times 9{\cdot}81 = 9{\cdot}52 \text{m s}^{-2} .$$

It thus follows that for near-Earth motions we are justified in considering the acceleration due to gravity to be constant ($g = 9{\cdot}81 \text{m s}^{-2}$).

7.2 The Bodies of the Solar System

7.2.1 PLANETS

The discovery of Neptune in 1846 verified beyond doubt Newton's gravitational hypothesis. Perturbations in the predicted orbit of Uranus had been observed, and could only be explained by the influence of another planetary body. Calculations were made to predict the expected position of this unknown planet, as a result of which powerful telescopes were able to observe the new planet, Neptune. Pluto, the last of the known planets, was only discovered in 1930, again after a deliberate search instigated by its effects on the orbits of other planets. There has also been recent speculation about a tenth planet (Planet X), but its existence has yet to be verified.

A summary of solar system data is presented in table 7.1. The sun and the planets constitute over 99% of the material in the solar system, but in number they are only a minute fraction of the bodies in the solar system.

7.2.2 ASTEROIDS

Thousands of asteroids, or little planets, have been detected. The first discovered, at the beginning of the 19th century, was called Ceres and was soon followed by Palles, Juno and Vesta. The latter can at times be seen by the naked eye. By 1950, 1500 asteroids had been found, most of them in orbit between Mars and Jupiter. This has lead to speculation that the asteroids are either a disintegrated planet or a planet partially formed. [Inspection of the radii of the planetary orbits indicates that one might expect a planet between Mars and Jupiter.]

7.2.3 COMETS

The most spectacular bodies in the solar system, except possibly the planet Saturn, are the comets, which in the vicinity of the sun exhibit long and bright tails. Edward Halley (1656-1742) encouraged Newton in his gravitational investigations, and after the publication of the gravitational laws, Halley employed the theory to establish that a certain comet, observed in 1682, was on an elliptical orbit about the sun. He concluded that the comet had a periodic time of 76 years and so predicted its return in 1758. In fact the comet, later named after Halley, had been observed many times before 1682, being pictured for instance, in the Bayeux tapestry. It was also thought to be an omen signifying disaster for the Saxons in 1066 at the Battle of Hastings. Its next return is expected in 1986.

Comets of varying size and composition have since been observed and a number of theories have been put forward. One of the most acceptable is due to Whipple, who considers comets to be a collection of small, solid particles frozen together

Table 7.1 – Solar System Data

Planetary Body	Mean Distance from SUN (compared with the Earth)	Mass (compared with the Earth)	Volume (compared with the Earth)	Radius (compared with the Earth)	Period of Revolution about Sun (Years)	Period of Rotation about own axis (Days)	Eccentricity of Orbit	Number of Moons
SUN	–	333,434	1,301,813	109·187(?)	–	24·65	–	–
MERCURY	0·387	0·04	0·055	0·366	0·24	0·59	0·206	0
VENUS	0·723	0·83	0·904	0·960	0·61	30·0(?)	0·007	0
EARTH	1·000	1·00	1·000	1·000	1·00	1·00	0·017	1
MARS	1·524	0·11	0·150	0·273	1·90	1·03	0·093	2
JUPITER	5·204	318·0	1318·7	10·969	11·9	0·41	0·048	12
SATURN	9·5	95·0	767·2	9·036	29·5	0·42	0·056	9
URANUS	19·2	15·0	49·4	3·715	84·0	0·45	0·047	5
NEPTUNE	30·1	17·0	41·8	3·538	165·0	0·66	0·009	2
PLUTO	39·5	0·8(?)	1·07(?)	1·02(?)	248·0	?	0·249	0
EARTH'S MOON	–	0·012	0·020	0·273	–	27·32	–	–
	Earth = $149{\cdot}5 \times 10^6$ km	Earth = $5{\cdot}976 \times 10^{24}$ kg	Earth = $1{\cdot}08 \times 10^{12}$ km^3	Earth = 6368 km	Gravitational constant $\gamma = 6{\cdot}67 \times 10^{-11}\ m^3\ kg^{-1}\ s^2$			

in an ice of water, ammonia or methane. Nearing the sun, the matter melts and vapourises causing the long tail when in the vicinity of the sun.

7.2.4 METEORS

These are small particles that revolve round the sun and, on entering the Earth's atmosphere, become incandescent. Most of them are vapourised in the atmosphere, but some, known as meterorites, reach the ground. Meteor showers are associated with comet tails but many meteors appear to be independent of any shower.

7.2.5 INTERPLANETARY DUST

A very faint glow in the night sky is due to interplanetary dust, which is thought to originate either from distant stars or fine meteoric dust. The sun also continuously emits particles forming the *solar wind*. This 'wind' has great significance in its effect on the magnetic conditions above the atmosphere, and has been the subject of many satellite observations.

7.3　Geometry of Conic Sections

A conic is the locus of a point whose distances from a fixed point O and a fixed line DD' have a constant ratio, e, called the eccentricity. This is illustrated in Fig. 7.2, and from the definition, we have

$$ie \quad \frac{OP}{PM} = const. = e \qquad OP = ePM ; \qquad (7.4)$$

O is called the focus and DD' the directrix. We also define the semi-latus rectum, l, as the distance from the focus to the conic in the same direction as that of the directrix; i.e.

$$ON = l . \qquad (7.5)$$

Introducing polar coordinates (R,ϕ) as in Fig. 7.2, we note that

$$R = e(OS - R \cos \phi) . \qquad \text{since} \quad \cos\frac{\pi}{2} = 0$$

But when

$$\phi = \pi/2, R = l = ON$$

and so

$$l = e.OS ,$$

and eliminating the distance OS

$$R = e\left(\frac{l}{e} - R\cos\phi\right) ; \quad R + eR\cos\phi = l$$
$$R(1 + e\cos\phi) = l$$

gives

$$\boxed{R = l/(1 + e \cos \phi)} . \qquad (7.6)$$

This is the general polar equation of a conic and ϕ is called the true anomaly. The type of curve depends on whether e is greater than, equal to, or less than unity.

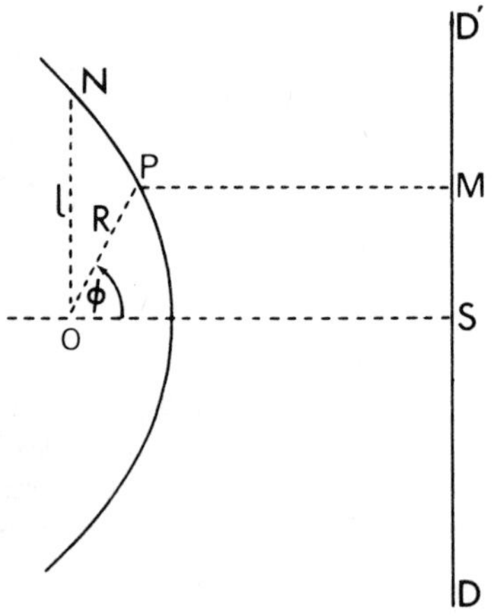

Fig. 7.2 Geometry of conic

We distinguish three types of conic as follows:

$$e > 1: \text{ hyperbola}$$
$$e = 1: \text{ parabola}$$
$$0 \leqslant e < 1: \text{ ellipse}$$
$$(e = 0 \text{ gives a circle}).$$

Since the *ellipse* is of most importance in orbit theory, we will consider this special case of a conic in some detail. Fig. 7.3 illustrates the geometry of the ellipse. The distances a and b are called the semi-major and semi-minor axes respectively. Z is the centre of the ellipse, A the apocentre and B the pericentre. Since

Since $R = \ell/(1 + e \cos \phi)$

Eq. (7.6)

at OB $\phi = 0°$
04 $\phi = 180°$
$\therefore \cos(180) = -1.$

$$2a = OB + OA$$
$$= \frac{l}{(1 + e)} + \frac{l}{(1 - e)}$$
$$= \frac{2l}{(1 - e^2)},$$

we have

$$\boxed{a = l/(1 - e^2)} \qquad . \qquad (7.7)$$

Also b is the maximum value of $y = R \sin \phi$ for varying ϕ, and using (7.6) it can be shown that

$$\boxed{b = l/(1 - e^2)^{1/2}} \quad . \tag{7.8}$$

Eliminating l from (7.7) and (7.8) gives

$$b^2 = a^2(1 - e^2) \ . \tag{7.9}$$

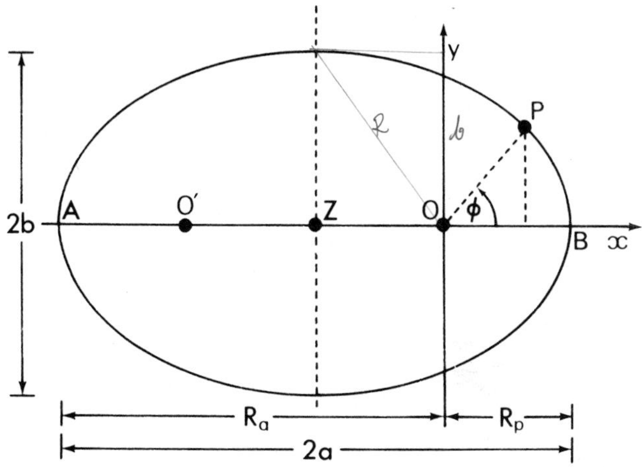

Fig. 7.3 Geometry of the ellipse

7.4 Planetary Motion

Suppose a planet of mass M is moving around the sun on a path $\mathscr{C}$ given by

$$\mathbf{r} = \mathbf{r}(t) \ . \tag{7.10}$$

Assuming Newton's gravitational law between the planet and the sun, and ignoring other interactions, the equation of motion for the planet is

$$M\ddot{\mathbf{r}} = -\frac{\gamma M M_s}{r^3}\mathbf{r} \ . \tag{7.11}$$

Here γ is the universal gravitational constant, and M_s is the mass of the sun. For simplicity we write $k = \gamma M_s$, so that (7.11) becomes

$$\boxed{\ddot{\mathbf{r}} = -\frac{k}{r^3}\mathbf{r}} \quad . \tag{7.12}$$

Suppose the planet is moving with velocity $\mathbf{v}$ at a particular instant. Since the only force acting is directed along the planet-sun line, there will be no component of force in a direction perpendicular to the plane containing this line and the velocity vector $\mathbf{v}$. It follows that the planet cannot acquire any linear momentum in a direction perpendicular to the defined plane, and must therefore continue to move in a plane through the sun. Choosing plane polar coordinates (R,ϕ) as in Fig. 7.4, we can write (7.12) as

$$\ddot{\mathbf{R}} = -\frac{k}{R^2}\mathbf{e}_R \,, \qquad (7.13)$$

$\mathbf{e}_R$ and $\mathbf{e}_\phi$ being unit tangents to the $R-$ and $\phi-$ coordinate lines. [See Appendix 3]. The axis from which ϕ is measured will be defined later.

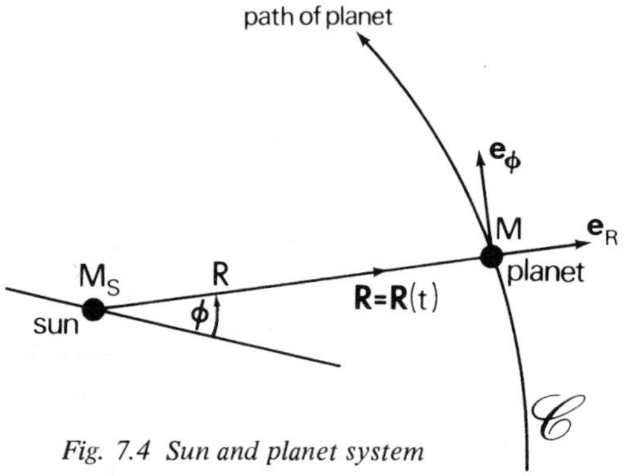

Fig. 7.4 Sun and planet system

Since $\mathbf{R} = R\mathbf{e}_R$, then

$$\dot{\mathbf{R}} = \dot{R}\mathbf{e}_R + R\dot{\phi}\mathbf{e}_\phi$$

$$\ddot{\mathbf{R}} = (\ddot{R} - R\dot{\phi}^2)\mathbf{e}_R + (2\dot{R}\dot{\phi} + R\ddot{\phi})\mathbf{e}_\phi \,.$$

Substituting in (7.13) gives

$$(\ddot{R} - R\dot{\phi}^2)\mathbf{e}_R + (2\dot{R}\dot{\phi} + R\ddot{\phi})\mathbf{e}_\phi = -\frac{k}{R^2}\mathbf{e}_R \,,$$

and equating coefficients of $\mathbf{e}_R$, $\mathbf{e}_\phi$

$$\ddot{R} - R\dot{\phi}^2 = -k/R^2 \qquad (7.14)$$

$$2\dot{R}\dot{\phi} + R\ddot{\phi} = 0 \,. \qquad (7.15)$$

Although this pair of simultaneous differential equations may not appear to be readily soluble, there are in fact many analytic methods available to solve them.

We will use a method of substitution. As expected the second equation can be integrated to give

$$R^2\dot{\phi} = \text{constant}$$

$$= H/M\,, \tag{7.16}$$

H being the magnitude of the constant angular momentum about the sun. Eliminating $\dot{\phi}$ from (7.14) gives

$$\ddot{R} - H^2/M^2R^3 = -\,k/R^2\,. \tag{7.17}$$

This is a second order non-linear differential equation and can be solved by substituting $p = 1/R$ and regarding p as a function of ϕ, rather than time t. Thus

$$\dot{R} = \frac{d}{dt}(1/p)$$

$$= -\frac{1}{p^2}\frac{dp}{d\phi}\dot{\phi}$$

$$= -R^2\dot{\phi}\frac{dp}{d\phi}$$

$$= -\frac{H}{M}\frac{dp}{d\phi}\,, \text{ using (7.16)}\,. \tag{7.18}$$

Hence $$\ddot{R} = -\frac{H}{M}\frac{d^2p}{d\phi^2}\dot{\phi}$$

$$= -p^2\frac{H^2}{M^2}\frac{d^2p}{d\phi^2}\,, \text{ again using (7.16)}\,.$$

Substituting into (7.17) and rearranging gives

$$\frac{d^2p}{d\phi^2} + p = k\frac{M^2}{H^2}\,.$$

Solving, $$p = \frac{1}{R} = k\frac{M^2}{H^2} + C\cos(\phi - \phi_0),$$

where C and ϕ_0 are arbitrary constants. Choosing the axis from which ϕ is measured so that $\phi_0 = 0$, the equation of the planet's orbit is

$$p = k\frac{M^2}{H^2} + C\cos\phi$$

or

$$\boxed{R = \frac{H^2/kM^2}{\left(1 + \dfrac{CH^2}{kM^2}\cos\phi\right)}} \quad . \tag{7.19}$$

Comparing with (7.6), (7.19) represents a conic with eccentricity CH^2/kM^2 and semi-latus rectum $l = H^2/kM^2$.

By observation, the paths of planets are closed and so the path $\mathscr{C}$ must be an ellipse (i.e. $e < 1$). This proves Kepler's first law. Table 7.1 gives the eccentricities of the planetary orbits and, with the possible exceptions of Mercury and Pluto, it is noted that the eccentricities are very small. This means that as a first approximation a circular orbit can be used to represent the path of a planet round the sun.

In (7.19) we still have one arbitrary constant C which determines the eccentricity of the orbit. In the following lemma we show that this constant is uniquely determined when given both the energy and angular momentum of the planet.

LEMMA 7.1:

If E denotes the total energy of the planet, then

$$\boxed{E = \frac{k^2M^2}{2H^2}(e^2 - 1)} \quad . \tag{7.20}$$

PROOF:

The total energy, E, is given by

$$E = K.E + P.E \; .$$

The only force acting, $\mathbf{F}$, is of the form

$$\mathbf{F} = -\frac{kM}{R^2}\mathbf{e}_R$$

$$= kM \operatorname{grad}\left(\frac{1}{R}\right) ;$$

and so the force is conservative and the $P.E.$ given by

$$P.E. = -kM/R \; .$$

Hence

$$E = \tfrac{1}{2}M(\dot{\mathbf{R}}.\dot{\mathbf{R}}) - kM/R$$

$$= \tfrac{1}{2}M(\dot{R}^2 + R^2\dot{\phi}^2) - kM/R$$

$$= \frac{H^2}{2M}\left[\left(\frac{dp}{d\phi}\right)^2 + p^2\right] - kMp, \text{ using (7.18) and (7.16);}$$

$$= \frac{H^2}{2M}\left[C^2\sin^2\phi + \left(k\frac{M^2}{H^2} + C\cos\phi\right)^2\right] - kM\left(k\frac{M^2}{H^2} + C\cos\phi\right),$$

from (7.19).

Rearranging gives $E = \dfrac{H^2C^2}{2M} - \dfrac{k^2M^3}{2H^2}$.

But $e = CH^2/kM^2$, so that

$$E = \frac{H^2}{2M}\frac{e^2k^2M^4}{H^4} - \frac{k^2M^2}{2H^2}$$

$$= \frac{k^2M^3}{2H^2}(e^2 - 1), \text{ as required.}$$

From this result we see that the orbit is completely determined from the two conservation constants H and E since

$$l = H^2/kM^2, \quad e = (1 + 2H^2E/k^2M^3)^{1/2}. \tag{7.21}$$

We also note that $E < 0$ for planetary motion since the orbit is an ellipse.

Returning to Kepler's laws, we prove the following lemma which is a statement of Kepler's second law:

LEMMA 7.2:

The line joining the sun to a given planet sweeps out equal areas in equal times.

PROOF:

In a small time δt, the increment in area swept out by the planet, δA, is given approximately by $\delta A = \frac{1}{2}R(R\delta\phi)$ (see Fig. 7.5). Hence

$$\frac{dA}{dt} = \frac{1}{2} R^2 \dot{\phi}$$

$$= \frac{1}{2}\frac{H}{M}, \text{ using (7.16)} \quad . \tag{7.22}$$

Since H is a constant, dA/dt is constant and this proves the lemma.

Kepler's third law can now easily be deduced. Integrating (7.22) we see that

$$A = \frac{1}{2}\frac{H}{M}(t - \tau)$$

assuming $A = 0$ at $t = \tau$. The orbital period T is given by $t = \tau + T$ and $A = \pi ab$. Hence

$$T = \frac{2M}{H}\pi ab \; .$$

Using (7.19), $l = H^2/kM^2$, so that

$$T = \frac{2\pi ab}{l^{1/2}k^{1/2}}$$

Using (7.8) to eliminate b, $T = \dfrac{2\pi a l^{1/2}}{(1 - e^2)^{1/2}k^{1/2}}$

and from (7.7), $$T = \frac{2\pi a^{3/2}}{k^{1/2}} \tag{7.23}$$

Hence $T^2 \propto a^3$, and this is Kepler's third law.

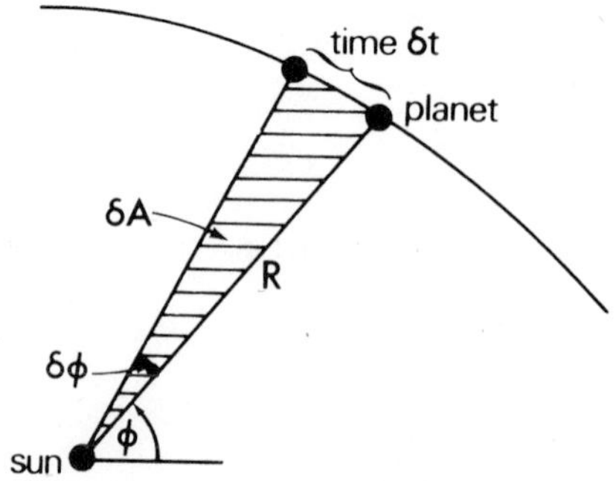

Fig. 7.5 Area swept out by planet

7.5 Satellite Orbits

When a satellite is in orbit round the Earth, the main force acting on the satellite is the attractive force of the Earth's gravitation. Provided the satellite does not orbit near the Moon, and stays in the vicinity of the Earth, we can write its equation of motion as

$$m\ddot{\mathbf{R}} = -\gamma\frac{mM_E}{R^3}\mathbf{R}$$

where m is the mass of the satellite, M_E the mass of the Earth and $\mathbf{R}$ the position vector of the satellite relative to the centre of the Earth. As in Ch. 7.4,

$$\ddot{\mathbf{R}} = -\frac{k}{R^2}\mathbf{e}_R \qquad (7.24)$$

where k now equals γM_E and not γM_s. The theory in Ch. 7.4 is equally valid for a satellite in orbit about the Earth, with γM_s replaced by γM_E, except that we cannot conclude Kepler's first and third laws unless the satellite is on a closed (i.e. elliptic) orbit.

Before considering some of the more important problems in satellite dynamics, we prove the following useful theorem:

THEOREM 7.1:

With the usual notation, when a satellite is in elliptic orbit, its speed v is given by

$$\boxed{v^2 = k\left(\frac{2}{R} - \frac{1}{a}\right)}, \qquad (7.25)$$

and when in parabolic orbit,

$$\boxed{v^2 = 2k/R}. \qquad (7.26)$$

PROOF:

From lemma 7.1,
$$E = \frac{k^2m^3}{2H^2}(e^2 - 1) .$$

For elliptic motion, (7.7) gives $e^2 - 1 = -l/a$, and from (7.19), $l = H^2/km^2$. Thus

$$E = \frac{k^2m^3}{2H^2}\left(-\frac{H^2}{km^2a}\right) = -\frac{km}{2a} .$$

But the energy equation is

$$E = \tfrac{1}{2}mv^2 - km/R \ ,$$

whence
$$\frac{1}{2}mv^2 - \frac{km}{R} = -\frac{km}{2a} \ ,$$

and the first result follows.

For parabolic motion, $e = 1$, and so $E = 0$. Thus $v^2 = 2k/R$.

7.6 The Apsidal Equation

A point on a satellite's orbit where its velocity is perpendicular to its position vector is called an *apse*. In Fig. 7.6, A and B are the apses.

Now at A or B, $\mathbf{e}_R \,.\, \dot{\mathbf{R}} = 0$ where $\mathring{\mathbf{R}} = \mathbf{v} = \mathring{R}\mathbf{e}_R + R\dot{\phi}\mathbf{e}_\phi$. Hence at an apse $\dot{R} = 0$, showing that the positions A and B correspond to maximum and minimum distances of the satellite from the Earth. B, the closest point on the satellite's path to the Earth, is called the *perigee* and A, the furthest point, is called the *apogee.* For satellites in orbit about the Sun, the corresponding distances are called the *perihelion* and *aphelion.*

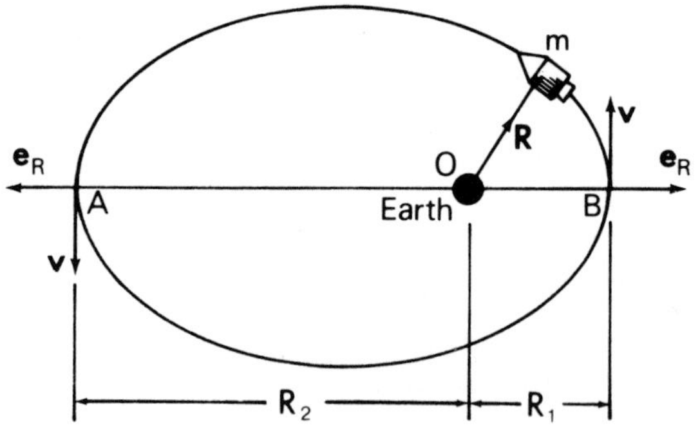

Fig. 7.6 Apsidal positions

LEMMA 7.3:

If OB $= R_1$, OA $= R_2$, then R_1 and R_2 are the roots of the equation

$$\boxed{R^2 + \frac{kmR}{E} - \frac{H^2}{2mE} = 0} \ . \qquad (7.27)$$

PROOF:

The energy equation is

$$E = \frac{1}{2}m(\dot{\mathbf{R}}.\dot{\mathbf{R}}) - \frac{km}{R}$$

$$= \frac{1}{2}m(\dot{R}^2 + R^2\dot{\phi}^2) - \frac{km}{R} .$$

But, from (7.16), $\dot{\phi}^2 = H^2/m^2R^4$, and substitution gives

$$E = \frac{1}{2}m\left(\dot{R}^2 + \frac{H^2}{m^2R^2}\right) - \frac{km}{R} .$$

At an apse, $\dot{R} = 0$, and so

$$E = \frac{1}{2}\frac{H^2}{mR^2} - \frac{km}{R} ,$$

from which (7.27) immediately follows.

Note: Equation (7.27) is normally referred to as the *apsidal equation.*

EXAMPLE:

A satellite is moving in a circular orbit of radius $2R_E$ about the Earth, radius R_E. At an instant of time, the direction of motion of the satellite is changed through an angle a towards the Earth, without a change in speed. Find the angle a in order that the satellite just touches the Earth.

SOLUTION:

From (7.25), on its circular motion with $a = 2R_E$ the satellite's speed is given by

$$v^2 = k/2R_E .$$

Its energy is thus

$$E = \frac{1}{2}mv^2 - \frac{mk}{R}$$

$$= \frac{mk}{4R_E} - \frac{mk}{2R_E}$$

i.e.

$$E = -mk/4R_E .$$

Now at some instant of time, the satellite's direction is changed through an angle α, thus changing H but not E. With reference to Fig. 7.7, the new angular momentum is given by

$$H = |\mathbf{H}| = m\,|\mathbf{R} \times \dot{\mathbf{R}}|$$
$$= m(2R_E)(k/2R_E)^{1/2}\sin(\alpha + \pi/2)$$
$$= m(2R_E k)^{1/2}\cos\alpha\ .$$

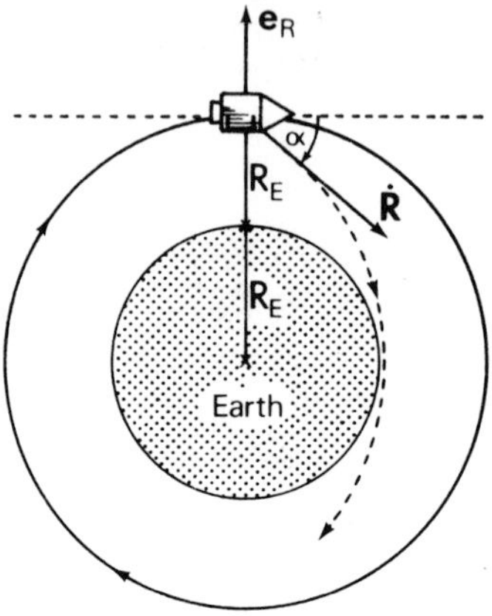

Fig. 7.7 Satellite path

The condition for the satellite to just touch the Earth is that the apsidal distance $R_1 = R_E$. Substituting for E and H in (7.27) gives the apsidal equation:

$$R^2 - 4R_E R + 4\cos^2\alpha\, R_E^{\,2} = 0\ .$$

Thus, the smaller apsidal distance is given by

$$\frac{R_1}{R_E} = 2 - (4 - 4\cos^2\alpha)^{1/2}$$
$$= 2 - 2\sin\alpha$$

and for $R_1 = R_E, \alpha = \sin^{-1}(1/2)$

$$\text{i.e.}\quad \alpha = 30^\circ\ .$$

7.7* Communication Satellites

Satellites open up a new approach to the problem of military and civil global communications. There are two distinct types of communication satellite systems:

(i) 'Passive' systems in which the satellites serve only as reflectors from which the signals are bounced from the transmitting ground station to the receiving ground station.

(ii) 'Active' systems in which satellites carry equipment to receive, amplify and retransmit signals. (See Fig. 7.8).

The first type is clearly extremely simple and hence reliable but the second type has more practical uses.

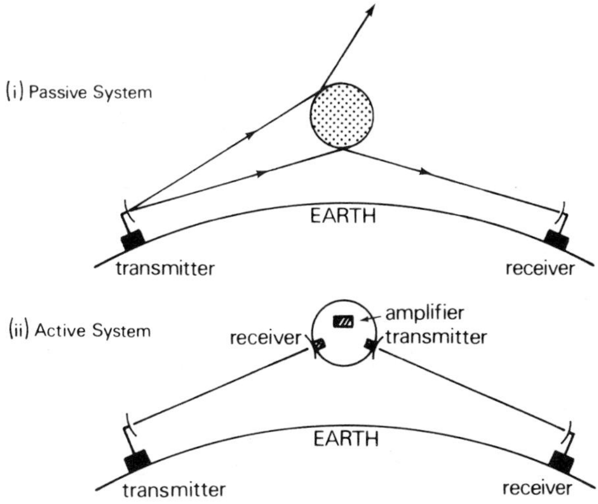

Fig. 7.8 Satellite communications systems

In both cases the satellites can either be moving or 'stationary' relative to the Earth. For civil communications it is an advantage to have a stationary satellite so that contact between two transmitting stations can be made at any time.

If a satellite is at rest relative to the Earth, then it must revolve round the Earth in a circular orbit with periodic time one day. But, from (7.23), the periodic time for a circular orbit of radius $(R_E + x)$ is

$$T = 2\pi(R_E + x)^{3/2}/k^{1/2}. \qquad (7.28)$$

This shows that there is a unique value of x, the height above the Earth, which ensures that a circular orbit is stationary relative to the Earth. Now one day is $8{\cdot}64 \times 10^4$ s and (7.28) gives

$$x = -R_E + (\gamma M_E)^{1/3}(8{\cdot}64 \times 10^4/2\pi)^{2/3}.$$

Substituting the values of R_E, γ and M_E gives

$$x = (-6368 + 42140)\text{km}$$

$$= 3{\cdot}58 \times 10^4 \text{ km} .$$

It is of interest to find out how many stationary satellites are needed so that every point on the equator is in view of at least one satellite. Assuming the satellites are placed in the appropriate circular orbit in the Earth's equatorial plane, a single satellite will cover an angle 2α on the Earth's surface, where from Fig. 7.9

$$\cos\alpha = R_E/(x + R_E)$$

$$= 6368/42140 .$$

Thus α is about 81°, and the angle covered by one satellite is approximately 162°. Clearly two satellites would not be sufficient for the purpose but three would certainly suffice.

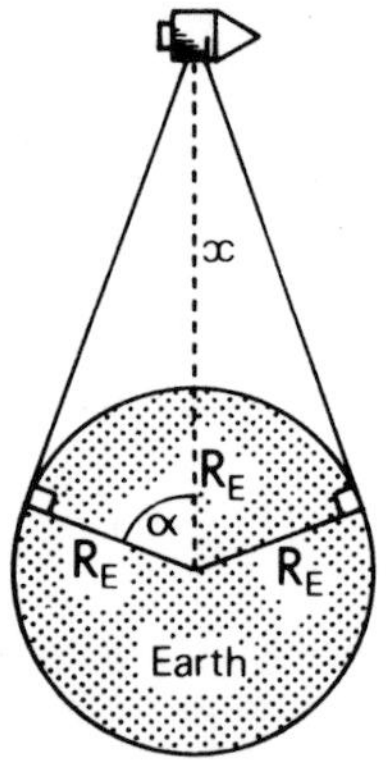

Fig. 7.9 Communication satellite

7.8 The Initial Value Problem

It is clearly of fundamental importance to be able to solve the initial value problem for a launched satellite. That is, given the initial position **R** and velocity $\mathbf{v} = \dot{\mathbf{R}}$ in some coordinate system, we wish to determine the nature of the satellite's orbit and its position after a time *t*. The first problem is quite straightforward and we will describe and illustrate one method for solving it. The second problem is much more difficult and beyond the scope of this introductory text.

Our initial conditions, often referred to as the 'cut-off' conditions as these are the conditions at rocket burnout, can be written as

$$R = R_o, \; v = V_o, \; \beta = \beta_o$$

where β, called the heading angle, is the angle between the normal to $\mathbf{R}$ and the velocity $\mathbf{v}$ (Fig. 7.10).

To determine the orbit completely, we must find the eccentricity e, the semi-latus rectum l, and the angle, say ϕ_0, between perigee and its initial position. With a coordinate system chosen so that ϕ is measured from perigee, the orbit equation is

$$R = \frac{l}{1 + e \cos \phi} . \tag{7.29}$$

Eccentricity: From (7.20),

$$\begin{aligned} e^2 &= 1 + 2H^2 E/k^2 m^3 \\ &= 1 + \frac{2H^2}{k^2 m^3}\left(\frac{1}{2} m V_0{}^2 - \frac{mk}{R_o}\right), \end{aligned}$$

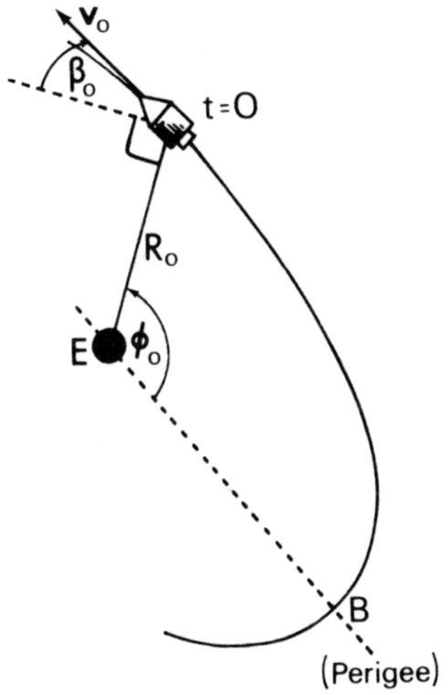

Fig. 7.10 Cut-off conditions

and $H = m \,|\, \mathbf{R} \times \dot{\mathbf{R}} \,| = mR_o V_o \cos \beta_o$, since E and H are constant throughout the motion and can therefore be evaluated initially.

Hence

$$e^2 = 1 + 2R_o^2V_o^2\cos^2\beta_o\left(\frac{1}{2}V_o^2 - \frac{k}{R_o}\right)\Big/k^2$$

$$= 1 + \cos^2\beta_o\left(\frac{R_o^2V_o^4}{k^2} - 2\frac{R_oV_o^2}{k}\right)$$

i.e.

$$\boxed{e^2 = 1 + \frac{R_oV_o^2}{k}\left(\frac{R_oV_o^2}{k} - 2\right)\cos^2\beta_o} \quad . \tag{7.30}$$

Semi-latus rectum: From (7.19),

$$l = H^2/m^2k = \frac{R_o^2V_o^2}{k}\cos^2\beta_o$$

i.e.

$$\boxed{l/R_o = \left(\frac{R_oV_o^2}{k}\right)\cos^2\beta_o} \quad . \tag{7.31}$$

Angle ϕ_o. From (7.29),

$$1 + e\cos\phi_o = l/R_o = \left(\frac{R_oV_o^2}{k}\right)\cos^2\beta_o$$

and so

$$\cos\phi_o = \left[\left(\frac{R_oV_o^2}{k}\right)\cos^2\beta_o - 1\right]\Big/e\ .$$

Using the value (7.30) for e and rearranging, it can be shown that

$$\boxed{\tan\phi_o = \frac{(R_oV_o^2/k)\cos\beta_o\sin\beta_o}{[(R_oV_o^2/k)\cos^2\beta_o - 1]}} \quad . \tag{7.32}$$

Note: It is interesting to note the dependence of e, l/R_o and ϕ_o on the heading angle β_o and on the factor $R_oV_o^2/k$.

EXAMPLE:

A satellite is launched in the Earth's equatorial plane at height $R_E/2$ above the Earth's surface with speed $\sqrt{k/R_E}$ and heading angle 30°. Determine the orbit and find the maximum and minimum heights of the satellite above the Earth's surface.

SOLUTION:

Now
$$\frac{R_o V_o^2}{k} = 3\frac{R_E}{2}\;\frac{k}{R_E}\;\frac{1}{k} = \frac{3}{2}\,,$$

and from (7.30),
$$e^2 = 1 + \frac{3}{2}\left(-\frac{1}{2}\right)\frac{3}{4} = \frac{7}{16}\;;$$

i.e. the eccentricity is $\sqrt{7}/4$, which shows that the orbit is an ellipse ($e < 1$). Also, from (7.31),

$$l = \frac{3R_E}{2} \times \frac{3}{2} \times \frac{3}{4} = \frac{27}{16}R_E\;;$$

and from (7.32),

$$\tan\phi_o = \frac{3}{2} \times \frac{\sqrt{3}}{2} \times \frac{1}{2}\bigg/\left[\frac{3}{2} \times \frac{3}{4} - 1\right] = 3\sqrt{3}\;;$$

$$\text{i.e.}\quad \phi_o = \tan^{-1}(3\sqrt{3})\;.$$

Thus the orbit has equation

$$R = \frac{27R_E}{16 + 4\sqrt{7}\cos\phi}$$

and initially $\phi = \tan^{-1}(3\sqrt{3})$. Now the apsidal distances are given by

$$R_1 = \frac{27R_E}{(16 + 4\sqrt{7})}\,,\quad R_2 = \frac{27R_E}{(16 - 4\sqrt{7})}$$

and so the maximum/minimum heights above the Earth's surface are

$$\left(\frac{11 + 4\sqrt{7}}{16 - 4\sqrt{7}}\right)R_E\,,\quad \left(\frac{11 - 4\sqrt{7}}{16 + 4\sqrt{7}}\right)R_E$$

respectively; i.e. $3{\cdot}984\,R_E$ and $0{\cdot}016\,R_E$.

7.9 Orbit Parameters

From (7.20), we see that $e \leqslant 1$ if and only if $E \leqslant 0$, and since

$$E = \tfrac{1}{2}mV_o^2 - mk/R_o\,,$$

$$\frac{2ER_o}{mk} = \frac{R_o V_o^2}{k} - 2\,.$$

Thus $e \leqslant 1$ if and only if $R_o V_o^2/k \leqslant 2$. The dependence of the type of orbit on the parameters e, E and $R_o V_o^2/k$ is illustrated below:

Orbit	e	E	$R_o V_o^2/k$
Ellipse	$e<1$	$E<0$	$R_o V_o^2/k < 2$
Parabola	$e=1$	$E=0$	$R_o V_o^2/k = 2$
Hyperbola	$e>1$	$E>0$	$R_o V_o^2/k > 2$

Note also that $e = 0$ gives a circle and $R_o V_o^2/k = 1$ is a necessary but not sufficient condition for a circular orbit. To see this last result, we note that for a circular orbit of radius R_o, the speed is constant and equal to its initial value V_o, so that from (7.25)

$$V_o^2 = k\left(\frac{2}{R_o} - \frac{1}{R_o}\right),$$

giving $R_o V_o^2/k = 1$. Hence this is a necessary but not sufficient condition for a circular orbit, for β_o need not be zero.

7.10* Transfer Orbits

In this section we consider the transfer of a satellite from one orbit to another, and in particular from Earth's orbit to Mars' orbit. We assume that the orbits of Earth and Mars are coplanar circles, centres at the sun, of radii R_{ES} and R_{MS} respectively, and we neglect the gravitational forces on the satellite due to Earth and Mars. After the initial firing at E (Fig. 7.11), the satellite moves in an orbit under the influence of the Sun until it reaches the orbit of Mars at M.

Any satellite fired from the Earth will have the velocity of the Earth in addition to its projected velocity. To use this property to its fullest extent, we assume the satellite leaves the Earth's orbit in a tangential direction to the Earth's orbit path. The two velocities are thus in the same direction, and will combine to give the greatest possible initial speed to the satellite.

If the equation of the orbit between E and M is

$$R = l/(1 + e\cos\phi) \tag{7.33}$$

then, since E is the perihelion of the orbit, we must have

$$R_{ES} = l/(1 + e)$$

$$R_{MS} = l/(1 + e\cos\phi_o),$$

ϕ_o being the value of ϕ when the satellite reaches the orbit of Mars.

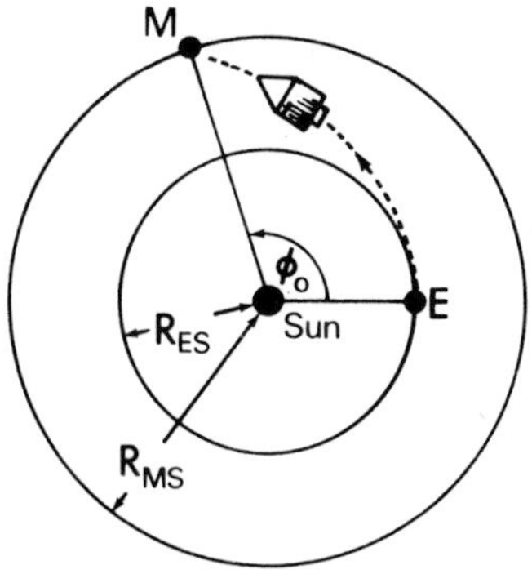

Fig. 7.11 Transfer orbit

Eliminating l from the above two equations gives

$$e = (R_{MS} - R_{ES})/(R_{ES} - R_{MS}\cos\phi_0) \tag{7.34}$$

and so
$$l = R_{MS}R_{ES}(1 - \cos\phi_0)/(R_{ES} - R_{MS}\cos\phi_0)\,. \tag{7.35}$$

This defines the orbit for a given value of ϕ_0. We choose this parameter in order to minimize the velocity increment given to the satellite at E.

From theorem 7.1, the velocity required at E is given by

$$v^2 = k\left[\frac{2}{R_{ES}} - \frac{(1 - e^2)}{l}\right]$$

$$= k\left[\frac{2}{R_{ES}} - \frac{(1 - e)}{R_{ES}}\right]$$

$$= \frac{k}{R_{ES}}(1 + e)$$

i.e.
$$v = [k(1 + e)/R_{ES}]^{1/2}\,.$$

Since the satellite leaves with the velocity of the Earth, $(k/R_{ES})^{1/2}$, the required increment, δv_1, at E is given by

$$\delta v_1 = \left(\frac{k}{R_{ES}}\right)^{1/2}[(1 + e)^{1/2} - 1]\,. \tag{7.36}$$

This increment is minimized when e is minimized, and from (7.34) e has a minimum value when $\phi_0 = \pi$. In this case the elliptic orbit traversed by the satellite has perihelion and aphelion distances equal to R_{ES} and R_{MS} respectively (Fig. 7.12). In this case

$$e = (R_{MS} - R_{ES})/(R_{MS} + R_{ES}) \tag{7.37}$$

with $$\delta v_1 = \left(\frac{k}{R_{ES}}\right)^{1/2} \left\{[2R_{MS}/(R_{MS} + R_{ES})]^{1/2} - 1\right\} . \tag{7.38}$$

This orbit is called the *Hohmann Transfer Orbit* and it minimizes the velocity that has to be given to the satellite.

If the satellite goes into the circular orbit of Mars at M, it will again require an increment in velocity, δv_2, say. This can easily be shown to be

$$\delta v_2 = \left(\frac{k}{R_{MS}}\right)^{1/2} \left\{1 - [2R_{ES}/(R_{MS} + R_{ES})]^{1/2}\right\}, \tag{7.39}$$

and the total velocity increment in order to place the satellite in the orbit of Mars is $\delta v_1 + \delta v_2$. Using the appropriate values of R_{ES}, R_{MS} and $k = \gamma M_s$,, we find that

$$\delta v_1 = 2{\cdot}8 \text{ km s}^{-1}, \ \delta v_2 = 2{\cdot}6 \text{ km s}^{-1} .$$

These are certainly attainable velocity increments. It is also of interest to find the journey time. From (7.23), the journey time τ is given by

$$\begin{aligned} \tau &= \pi a^{3/2}/k^{1/2} \\ &= \pi(R_{ES} + R_{MS})\, R_{MS})^{3/2}/2^{3/2}k^{1/2} , \end{aligned}$$

and substituting the appropriate values gives $\tau \approx 8$ months.

Although the Hohmann transfer orbit is a minimum velocity orbit, it does have many disadvantages. For instance, the journey time is very long for a manned space flight and also the relative positions of Earth and Mars are only correct for this transfer orbit every 780 days! Furthermore, for a return trip via a Hohmann transfer orbit, one would have to wait on Mars until the relative planetary positions were again correct.

The method above can also be applied to transfer orbits between confocal elliptic orbits but if the ratio R_{MS}/R_{ES} for other planets becomes large there are more economical paths to follow. One method is to move on an ellipse to a point C, beyond the target orbit (Fig. 7.12), followed by a return ellipse to B. This is called a 'bi-elliptic' orbit. A limiting case of the bi-elliptic orbit is a parabolic path from E to infinity, followed by a return to B on another parabolic path. Using theorem 7.1, the first increment in velocity for this path is

$$\delta v_1 = \sqrt{\frac{k}{R_{ES}}}(\sqrt{2} - 1)$$

and similarly $$\delta v_2 = \sqrt{\frac{k}{R_{MS}}}(\sqrt{2} - 1) .$$

If v_c is the original circular velocity and with $\alpha = R_{MS}/R_{ES}$,

$$\frac{\delta v_1 + \delta v_2}{v_c} = (\sqrt{2} - 1)(1 + \alpha^{-\frac{1}{2}})\,. \tag{7.40}$$

But for the Hohmann transfer orbit, from (7.38) and (7.39),

$$\frac{\delta v_1 + \delta v_2}{v_c} = \frac{(\alpha - 1)}{\alpha^{\frac{1}{2}}}\left(\frac{2}{1 + \alpha}\right)^{\frac{1}{2}} + \alpha^{-\frac{1}{2}} - 1\,. \tag{7.41}$$

Now when $\alpha = 25$, expression (7.40) has value 0·497 whereas (7.41) has value 0·531, showing that the 'bi-parabolic' transfer orbit requires a smaller velocity increment than the Hohmann transfer orbit. One should note that this paradox underlines the fact that in optimization problems it is necessary to describe *very carefully* the admissible paths. For treating such optimization problems the techniques of dynamic programming and the Pontryagin principle are of great value, but beyond the scope of this book.

Bi-elliptic orbits are in fact of little practical use for, although they are very economical in energy required, they are far too slow.

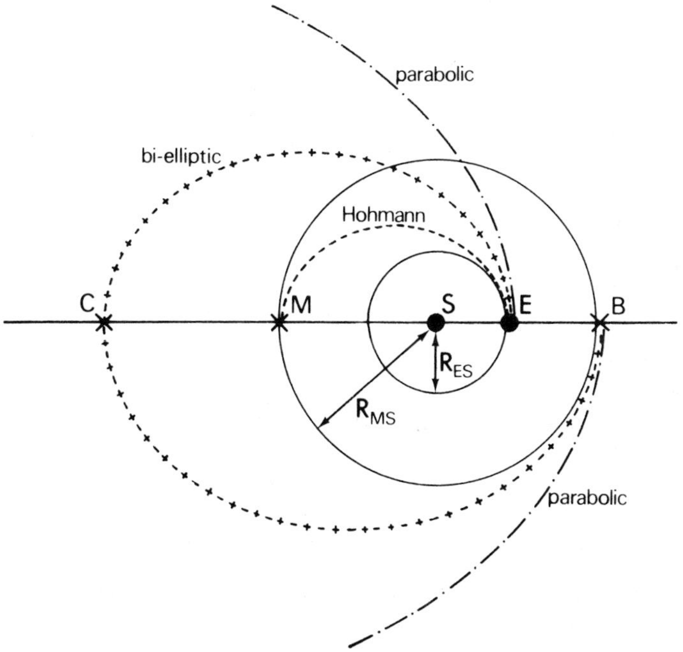

Fig. 7.12 Possible transfer orbits

7.11* Interception and Rendezvous

The interception and rendezvous of satellites in orbit is a very complex mathematical problem leading to transcendental equations that can only be solved by numerical techniques. In this section we will consider one of the simplest of this class of problems.

Two satellites, A and B, move initially in the same circular orbit, radius R_o, about the Earth, one leading the other by a specified angle ϕ_o. Suppose satellite A pursues satellite B, and further we suppose that satellite A intercepts B at position C, where ϕ_1 is the angle between B and C, by travelling on an elliptic path from A to C, whilst B continues to travel on its circular path to C. This is illustrated in Fig. 7.13.

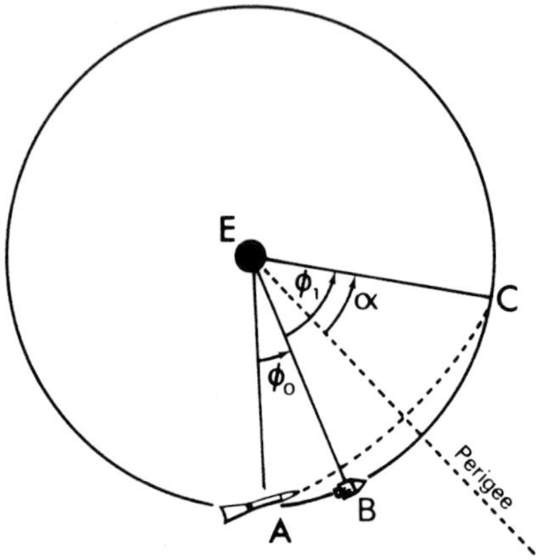

Fig. 7.13 Interception on a circular orbit

The perigee for the new orbit must bisect the angle $\phi_o + \phi_1$, and if the equation of the path is

$$R = l/(1 + e\cos\phi)\ ,,$$

then at C, $R_o = l/(1 + e\cos\alpha)$ where $\alpha = (\phi_o + \phi_1)/2$ and is a known angle. But the time of motion, τ say, for the elliptic path AC for interception at C is the same as the time for the circular path BC. Thus, using (7.23),

$$\tau = 2\pi\frac{R_o^{3/2}}{k^{1/2}}\frac{a}{2\pi} = \alpha\frac{R_o^{3/2}}{k^{1/2}} \qquad (7.42)$$

Hence the time from perigee to C for satellite A is $\alpha R_o^{3/2}/2k^{1/2}$: But from problem 7.13, we have

$$t = \frac{a^{3/2}}{k^{1/2}}\left[2\tan^{-1}\left\{\left(\frac{1-e}{1+e}\right)^{1/2}\tan\left(\frac{1}{2}\phi\right)\right\} - \frac{e(1-e^2)^{1/2}\sin\phi}{(1+e\cos\phi)}\right] \qquad (7.43)$$

where
$$a = \frac{l}{(1 - e^2)} = R_o \frac{(1 + e\cos\alpha)}{(1 - e^2)} . \tag{7.44}$$

Substituting for t and a, and with $\phi = \alpha$, we have

$$(1-e^2)^{3/2}\alpha = (1 + e\cos\alpha)^{3/2}\left[2\tan^{-1}\left\{\left(\frac{1-e}{1+e}\right)^{1/2}\tan\left(\frac{1}{2}\alpha\right)\right\} - \frac{e(1-e^2)^{1/2}\sin\alpha}{(1 + e\cos\alpha)}\right] \tag{7.45}$$

This is a transcendental equation for α, which clearly has to be solved numerically.

7.12* Ballistic Trajectories

The elliptic orbit equations developed in the preceeding sections have direct application to ballistic missile trajectories, terminating on the Earth's surface. Fig. 7.14 shows the geometry of a ballistic trajectory, L being the launch point and T the target point. It is assumed that the missle, given an initial velocity V_o with heading angle β_o at L, moves in space under the gravitational influence of the Earth until it reaches its target at T. In practice the missile will be given small course corrections during its flight.

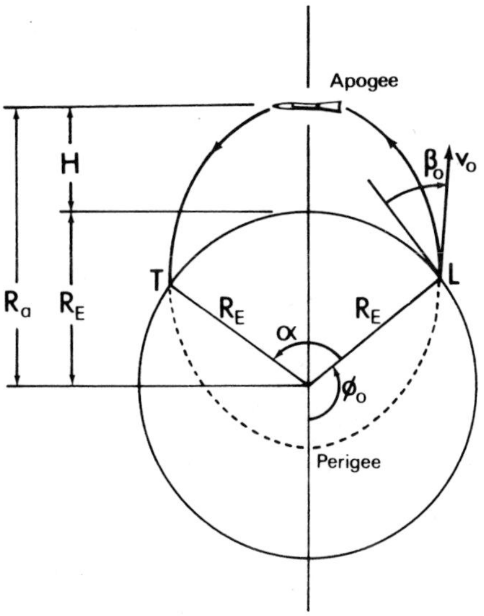

Fig. 7.14 Ballistic trajectory

Our interest here lies in determining the range αR_E and height H. The initial conditions are

$$R = R_E,\ v = V_o,\ \beta = \beta_o\ ;$$

and from (7.30),

$$e^2 = 1 + \frac{R_E V_o^2}{k}\left(\frac{R_E V_o^2}{k} - 2\right)\cos^2\beta_o\ . \qquad (7.46)$$

Since $\phi_o + \alpha/2 = \pi$, $\tan\phi_o = -\tan(\alpha/2)$, and from (7.32)

$$\tan(\alpha/2) = \frac{(R_E V_o^2/k)\sin\beta_o\cos\beta_o}{1-(R_E V_o^2/k)\cos^2\beta_o} \qquad (7.47)$$

As expected this result shows that the range depends on $(R_E V_o^2/k)$ and β_o. For $0 < \alpha < \pi$, maximum range for a given value of $R_E V_o^2/k$ corresponds to a maximum value of $\tan(\alpha/2)$, and this occurs when

$$(2\cos^2\beta_o - 1)[(R_E V_o^2/k)\cos^2\beta_o - 1] + 2(R_E V_o^2/k)\sin^2\beta_o\cos^2\beta_o = 0\ .$$

Solving,
$$\cos^2\beta_o = 1/(2 - R_E V_o^2/k)\ , \qquad (7.48)$$

provided $R_E V_o^2/k < 1$. This expression gives the optimum value of β_o for given velocity V_o in order to obtain maximum range. Fig. 7.15 shows the range α plotted against β_o for given values of $R_E V_o^2/k$.

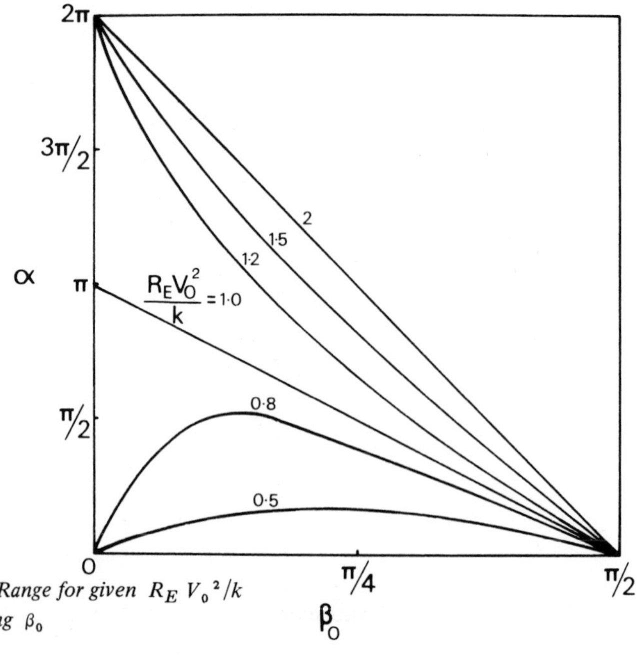

Fig. 7.15 Range for given $R_E V_o^2/k$ qnd varying β_o

From (7.48), $$\sin^2 \beta_o = [1 - (R_E V_o^2/k)]/[2 - (R_E V_o^2/k)] \tag{7.49}$$

and substitution in (7.47) gives

$$\tan(\alpha/2) = (R_E V_o^2/k)/2[1 - (R_E V_o^2/k)]^{1/2}. \tag{7.50}$$

In the general case, we see that

$$H + R_E = R_a ,$$

where R_a is the apogee distance. But

$$R_a = a(1 + e)$$

and from the equation for the ellipse, at L or T.

$$R_E = \frac{a(1 - e^2)}{1 + e\cos\phi_o}$$

$$= \frac{a(1 - e^2)}{1 - e\cos(a/2)} .$$

Hence $$H = \frac{[1 - e\cos(\alpha/2)]}{(1 - e)} R_E - R_E$$

i.e. $$H = \frac{e[1 - \cos(\alpha/2)]}{(1 - e)} R_E . \tag{7.51}$$

Worked Examples

EXAMPLE 7.1:

Compare the acceleration due to gravitational attraction experienced on the surface of the following planets with that experienced on the earth's surface:

(i) Moon, (ii) Jupiter, (iii) Uranus.

SOLUTION:

From equation (7.2), we have

$$g_P = \gamma \frac{M_P}{R_P^2} ,$$

where g_P is the acceleration on planet P, with mass M_P and radius R_P. Thus if suffix E denotes the earth,

$$\frac{g_P}{g_E} = \left(\frac{M_P}{M_E}\right)\left(\frac{R_E}{R_P}\right)^2 .$$

Using this formula with the data given in Ch. 7.2 gives

(i) Moon: $g_{moon} \approx 0{\cdot}16 g_E$

(ii) Jupiter: $g_{jupiter} \approx 2{\cdot}64 g_E$

(iii) Uranus: $g_{uranus} \approx 1{\cdot}09 g_E$

EXAMPLE 7.2:

Assuming Kepler's second law for the motion of a planet, prove that the force acting on the planet must be in the radial direction.

SOLUTION:

The increment in area swept out by the planet in a small time δt is given (Fig. 7.5) by

$$\delta A = \tfrac{1}{2} R^2 \delta \phi$$

where $\delta\phi$ is the angle through which the radius vector turns in time δt. In the limit, as $\delta t \to 0$, we have

$$\frac{dA}{dt} = \tfrac{1}{2} R^2 \dot{\phi} .$$

Now Kepler's second law implies that dA/dt is constant (i.e. area swept out is proportional to time) and hence we have

$$d^2 A/dt^2 = 0 ;$$

i.e.

$$\frac{d}{dt}\left(\frac{1}{2} R^2 \dot{\phi}\right) = 0 ,$$

which gives

$$R\dot{R}\dot{\phi} + \tfrac{1}{2} R^2 \ddot{\phi} = 0 ;$$

i.e.

$$2\dot{R}\dot{\phi} + R\ddot{\phi} = 0 .$$

But, in terms of plane polar coordinates, the planet's acceleration is given by

$$\ddot{\mathbf{R}} = (\ddot{R} - R\dot{\phi}^2)\mathbf{e}_R + (2\dot{R}\dot{\phi} + R\ddot{\phi})\mathbf{e}_\phi .$$

Thus $$\ddot{\mathbf{R}} = (\ddot{R} - R\dot{\phi}^2)\mathbf{e}_R$$

and, by Newton's second law, the force $\mathbf{F}$ acting is given by

$$\mathbf{F} = m\ddot{\mathbf{R}} = m(\ddot{R} - R\dot{\phi}^2)\mathbf{e}_R$$

EXAMPLE 7.3:

A particle of mass m is projected from a point distant R_0 from the origin with a velocity of magnitude V_0 in a direction perpendicular to the radius vector. Assuming that the particle moves under an inverse cube law such that it experiences a force directed towards the origin and of magnitude λ/R^3 per unit mass, where λ is a constant and R is the radial distance, determine its orbit for the case when $R_0 V_0 > \lambda^{1/2}$.

SOLUTION:

Using plane polar coordinates (R, ϕ) so that the initial radius vector is given by $\phi = 0$, the equation of motion is

$$m\ddot{\mathbf{r}} = m(\ddot{R} - R\dot{\phi}^2)\mathbf{e}_R + m(2\dot{R}\dot{\phi} + R\ddot{\phi}^2)\mathbf{e}_\phi = -\frac{\lambda m}{R^3}\mathbf{e}_R$$

Thus $$2\dot{R}\dot{\phi} + R\ddot{\phi}^2 = 0$$

i.e. $$\frac{1}{R}\frac{d}{dt}(R^2\dot{\phi}) = 0 ,$$

and hence $h = R^2\dot{\phi}$ is a constant. We also have

$$\ddot{R} - R\dot{\phi}^2 = -\frac{\lambda}{R^3} ,$$

whence, eliminating $\dot{\phi}$, we obtain

$$\ddot{R} - \frac{h^2}{R^3} = -\frac{\lambda}{R^3} .$$

This is the analogous equation to (7.17), and we can use the same method of solution; i.e. substitute $p = 1/R$ and solve for p as a function of ϕ. Thus

$$\dot{R} = \frac{d}{dt}(1/p) = \frac{d}{d\phi}(1/p)\frac{d\phi}{dt}$$

$$= -R^2\dot{\phi}\frac{dp}{d\phi}$$

$$= -h\frac{dp}{d\phi},$$

and

$$\ddot{R} = -\frac{d^2p}{d\phi^2}\dot{\phi}$$

$$= -h^2p^2\frac{d^2p}{d\phi^2}.$$

Substituting into the differential equation,

$$\frac{d^2p}{d\phi^2} + \left(1 - \frac{\lambda}{h^2}\right)p = 0.$$

Clearly the form of the solution depends on the sign of $(1 - \lambda/h^2)$, and the value of h is found from the initial conditions.

Now, initially, $\mathbf{v} = \dot{R}\mathbf{e}_R + R\dot{\phi}\mathbf{e}_\phi = V_0\mathbf{e}_\phi$ i.e. initially $\dot{\phi} = V_0/R_0$ and $\dot{R} = 0$, so that $h = R_0{}^2(V_0/R_0) = R_0V_0$. Thus if $R_0V_0 > \lambda^{1/2}$, then $h^2 > \lambda$ and $1 - \lambda/h^2 > 0$. Writing $\alpha^2 = 1 - \lambda/h^2$, we have the solution,

$$p = A\sin(\alpha\phi + \beta),$$

where A and β are constants. Applying the initial conditions $p = 1/R_0$ when $\phi = 0$ gives

$$\frac{1}{R_0} = A\sin\beta$$

and $\dot{R} = -h\,dp/d\phi = 0$ when $\phi = 0$ gives

$$0 = A\alpha\cos\beta.$$

Thus $\beta = \pi/2$ and $A = 1/R_0$, giving

$$p = \frac{1}{R_0}\sin(\alpha\phi + \pi/2).$$

The orbit equation is then $R\cos(\alpha\phi) = R_0$.

EXAMPLE 7.4:

Defining $I = \frac{1}{2}mR^2$, show that for motion under an inverse square law of force,

$$\frac{d^2 I}{dt^2} = T + E \ ,$$

where T is the kinetic energy, and E the total energy.

SOLUTION:

Since $I = \frac{1}{2}mR^2$, we have $$\frac{dI}{dt} = mR\dot{R}$$

and $$\frac{d^2 I}{dt^2} = m\dot{R}^2 + mR\ddot{R} \ .$$

But from (7.14), $\ddot{R} = R\dot{\phi}^2 - k/R^2$, so that

$$\frac{d^2 I}{dt^2} = m\dot{R}^2 + mR^2\dot{\phi}^2 - mk/R$$

$$= 2T + V \ ,$$

where $T = \frac{1}{2}m(\dot{R}^2 + R^2\dot{\phi}^2)$ is the kinetic energy, and $V = -mk/R$ (lemma 7.1) is the potential energy. But the total energy, E, is given by $E = T + V$, and hence the result follows.

EXAMPLE 7.5:

A satellite is moving uniformly in a circular orbit of radius $4a$ about the centre of the earth, O, where a is the radius of the earth. Retro-rockets are fired to reduce the speed of the satellite, without a change in direction, so that it subsequently moves in an elliptic orbit, the least distance from O in the ensuing motion being $2a$. Show that the satellite's speed must be reduced in the ratio $\sqrt{2} : \sqrt{3}$.

SOLUTION:

For circular motion, of radius d, from (7.25), we have $v^2 = k/d$, so that just before the retro-rockets are fired

$$v_0^2 = k/4a \ .$$

Suppose that the speed is changed to λv_0. The energy constant for the subsequent motion is

$$E = K.E. + P.E.$$

$$= \frac{1}{2}m\lambda^2\left(\frac{k}{4a}\right) - \frac{mk}{4a},$$

evaluating initially, and since $P.E. = -mk/R$. Thus

$$E = \frac{mk}{8a}(\lambda^2 - 2).$$

Also $H = |\mathbf{H}| = m\,|\mathbf{R} \times \dot{\mathbf{R}}|$, and evaluating this constant initially gives

$$H = m(4a)\lambda\,(k/4a)^{1/2} = m\lambda(4ak)^{1/2}.$$

We can now use the apsidal equation, (7.27), to determine the satellite's least distance from O. Substituting for E and H in (7.27) gives

$$R^2 + \frac{8aR}{(\lambda^2 - 2)} - \frac{16\lambda^2 a^2}{(\lambda^2 - 2)} = 0.$$

We know that the apogee distance must be $R = 4a$, and so this must be one root. Thus factorising,

$$(R - 4a)\left\{R + \frac{4\lambda^2 a}{(\lambda^2 - 2)}\right\} = 0,$$

giving the perigee distance as $4\lambda^2 a/(2 - \lambda^2)$. We require this to be equal to $2a$; i.e.

$$\frac{4\lambda^2 a}{2 - \lambda^2} = 2a,$$

giving $3\lambda^2 = 2$; i.e. $\lambda = \sqrt{2}/\sqrt{3}$. This is the required result.

EXAMPLE 7.6:

Show that, for a satellite moving on an elliptic orbit,

$$e = \frac{R_p {v_p}^2}{k} - 1$$

where R_p and v_p are the perigee values of R and v respectively.

SOLUTION:

For an elliptic orbit $R = l/(1 + e\cos\phi)$, and at perigee $\phi = 0$. Thus $R_p = l/(1 + e)$, and using (7.7),

$$R_p = a(1 - e) \; .$$

Now, from theorem 7.1,

$$v^2 = k\left(\frac{2}{R} - \frac{1}{a}\right)$$

$$= k\left\{\frac{2}{R} - \frac{(1 - e)}{R_p}\right\}$$

Evaluating at perigee gives

$$v_p{}^2 = \frac{k}{R_p}(1 + e)$$

or

$$e = \frac{R_p v_p{}^2}{k} - 1 \; .$$

EXAMPLE 7.7:

At perigee on its first elliptic orbit of Earth, a satellite experiences a drag due to Earth's upper atmosphere which has the effect of reducing its speed from v_p to $v_p(1 - \epsilon)$, without a change in direction, where ϵ is a small positive constant. Show, using Worked Example 7.6, that the eccentricity of its new orbit about the Earth is given approximately by $e - 2\epsilon(1 + e)$, where e is the eccentricity of the original orbit.

Deduce the eccentricity after n perigee passes. If $e = 0{\cdot}9$ and $\epsilon = 0{\cdot}01$, how many perigee passes does the satellite make before its path is approximately circular?

SOLUTION:

If R_p, v_p are perigee distance and speed then, from Example 7.6,

$$e = \frac{R_p v_p{}^2}{k} - 1 \; .$$

But at perigee on its first orbit the speed is reduced to $(1 - \epsilon)v_p$ and, again using Example 7.6, the new orbit's eccentricity is given by

$$e_1 = R_p(1 - \epsilon)^2 v_p{}^2/k - 1$$

$$= \frac{R_p v_p{}^2}{k} - 1 - 2\frac{\epsilon R_p v_p{}^2}{k} + \text{terms of order } \epsilon^2$$

i.e.

$$e_1 \approx e - 2\epsilon(1 + e) ,$$

using the formula for e.

Clearly, the eccentricity of the orbit after the satellite's second perigee pass is given by

$$e_2 \approx e_1 - 2\epsilon(1 + e_1)$$

$$= e - 2\epsilon(1 + e) - 2\epsilon(1 + e) + \text{terms of order } \epsilon^2$$

i.e.

$$e_2 \approx e - 4\epsilon(1 + e) .$$

Thus after n perigee passes

$$e_n \approx e - 2n\epsilon(1 + e) .$$

For a circular path $e_n \approx 0$ giving $e \approx 2n\epsilon(1 + e)$,

$$\text{or} \quad n \approx e/2\epsilon(1 + e) .$$

With $e = 0{\cdot}9$ and $\epsilon = 0{\cdot}01$, this gives $n \approx 24$. Note that as the orbit approaches a circle, the satellite will begin to experience drag due to the upper atmosphere throughout its orbit, and the model would not then be very realistic, so that the number of perigee passes, 24, is only a rough order of magnitude.

Problems

7.1 Determine, in terms of g, the gravitational force experienced by a particle of mass m on the surface of Mars.

7.2 Show that the equation of the ellipse

$$R = l/(1 + e \cos \phi)$$

in terms of plane polar coordinates (R, ϕ) with origin at one focus, becomes

$$\frac{x^2}{a^2} + \frac{y^2}{b^2} = 1 ,$$

where x and y are rectangular Cartesian coordinates with origin at the centre of the ellipse and axes orientated so that the x-axis passes through the focus.

7.3 With the usual notation for an ellipse, show that

$$R_p = a(1 - e), \; R_a = a(1 + e) \; .$$

Also prove that

(i) $a = (b^2 + R_a{}^2)/2R_a = (b^2 + R_p{}^2)/2R_p$;

(ii) $e = (R_a - R_p)/(R_a + R_p)$.

7.4 Show that a circular path, $R = a$, is a solution of the equation of planetary motion

$$\ddot{\mathbf{R}} = -\frac{k}{R^3}\mathbf{R} \; ,$$

provided the angular momentum about the origin has a specified value. Find this value.

7.5 Halley's comet moves in an elliptic orbit about the sun with period 76 years. Find its semi-major axis and, if $e = 0{\cdot}97$, determine its aphelion and perihelion distances from the sun, in terms of the semi-major axis of the Earth's orbit about the sun.

7.6 A satellite is moving in a circular orbit of radius αR_E $(\alpha > 1)$ about the Earth, radius R_E. At an instant of time the velocity of the satellite, $\mathbf{v}$, is suddenly changed to $\lambda\mathbf{v}$ (λ constant). Find λ such that the satellite just touches the Earth.

7.7 A satellite, moving in a circular orbit of radius $2R_E$ about the Earth, changes the direction of its velocity through an angle α, its magnitude remaining unchanged. Find the value of α so that its nearest approach to the Earth is $R_E/2$ from the Earth's surface.

7.8* Communication satellites are placed in a circular orbit in the equatorial plane of Mars, such that they remain stationary with respect to the surface of Mars. What is the minimum number of such satellites required for every point on the equator to be in view of at least one satellite.

7.9 If the initial conditions for a satellite at rocket burnout are

$$\beta = \beta_o, \; R = R_o, \; v = V_o \; ,$$

show that the perigee distance is

$$R_p = R_o[1 - \{1 - K(2 - K)\cos^2\beta_o\}^{1/2}]/(2 - K)$$

and apogee distance

$$R_a = R_o[1 + \{1 - K(2 - K)\cos^2\beta_o\}^{1/2}]/(2 - K)\ ,$$

where $K = R_o V_o^2/k$.

7.10 A satellite is launched parallel to the Earth's surface at $R_o = (1{\cdot}1)R_E$ and with $R_o V_o^2/k = 1{\cdot}2$. Determine the apogee distance and the ratio of the apogee to perigee heights above the Earth's surface.

7.11 It is intended that a satellite is to be placed in a circular orbit $R = a$. The conditions at rocket burnout should be

$$R = a,\ v = \sqrt{k/a},\ \beta = 0\ ;$$

but due to an error in the rocket programme, the initial conditons are

$$R = a,\ v = \sqrt{k/a},\ \beta = \alpha\ .$$

Assess the effect of the heading angle error α on the perigee and apogee heights.

7.12 A satellite is moving in an elliptic orbit round the Earth. With the usual notation, show that the rate of change of the true anomaly is given by

$$\frac{d\phi}{dt} = \frac{k^{1/2}(1 + e\cos\phi)^2}{a^{3/2}(1 - e^2)^{3/2}}\ ,\ 0 \leqslant e \leqslant 1\ .$$

Hence show that the time of flight from perigee is given by

$$t = \frac{a^{3/2}(1 - e^2)^{3/2}}{k^{1/2}} \int_o^\phi \frac{d\phi}{(1 + e\cos\phi)^2}\ .$$

7.13 Show that

$$\int_o^\phi \frac{d\phi}{(1 + e\cos\phi)^2} = \frac{1}{(1 - e^2)^{3/2}}\left[2\tan^{-1}\left\{\left(\frac{1 - e}{1 + e}\right)^{1/2}\tan\left(\frac{1}{2}\phi\right)\right\} - \frac{e(1 - e^2)^{1/2}\sin\phi}{(1 + e\cos\phi)}\right]$$

and from problem 7.12 deduce that

$$t = \frac{a^{3/2}}{k^{1/2}}\left[2\tan^{-1}\left\{\left(\frac{1 - e}{1 + e}\right)^{1/2}\tan\left(\frac{1}{2}\phi\right)\right\} - \frac{e(1 - e^2)^{1/2}\sin\phi}{(1 + e\cos\phi)}\right].$$

$$\left[\text{Hint: } \frac{d}{d\phi}\left(\frac{e\sin\phi}{1 + e\cos\phi}\right) = \frac{1}{1 + e\cos\phi} - \frac{(1 - e^2)}{(1 + e\cos\phi)^2}\ .\right]$$

7.14 A satellite is moving in a circular orbit with speed v at a height h above the Earth's surface. Retro-rockets are fired which reduce the satellite's speed to $\lambda v (0 < \lambda < 1)$. Show that the satellite will collide with the Earth if

$$\lambda^2 \leqslant 2R_E/(2R_E + h) \ .$$

Also show that the impact velocity v_c is given by

$$v_c^{\,2} = k(2h + R_E\lambda^2)/(R_E + h)R_E \ ,$$

where $k = \gamma M_E$.

7.15* A rocket is fired from the vicinity of the Earth in the direction that the Earth is moving in its orbit round the sun. Find the minimum speed with which the rocket must be fired if it is to reach the orbit of Saturn on the Hohmann transfer orbit. (Assume Earth and Saturn move in coplanar circular orbits centred at the sun). Find the time of flight and the initial relative positions of the planets for the flight to take place.

7.16* A ballistic missile requires a range of $7{\cdot}5 \times 10^3$ km. Determine the optimum heading angle, β_o, initial velocity, V_o, and maximum height, H, reached by the missile.

Chapter 8

Stability

8.1 Intuitive Ideas

If a particle moves over a series of 'hills' and 'valleys' as in Fig. 8.1, we know intuitively that the particle can remain at rest at the highest and lowest points of the hills and valleys respectively; i.e. at points A, B, C in Fig. 8.1. We say that the particle is in equilibrium at points A, B or C, and call such points *equilibrium positions* of the particle. Suppose the particle is at rest at position A, and consider its motion when it is displaced slightly. Clearly the particle will move away from A with increasing momentum. This type of equilibrium position is termed *unstable.* Conversely, if the particle is slightly pushed away from B, it will return to B and oscillate about this equilibrium position. This type of equilibrium position is termed *stable.*

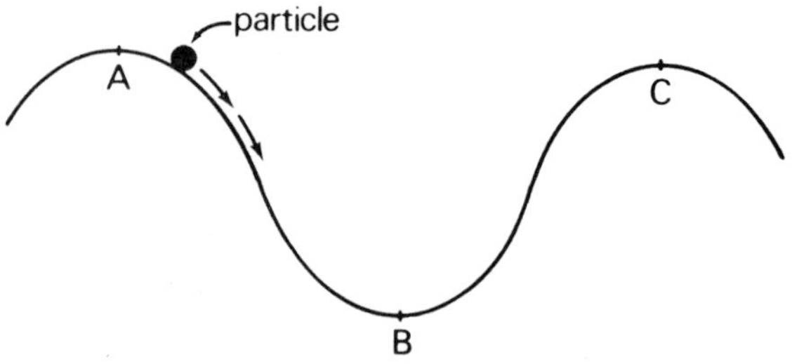

Fig. 8.1 Particle moving over hills and valleys

So far we have considered stability from a physical viewpoint only. Let us now investigate the mathematics of the problem. For instance, for motion near the equilibrium position A, assuming the particle and surface are perfectly smooth and the surface approximates to an arc of a circle, of radius a, at A, the equation of motion (in the usual notation, shown in Fig. 8.2) is

$$m\ddot{\mathbf{R}} = N\mathbf{e}_R - mg\cos\phi\,\mathbf{e}_R + mg\sin\phi\,\mathbf{e}_\phi .$$

But

$$\ddot{\mathbf{R}} = -a\dot{\phi}^2\,\mathbf{e}_R + a\ddot{\phi}\,\mathbf{e}_\phi ,$$

and taking components in the $\mathbf{e}_\phi$ direction gives

$$\ddot{\phi} = \frac{mg}{a} \sin \phi \,. \tag{8.1}$$

Provided ϕ is small, we have, to a first approximation,

$$\ddot{\phi} - \left(\frac{mg}{a}\right)\phi = 0\,. \tag{8.2}$$

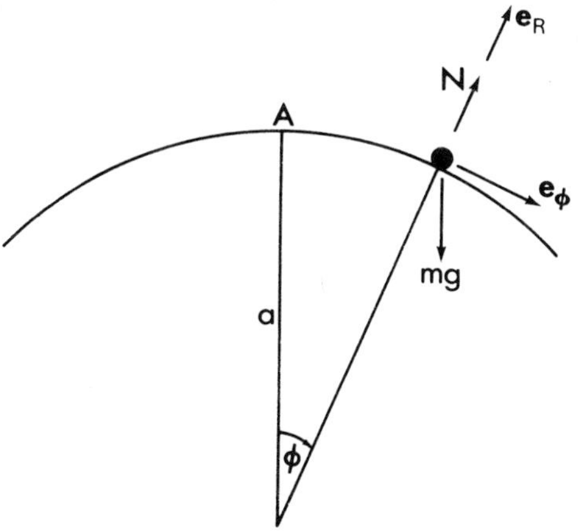

Fig. 8.2 Motion near equilibrium position

Solving (8.2) we obtain

$$\phi = A\,\mathrm{e}^{(mg/a)^{1/2}t} + B\,\mathrm{e}^{-(mg/a)^{1/2}t}$$

and so as time increases, ϕ increases indefinitely according to this solution. If the particle was at a lowest point, say B in Fig. 8.1, it is easily shown that the equation of motion for small ϕ is of the form

$$\ddot{\phi} + \left(\frac{mg}{a}\right)\phi = 0\,. \tag{8.3}$$

where ϕ is measured from the downward vertical. This equation has solution

$$\phi = A \cos\,[(mg/a)^{1/2}t] + B \sin\,[(mg/a)^{1/2}t] \tag{8.4}$$

and in this case ϕ is bounded for all t.

The above example shows that there is indeed a link between the classification of stability and the differential equation governing small displacements away from an equilibrium position. In the next section we make a precise mathematical definition of stability.

8.2 General Stability Theory

In this section we deal first with the differential equation

$$\boxed{\ddot{x} = f(x).} \tag{8.5}$$

Physically we can interpret this differential equation as representing the equation of motion of a particle moving on a straight line under a force $f(x)$ per unit mass. Equilibrium positions are given by $\ddot{x} = 0$. Suppose $x = x_0$ is an equilibrium position, so that

$$f(x_0) = 0. \tag{8.6}$$

Consider a small displacement away from the position $x = x_0$. If $x = x_0 + \xi$, where ξ is small, then substitution in (8.5) gives

$$\begin{aligned} \ddot{x} = \ddot{\xi} &= f(x_0 + \xi) \\ &= f(x_0) + \xi f'(x_0) + \tfrac{1}{2}\xi^2 f''(x_o) + \ldots , \end{aligned}$$

using Taylor's theorem. Since ξ is small, and x_0 is an equilibrium position

$$\boxed{\ddot{\xi} - f'(x_0)\,\xi \approx 0.} \tag{8.7}$$

If $f'(x_0) < 0$, say $f'(x_0) = -a^2$ where a is a positive constant, then the solution for ξ is

$$\xi = A\sin(at) + B\cos(at),$$

where A and B are arbitrary constants.

Clearly $|\xi| \leqslant |A| + |B|$ and so ξ is bounded. Alternatively, if $f'(x_0) > 0$, say $f'(x_0) = b^2$ where b is a positive constant, then

$$\xi = Ce^{bt} + De^{-bt},$$

and ξ is in general unbounded since e^{bt} increases without limit as t increases. Similarly, if $f'(x_0) = 0$, ξ is unbounded. Consequently we make the following definition:

DEFINITION:

The equilibrium position, $x = x_0$, of the differential equation

$$\ddot{x} = f(x)$$

is said to be *stable* if $f'(x_o) < 0$ and *unstable* if $f'(x_o) \geqslant 0$.

For example, consider the problem in Ch. 8.1 when the particle is at one of its highest points. The differential equation for its motion is

$$\ddot{\phi} = \frac{mg}{a} \sin \phi$$

and the equilibrium position is $\phi = 0$. In this case

$$f(\phi) = \frac{mg}{a} \sin \phi \, ,$$

giving

$$f'(\phi) = \frac{mg}{a} \cos \phi \, .$$

Thus $f'(0) = mg/a > 0$ and, by our definition, the equilibrium position is unstable.

The definition of stability introduced above can easily be extended in order to define stability of a solution of any differential equation. For instance, let $x = x_o(t)$ be a solution of the differential equation

$$\ddot{x} = f(x, \dot{x}, t). \tag{8.8}$$

Substituting $x = x_o + \xi$ in (8.8), where ξ is assumed small, results in a differential equation of the form

$$\ddot{\xi} = g(\xi, \dot{\xi}, t). \tag{8.9}$$

DEFINITION:

If the solution of (8.9) for ξ is bounded for all time t, the solution $x = x_o(t)$ of (8.8) is said to be *stable*. If ξ is not bounded for all time t, $x = x_o(t)$ is said to be an *unstable* solution of (8.8).

Clearly the definition can be generalized to the stability of solutions of any differential equations, including those of first order.

As an example, consider the stability of the equilibrium solutions $x = \pm \pi/2$ of the differential equation

$$\ddot{x} + \cos x = 0 \, .$$

If $x = \pi/2 + \xi$,

$$\ddot{\xi} + \cos\left(\frac{\pi}{2} + \xi\right) = 0 \, ;$$

i.e.
$$\ddot{\xi} - \sin\xi = 0,$$

and for small ξ,
$$\ddot{\xi} - \xi \approx 0.$$

Hence
$$\xi = A\,e^{t} + B\,e^{-t},$$

showing that ξ is not bounded, and so $x = \pi/2$ is an unstable solution.

But if $x = -\pi/2 + \xi$,
$$\ddot{\xi} + \xi \approx 0$$

giving
$$\xi = A\sin t + B\cos t.$$

Hence $|\xi| \leqslant |A| + |B|$, ξ is bounded for all time, and $x = -\pi/2$ is a stable solution.

As a second example, consider the stability of the solution $x = e^{-t}$ of the differential equation
$$\ddot{x} - \dot{x} - 2x = 0.$$

If $x = e^{-t} + \xi$,
$$(e^{-t} + \ddot{\xi}) - (-e^{-t} + \dot{\xi}) - 2(e^{-t} + \xi) = 0.$$

i.e.
$$\ddot{\xi} - \dot{\xi} - 2\xi = 0,$$

and solving,
$$\xi = A\,e^{-t} + B\,e^{2t}$$

Hence ξ is not bounded, and so $x = e^{-t}$ is an unstable solution.

8.3 Stability and Potential Energy

We now return to the physical interpretation of stability, and suppose a particle of mass m has equation of motion
$$m\ddot{x} = mf(x). \tag{8.10}$$

We have defined an equilibrium solution $x = x_0$ to be stable if $f'(x_0) < 0$ and unstable if $f'(x_0) \geqslant 0$. From Ch. 6.5, the potential energy, $V(x)$, is defined as
$$V(x) = -m\int_{x_0} f(x)\,dx, \tag{8.11}$$

x_0 being the zero potential energy level. Differentiating (8.11) gives
$$V'(x) = -mf(x)$$

and, differentiating again,

$$V''(x) = -mf'(x).$$

At an equilibrium position, $x = x_o$, we see that $V'(x_o) = 0$. This means that the potential energy has a stationary value at an equilibrium position. Now for *stable* equilibrium, $f'(x_o) < 0$ giving $V''(x_o) > 0$. This shows that the potential energy is a *minimum* at a stable equilibrium position. Similarly for an unstable equilibrium position, the potential energy has a maximum or inflection value.

We can soon see the physical significance of the potential energy extremum values and stability criteria. For the equation of motion (8.10), energy conservation gives

$$\text{Kinetic Energy} + \text{Potential Energy} = \text{constant}.$$

At an equilibrium position, say $x = x_o$,

$$\text{P.E.} = V(x_o),$$

$$\text{K.E.} = 0\,.$$

If the particle is slightly displaced from $x = x_o$ by giving it a small amount of kinetic energy, say T_o, then, denoting the K.E. by T, we see that

$$T + V(x) = T_o + V(x_o)$$

i.e.

$$T = T_o + V(x_o) - V(x). \tag{8.12}$$

If $V(x_o)$ is a maximum, $V(x_o) - V(x) > 0$, and T must therefore increase. This means that the speed of the particle must increase, corresponding to our definition of instability. Similarly if $V(x_o)$ is a minimum, $V(x_o) - V(x) < 0$, and T is decreasing, corresponding to stable equilibrium.

8.4 Stability of Circular Orbits in Central Orbit Theory

If the force acting on a particle of mass m is directed towards a fixed point and is a function only of its distance away from the fixed point, then we say that the particle is moving under a *central force*. The inverse square law of attraction is an example of a central force. In this section we first show that a circular orbit is always a possible orbit for a particle moving under any central force, and we investigate the stability of this solution.

The motion is illustrated in Fig. 8.3, the fixed point being O, and the distance of the particle from O being R. If the force towards O is $f(R)$ per unit mass, then the equation of motion is

$$m\ddot{\mathbf{R}} = -mf(R)\mathbf{e}_R\,. \tag{8.13}$$

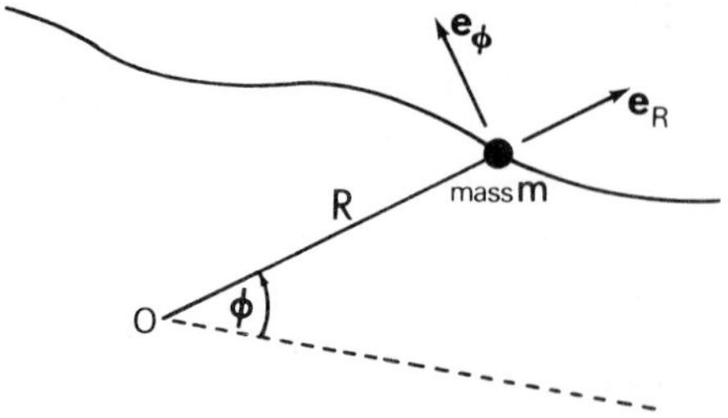

Fig. 8.3 Central force orbit

But

$$\ddot{\mathbf{R}} = (\ddot{R} - R\dot{\phi}^2)\mathbf{e}_R + (R\ddot{\phi} + 2\dot{R}\dot{\phi})\mathbf{e}_\phi ,$$

and equating $\mathbf{e}_R$ and $\mathbf{e}_\phi$ components,

$$\ddot{R} - R\dot{\phi}^2 = f(R) , \tag{8.14}$$

$$R\ddot{\phi} + 2\dot{R}\dot{\phi} = 0 . \tag{8.15}$$

From (8.15),

$$\frac{d}{dt}(R^2\dot{\phi}) = 0 ,$$

and integrating

$$R^2\dot{\phi} = h , \tag{8.16}$$

where h is a constant. This is just conservation of angular momentum about O. Eliminating $\dot{\phi}$ from (8.14) and (8.16) gives a second order non-linear differential equation

$$\ddot{R} = \frac{h^2}{R^3} - f(R) . \tag{8.17}$$

This is the differential equation of 'central orbits', and we note that it is of the form (8.10).

It is not possible to solve (8.17) completely unless we know the exact form of $f(R)$, and for most functions $f(R)$ the differential equation does not possess an analytic solution. But there is always one solution; for if

$$R = a , \tag{8.18}$$

that is, a circular orbit of radius a, equation (8.17) is satisfied provided

$$h^2 = a^3 f(a) . \tag{8.19}$$

This shows that a circular orbit of radius a is always a possible path for the particle provided h has the value given in (8.19). Using (8.16), this implies that the angular velocity is given by

$$\dot{\phi} = \{f(a)/a\}^{1/2}. \tag{8.20}$$

We will now investigate the stability of this circular motion. We have seen that $R = a$ is a solution of (8.17), and substituting $R = a + \xi$, where ξ is small, and $h^2 = a^3 f(a)$ in (8.17) gives

$$\ddot{\xi} = \frac{a^3 f(a)}{(a+\xi)^3} - f(a+\xi)\ .$$

Using Taylor's theorem for f we see that

$$\begin{aligned}\ddot{\xi} &= f(a)\,(1+\xi/a)^{-3} - \{f(a) + \xi f'(a) + \ldots\} \\ &= -\left\{\frac{3}{a} f\,(a) + f'(a)\right\}\xi + \mathrm{O}(\xi^2).\end{aligned}$$

For small ξ,

$$\ddot{\xi} + \left\{\frac{3}{a} f(a) + f'(a)\right\}\xi = 0$$

and we see that ξ is bounded provided

$$\frac{3}{a} f(a) + f'(a) > 0\,. \tag{8.21}$$

This means that a circular orbit is a stable solution of (8.17) provided (8.21) holds.

As an example, consider an inverse law of attraction of the form

$$f(R) = k/R^s, \tag{8.22}$$

where s is a constant. From (8.21), a circular orbit is stable if

$$3 - s > 0$$

Thus for $s < 3$, we have stable circular motion, which of course includes the case $s = 2$, the inverse square law of attraction. The result indicates that if a satellite, moving on a circular orbit about the Earth, is slightly displaced in its motion it will tend to continue moving near its circular orbit and not move quickly away.

Worked Examples

EXAMPLE 8.1:

A particle of mass m moves in a conservative field of force in such a way that its position at any time can be expressed in terms of a single coordinate, q, at that time;
i.e.

$$x = x(q), \quad y = y(q), \quad z = z(q).$$

Show that the conservation of energy can be expressed in the form

$$\frac{dF}{dq}\dot{q}^2 + 2F\ddot{q} + \frac{dV}{dq} = 0,$$

where $V = V(q)$ is the potential energy and the kinetic energy is given by $T = F(q)\dot{q}^2$.
Hence deduce that stationary equilibrium positions, $q = q_0$, satisfy $\frac{dV}{dq} = 0$,

SOLUTION:

Since the particle's position vector is of the form

$$\mathbf{r} = (x(q), y(q), z(q)),$$

its velocity can be written as

$$\mathbf{v} = \frac{d\mathbf{r}}{dq}\dot{q},$$

so that

$$T = \tfrac{1}{2}m\mathbf{v}.\mathbf{v} = \tfrac{1}{2}m\left(\frac{d\mathbf{r}}{dq}\cdot\frac{d\mathbf{r}}{dq}\right)\dot{q}^2$$

Defining $F(q) = \frac{1}{2}m(d\mathbf{r}/dq.d\mathbf{r}/dq)$, we have $T = F(q)\dot{q}^2$ and the conservation of energy takes the form

$$F(q)\dot{q}^2 + V(q) = \text{constant},$$

where $V(q)$ is the potential energy.

Regarding the left hand side as a function of q, where $q = q(t)$, and differentiating we obtain

$$\frac{d}{dq}[F(q)\dot{q}^2 + V(q)] = 0$$

i.e.
$$\frac{dF}{dq}\dot{q}^2 + F\frac{d}{dt}(\dot{q}^2)\frac{dt}{dq} + \frac{dV}{dq} = 0$$

or
$$\frac{dF}{dq}\dot{q}^2 + 2F\ddot{q} + \frac{dV}{dq} = 0\,.$$

At a stationary equilibrium position $q = q_0$ we have $\mathbf{v} = \dot{\mathbf{v}} = \mathbf{0}$. Now $\mathbf{v} = \mathbf{0}$ implies that $\dot{q} = 0$. Also,,

$$\dot{\mathbf{v}} = \frac{d^2\mathbf{r}}{dt^2} = \frac{d}{dt}\left(\frac{d\mathbf{r}}{dq}\dot{q}\right) = \frac{d^2\mathbf{r}}{dq^2}\dot{q}^2 + \frac{d\mathbf{r}}{dq}\ddot{q}$$

so that, at equilibrium, $q = 0$. But the energy conservation equation above holds for all time and hence in particular at equilibrium. Thus, at equilibrium, when $\dot{q} = \ddot{q} = 0$, we must have

$$\frac{dV}{dq} = 0\,.$$

Note: Since the particle's position is expressible in terms of one coordinate, q, we say that the particle has *one degree of freedom.* Following the analysis in Ch. 8.3, the equilibrium position will be stable if V has a minimum value at $q = q_0$ and unstable otherwise.

EXAMPLE 8.2:

A simple pendulum consists of a weightless rod of length l and a bob of mass m. Show that there are two positions of equilibrium, one of which is stable.

SOLUTION:

If ϕ is the angle made by the pendulum with the downward vertical, as in Fig. 8.4, the potential energy is given by

$$V = -\,mgl\cos\phi\,,$$

where zero potential energy level is taken as the horizontal through O.

For equilibrium (from Worked Example 8.1) $dV/d\phi = 0$, which gives $\sin\phi = 0$. Thus $\phi = 0$ and $\phi = \pi$ are positions of equilibrium. To investigate stability we consider the second derivative of V. Now

$$\frac{d^2V}{d\phi^2} = mg\,l\cos\phi\,.$$

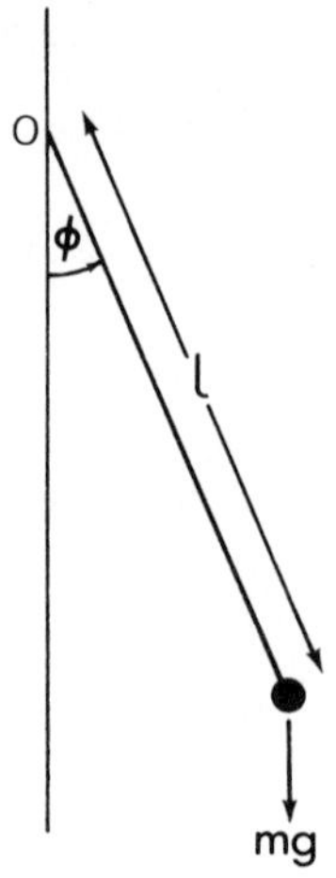

Fig. 8.4 Simple pendulum

Hence, at $\phi = 0$, $d^2V/d\phi^2$ is positive so that the equilibrium position $\phi = 0$ is stable. At $\phi = \pi$, $d^2V/d\phi^2$ is negative, and we have an unstable equilibrium position.

EXAMPLE 8.3:

A bead of mass m can slide freely on a vertical circular wire of radius a. Attached to the bead is a spring of natural length b and modulus λ, the other end of the spring being attached to the highest point of the wire.

Find the possible equilibrium positions and investigate their stability when $b = a$ and

$$\text{(i) } \lambda = mg; \qquad \text{(ii) } \lambda = 3mg.$$

SOLUTION:

The configuration is illustrated in Fig. 8.5, ϕ being the angle made by the spring with the downward vertical.

To find equilibrium positions, we must find the total potential energy, which has contributions due to the extension of the spring and to the position of the bead in the gravitational field. Thus, since the spring's length is $2a\cos\phi$, its potential energy is $\lambda(2a\cos\phi - b)^2/(2b)$, and we obtain

$$V = \lambda(2a\cos\phi - b)^2/(2b) - mg(2a\cos\phi)\cos\phi\,.$$

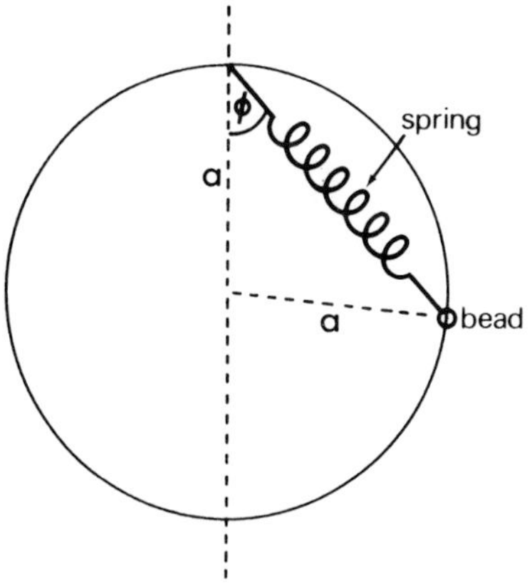

Fig. 8.5 Stability of bead on vertical wire

For equilibrium $dV/d\phi = 0$, giving

$$\frac{dV}{d\phi} = \frac{\lambda}{b}(2a\cos\phi - b)(-2a\sin\phi) + 4mga\cos\phi\sin\phi = 0 .$$

Thus $$2a\sin\phi\left(\lambda - \frac{2a\lambda}{b}\cos\phi + 2mg\cos\phi\right) = 0 ,$$

so that either $\sin\phi = 0$ or $\cos\phi = \lambda b/2(a\lambda - mgb)$. Now if $\sin\phi = 0$, the only realistic solution is $\phi = 0$, so that the bead is at the lowest point of the wire. A second solution is possible, namely

$$\phi = \cos^{-1}\left[\frac{\lambda b}{2(a\lambda - mgb)}\right],$$

provided $0 < \phi < \pi/2$, which means $1 > \cos\phi > 0$;

i.e. $$0 < \frac{\lambda b}{2(a\lambda - mgb)} < 1.$$

Thus we require $a\lambda > mgb$ and $a\lambda > (mg + \lambda/2)b$, and since the second inequality implies the first, the second solution exists provided

$$a\lambda > (mg + \lambda/2)\,b.$$

(i) $b = a, \quad \lambda = mg$

The inequality above is not satisfied, so that $\phi = 0$ is the only equilibrium position. But, when $b = a$ and $\lambda = mg$, we have

$$\frac{dV}{d\phi} = 2mga\sin\phi$$

and
$$\frac{d^2V}{d\phi^2} = 2mga\cos\phi\,.$$

Thus, when $\phi = 0$, we have $d^2V/d\phi^2 = 2mga > 0$, and so V has a minimum at $\phi = 0$. Hence $\phi = 0$ is a STABLE equilibrium position.

(ii) $b = a,\ \lambda = 3mg$

The inequality is satisfied and we have two equilibrium positions, $\phi = 0$ and $\phi = \cos^{-1}(3/4)$. With $b = a$ and $\lambda = 3mg$, we have

$$\frac{dV}{d\phi} = 2mga\ (3\sin\phi - 2\sin 2\phi)$$

and

$$\frac{d^2V}{d\phi^2} = 2mga\ (3\cos\phi - 4\cos 2\phi)\,.$$

When $\phi = 0$, we have $d^2V/d\phi^2 = -2mga < 0$, giving an UNSTABLE equilibrium position at $\phi = 0$. Now when $\phi = \cos^{-1}(3/4)$, we obtain $d^2V/d\phi^2 = 7mga/2 > 0$, showing that this is a STABLE position.

Problems

8.1 The equation of a conical pendulum is

$$\ddot{\theta} - \frac{h^2\cos\theta}{\sin^3\theta} + \frac{g\sin\theta}{a} = 0\,,$$

where θ is the angle made with the downward vertical and h and a are positive constants. Show that the pendulum bob can move in a horizontal circle at a constant angle α, where

$$g\sin^4\alpha = ah^2\cos\alpha.$$

Show that this is a position of stable equilibrium.

8.2 An electron of mass m moves along a straight line subject to a conservative field of force which has potential energy

$$V = cx/(a^2 + x^2),$$

where a, c are positive constants and x is the displacement of the electron from a fixed point on the line.

Find the position of stable equilibrium and determine the period† of small oscillations about this position.

† See Ch. 9.2 for definition of period.

8.3 A simple pendulum consists of a bob of mass m suspended by a light rod of length a from a fixed smooth pivot A. One end of a spring, modulus λ and natural length b, is attached to a small wheel which runs in a smooth horizontal groove passing through A. The spring remains vertical and may be assumed to obey Hooke's law.

Show that, providing $mg + \lambda \leqslant a\lambda/b$, there are three distinct positions of equilibrium, and discuss their stability.

8.4 Investigate the stability of the solution $x_0(t)$ of the following differential equations:

(i) $t\dfrac{dx}{dt} - 2x = t + 1, \qquad x_0 = 8t^2 - t - \frac{1}{2}$

(ii) $\dfrac{dx}{dt} + tx - t^3 = 0, \qquad x_0 = t^2 - 2 + 2e^{-t^2/2};$

(iii) $t\dfrac{dx}{dt} + 3x = x^2t^2, \qquad x_0 = 1/t^2(1+t).$

8.5 Investigate the stability of the solution $x_0 = (1 + e^{-2t^2})^{1/4}$ of the differential equation

$$\frac{dx}{dt} + tx = tx^{-3}.$$

8.6 Investigate the stability of a circular orbit of a particle of mass m moving under a law of force mk/R^3, directed towards the origin, where k is a constant and (R, ϕ) are plane polar coordinates.

8.7 A particle is moving under a central force $f(R)$ per unit mass. Write down the conditions for stability of a circular orbit of radius a.

(i) If f is given by $f(R) = \mu e^{-\lambda R}/R^n$, μ being a positive constant, λ and n being constants, prove that a circular orbit is stable provided

$$a\lambda + n < 3;$$

(ii) If f is given by $f(R) = k/R^2 + k'/R^3$, where k and k' are positive constants, show that a circular orbit is stable.

Chapter 9

Oscillations and Vibrations

9.1 Introduction

We have previously seen that the equation of motion for a single particle dynamical problem is of the form

$$\ddot{\mathbf{r}} = \mathbf{F}\,, \tag{9.1}$$

$\mathbf{r}$ being the position vector of the particle and $\mathbf{F}$ a vector function. If we consider the special case of motion taking place in one direction only, the equation of motion can be written as

$$\boxed{\ddot{x} = f(x, \dot{x}, \ddot{x}, \ldots, t)}\,, \tag{9.2}$$

f being a function of x, its derivatives, and time t. This is a differential equation which, unfortunately, does not in general have an analytic solution. In this chapter, we solve the equation for some special cases of the function f.

Equation (9.2) can represent many physical situations, ranging from a satellite in motion round the Earth to the wave motions of crystal lattices. In particular it will be shown that if $f = -\omega^2 x$, then x will oscillate with time; and if $f = -\omega^2 x - k\dot{x}$, x will either oscillate with decreasing amplitude or tend to zero without oscillation.

Vibrations of this type are of great importance in the dynamics of moving machinery. For whenever a system has an oscillating or varying external force, a vibrational problem will arise. The force might cause the system to oscillate slightly; for instance a stationary bus with its engine running; or the vibrations might increase in strength as time elapses, causing damage or perhaps complete ruin of the system. This phenomenon is called *resonance*, and will be discussed in Ch. 9.8.

9.2 Simple Harmonic Motion

If a particle of mass m moves in a straight line under a force per unit mass of $-\omega^2 x$ (ω being a real positive constant and x denoting the distance along the line measured from a fixed point O on the line) then its equation of motion is

$$\boxed{\ddot{x} = -\omega^2 x} \quad . \tag{9.3}$$

This differential equation has the general solution

$$\boxed{x = A\cos(\omega t) + B\sin(\omega t)} \quad , \tag{9.4}$$

A and B being arbitrary constants. Alternatively, 9.4 can be written as

$$x = C\cos(\omega t + \alpha) \tag{9.5}$$

where C and α are now the arbitrary constants. The two forms of the solution for x are equivalent, and it can easily be seen that

$$A = C\cos\alpha, \quad B = -C\sin\alpha.$$

Using solution (9.5), at $t = 0$, the particle is at position $x = C\cos\alpha$, and at time $t = 2\pi/\omega$, again $x = C\cos\alpha$. In fact we see that each value of x is repeated at intervals of $2\pi/\omega$. In other words, we say that $x(t)$ has *period* $2\pi/\omega$, The motion described by (9.3) is called simple harmonic motion (S.H.M.). In general, we have the definition:

DEFINITION:

A function $g(t)$ is said to be a periodic function of t with period l if

$$\boxed{g(t + l) = g(t)} \quad , \tag{9.6}$$

for all t. Thus $\sin t$ has period 2π; $\sin(kt)$ has period $2\pi/k$.

If the particle is initially (i.e. at $t = 0$) at the position $x = a$, and is then released so that $x(0) = (0)$, the arbitrary constants C and α are determined from the conditions

$$x = a, \quad t = 0; \quad \text{and} \quad \dot{x} = 0, \quad t = 0.$$

The first gives $a = C\cos\alpha$, and the second gives $0 = \omega\sin\alpha$. Since $\omega \neq 0$, $\alpha = 0$ and the complete solution is

$$x = a\cos(\omega t). \tag{9.7}$$

This is illustrated in Fig. 9.1. The distance a is known as the *amplitude* of the motion, and as we have already noted the period is $2\pi/\omega$.

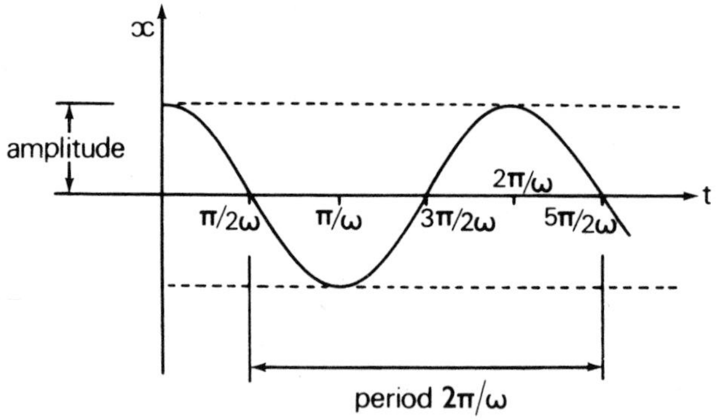

Fig. 9.1 Simple harmonic motion

For motion in general, when equation (9.5) is the governing equation, C is the *amplitude*, α is known as the *phase angle* and the *frequency* is defined as (2π/period) which is clearly equal to ω.

9.3 Simple Pendulum

One practical example of S.H.M. is the small oscillations of a simple pendulum. For our model of a simple pendulum, we attach a particle of mass m to one end of an inextensible light string of length a. The other end of the string is fixed at some point O and the particle is allowed to move freely in a vertical plane through O. The model is illustrated in Fig. 9.2. Here (R, ϕ) are plane polar coordinates, ϕ being measured from the vertical.

If T is the tension in the string, the equation of motion of the particle is

$$m\ddot{\mathbf{R}} = m\mathbf{g} - T\mathbf{e}_R \,, \tag{9.8}$$

where $\mathbf{g} = (g, 0, 0)$ in terms of xyz-axes as shown in Fig. 9.2. From (2.15), with $R = a$

$$\ddot{\mathbf{R}} = -a\dot{\phi}^2\mathbf{e}_R + a\ddot{\phi}\mathbf{e}_\phi \,;$$

and substituting in (9.8) gives

$$-ma\dot{\phi}^2\mathbf{e}_R + ma\ddot{\phi}\mathbf{e}_\phi = m(g\cos\phi\,\mathbf{e}_R - g\sin\phi\,\mathbf{e}_\phi) - T\mathbf{e}_R \,.$$

Taking components in the $\mathbf{e}_R$ and e_ϕ directions gives

$$ma\dot{\phi}^2 = -mg\cos\phi + T \tag{9.9}$$

$$ma\ddot{\phi} = -mg\sin\phi \,. \tag{9.10}$$

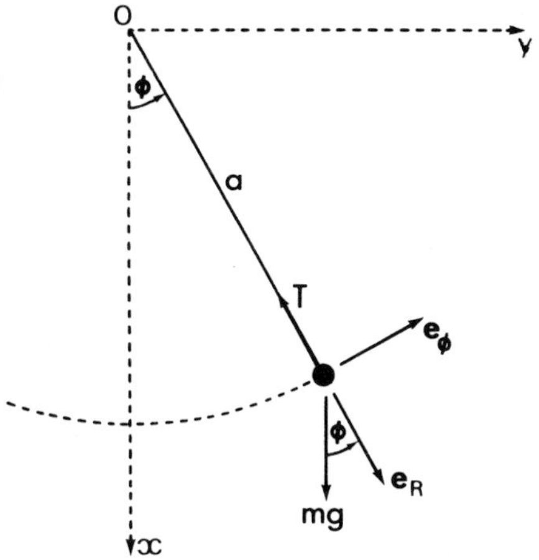

Fig. 9.2 Simple pendulum

Equation (9.10) can readily be integrated to give

$$\tfrac{1}{2}\,a\dot{\phi}^2 \;=\; g\cos\phi \,+\, A\,, \tag{9.11}$$

A being a constant dependent on the initial conditions for the pendulum. In order to find $\phi(t)$, we note that

$$\frac{dt}{d\phi} \;=\; \left[\frac{2}{a}\,(g\cos\phi + A)\right]^{-\frac{1}{2}}$$

giving

$$t \;=\; \int \frac{d\phi}{\left[\dfrac{2}{a}\,(g\cos\phi + A)\right]^{\frac{1}{2}}} \,+\, B. \tag{9.12}$$

In general we cannot perform the integration in (9.12), although substitution of (9.11) in (9.9) determines the tension T. But if we assume the movement of the pendulum away from the vertical position is always *small*, then the angle ϕ is small, and (9.10) reduces to

$$\ddot{\phi} \;=\; -\left(\frac{g}{a}\right)\phi\,. \tag{9.13}$$

Comparing with (9.3), we see that the motion is S.H.M. about $\phi = 0$ with period $2\pi\sqrt{a/g}$. This leads us to the concept of *simple equivalent pendulum* (S.E.P.). If the differential equation of motion of a particle (or system) is given by

$$\ddot{x} \;=\; -\,\omega^2 x\,,$$

then we define the length of the S.E.P., l, by

$$2\pi\sqrt{\frac{l}{g}} = \frac{2\pi}{\omega}$$

giving
$$l = g/\omega^2. \tag{9.14}$$

Thus a particle performing harmonic oscillations with frequency ω has the *same period* of oscillation as a simple pendulum of length g/ω^2, oscillating about the vertical.

One interesting point is that the frequency of oscillation of the simple pendulum depends only on the length, a, of the pendulum and g, and is independent of the mass of the particle.

9.4 Springs: Hooke's Law

We have already noted that under most conditions, to a good approximation, a stretched spring obeys Hooke's law, which can be expressed mathematically by

$$\boxed{T = \lambda y/a}\quad , \tag{9.15}$$

T being the tension in the spring, a the natural length, y the extension of the spring and λ a constant, called the modulus of elasticity.

Suppose a particle of mass m is attached to one end of a spring (natural length a and modulus of elasticity λ), which is suspended vertically by its other end. This is illustrated in Fig. 9.3. If T_o is the tension and d the extension in the spring when the system is in equilibrium, (9.15) shows that

$$mg = T_o = \lambda d/a\ .$$

Hence the extension is given by

$$d = mag/\lambda. \tag{9.16}$$

The particle is then pulled down a further distance and released. If x denotes its distance from the equilibrium position in the subsequent motion, its equation of motion is

$$m\ddot{x} = mg - T \tag{9.17}$$

where T is now given by
$$T = \lambda(x + d)/a.$$

Hence
$$m\ddot{x} = mg - \frac{\lambda x}{a} - \frac{\lambda d}{a},$$

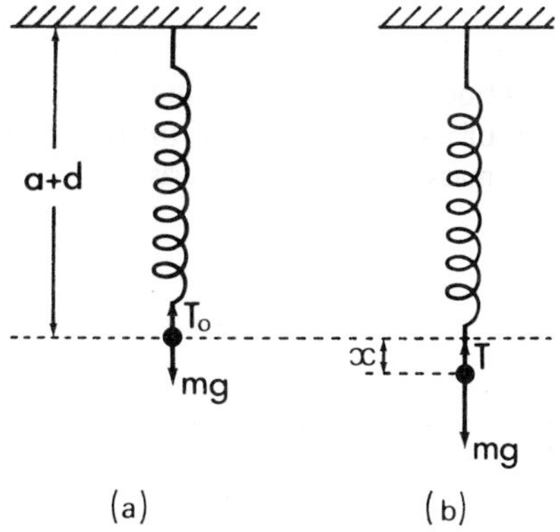

Fig. 9.3 Spring and mass system: (a) In equilibrium, (b) In general motion

and eliminating d from (9.16),

$$\ddot{x} = -\left(\frac{\lambda}{am}\right)x. \tag{9.18}$$

This shows that the particle will oscillate about the equilibrium position with period $2\pi\sqrt{ma/\lambda}$. Note that unlike the simple pendulum, the period of oscillation increases with increasing particle mass.

A more interesting problem is to consider two springs with two masses, suspended as in Fig. 9.4. Clearly if the second mass, B, is pulled vertically downwards from

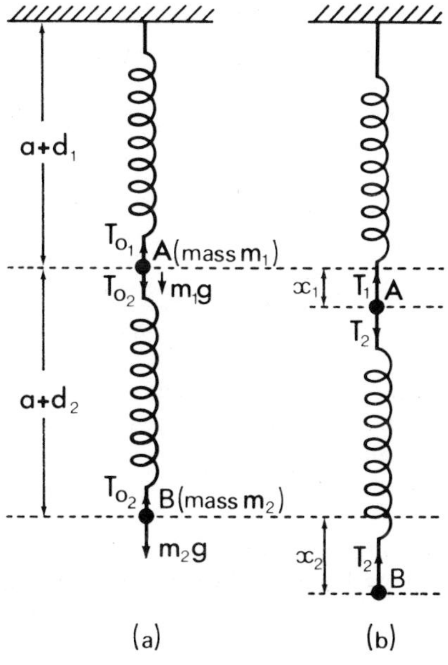

Fig. 9.4 Two mass-spring system: (a) In equilibrium, (b) In general motion

equilibrium keeping A fixed and both masses then released, we expect the motion of B to affect the mass A. To find out exactly what this effect is, we must consider the masses in extended positions away from equilibrium, solve the equations of motion and apply the initial conditions. Now if the mass of A is m_1 and the mass of B is m_2, and we assume the springs to be identical, then the equilibrium conditions are

$$T_{o_1} - T_{o_2} = m_1 g ,$$

$$T_{o_2} = m_2 g ,$$

where T_{o_1} and T_{o_2} are the tensions in the two springs. If d_1 and d_2 are the respective extensions in equilibrium, Hooke's law gives us

$$T_{o_1} = \lambda d_1/a ,$$

$$T_{o_2} = \lambda d_2/a .$$

Hence

$$\lambda(d_1 - d_2)/a = m_1 g , \tag{9.19}$$

$$\lambda d_2/a = m_2 g . \tag{9.20}$$

Now consider a general position during the motion of the system. If x_1 and x_2 are the respective displacements of the masses from their equilibrium positions,

$$m_1\ddot{x}_1 = -T_1 + T_2 + m_1 g , \tag{9.21}$$

$$m_2\ddot{x}_2 = -T_2 + m_2 g , \tag{9.22}$$

where T_1 and T_2 are now given by

$$T_1 = \lambda(d_1 + x_1)/a ,$$

$$T_2 = \lambda(d_2 + x_2 - x_1)/a .$$

Substituting T_1 and T_2 in (9.21) and (9.22), and eliminating d_1 and d_2 using (9.19) and (9.20), gives

$$\ddot{x}_1 = -\left(\frac{2\lambda}{am_1}\right)x_1 + \left(\frac{\lambda}{am_1}\right)x_2 , \tag{9.23}$$

$$\ddot{x}_2 = \left(\frac{\lambda}{am_2}\right)x_1 - \left(\frac{\lambda}{am_2}\right)x_2 . \tag{9.24}$$

We thus have a pair of coupled simultaneous differential equations. One method of solving them is to eliminate either x_1 or x_2; for instance by substituting the value for x_2 from (9.23) into (9.24). This leads to a fourth order linear differential equation, which can be solved.

A more sophisticated method is to assume a solution for x_1 and x_2, and then by direct substitution verify its existence. For instance, since we expect some sort of S.H.M., assume

$$x_i = A_i \cos(\omega t + \epsilon) \quad (i = 1, 2) , \tag{9.25}$$

where A_1, A_2, ω and ϵ are constants to be determined. Substituting in (9.23) gives

$$-A_1\omega^2\cos(\omega t+\epsilon) = -\left(\frac{2\lambda}{am_1}\right)A_1\cos(\omega t+\epsilon) + \left(\frac{\lambda}{am_1}\right)A_2\cos(\omega t+\epsilon).$$

Hence, writing $\beta_i = \lambda/am_i$ $(i = 1, 2)$,

$$A_1(\omega^2-2\beta_1) + A_2\beta_1 = 0\,; \qquad (9.26)$$

and, similarly, (9.24) reduces to

$$A_1\beta_2 + A_2(\omega^2-\beta_2) = 0\,. \qquad (9.27)$$

Eliminating the ratio A_1/A_2 from (9.26) and (9.27) gives the condition for the two equations to be consistent:

$$\begin{vmatrix} \omega^2-2\beta_1 & \beta_1 \\ \beta_2 & \omega^2-\beta_2 \end{vmatrix} = 0\,,$$

which gives

$$\omega^4 - (2\beta_1+\beta_2)\omega^2 + \beta_1\beta_2 = 0\,. \qquad (9.28)$$

Thus, if (9.25) is a solution of the differential equations, the frequency ω must satisfy (9.28). But solving (9.28) gives

$$\omega^2 = [(2\beta_1+\beta_2) \pm (4\beta_1{}^2+\beta_2{}^2)^{1/2}]/2\,,$$

and so there are two positive real values of ω, namely

$$\omega_+ = \{[(2\beta_1+\beta_2) + (4\beta_1^2+\beta_2^2)^{1/2}]/2\}^{1/2}, \qquad (9.29)$$

$$\omega_- = \{[(2\beta_1+\beta_2) - (4\beta_1^2+\beta_2^2)^{1/2}]/2\}^{1/2}. \qquad (9.30)$$

This shows that there are two possible values for the frequency, ω_+ or ω_-. The solutions corresponding to $\omega = \omega_+$ and $\omega = \omega_-$ are referred to as the *normal modes of oscillation* of the system.

Now when $\omega = \omega_+$, (9.26) gives

$$A_2^+ = A_1^+\left(2-\frac{\omega_+^2}{\beta_1}\right)$$

and the solution is

$$\left.\begin{aligned} x_1 &= A_1^+\cos(\omega_+t+\epsilon_+)\,, \\ x_2 &= A_1^+\left(2-\frac{\omega_+^2}{\beta_1}\right)\cos(\omega_+t+\epsilon_+)\,. \end{aligned}\right\} \qquad (9.31)$$

Hence, in this particular case

$$\frac{x_1}{x_2} = \frac{1}{(2-\omega_+^2/\beta_1)}$$

and, for example, when $m_1 = m_2 = m$, say, $\omega_+^2 = (3+\sqrt{5})\beta_1/2$, giving

$$\frac{x_1}{x_2} = \frac{2}{(1-\sqrt{5})} = \frac{-(1+\sqrt{5})}{2}.$$

This shows that in this mode of oscillation both masses are performing S.H.M. with the same period but in *opposing* directions. Also the amplitude of A's motion is larger than B's.

Similarly, when $\omega = \omega_-$, the solution is

$$\left.\begin{aligned} x_1 &= A_1^- \cos(\omega_- t + \epsilon_-)\,, \\ x_2 &= A_1^- \left(2 - \frac{\omega_-^2}{\beta_1}\right) \cos(\omega_- t + \epsilon_-)\,; \end{aligned}\right\} \qquad (9.32)$$

and in particular, if $m_1 = m_2 = m$, $\omega_-^2 = (3-\sqrt{5})\beta_1/2$, giving

$$\frac{x_1}{x_2} = \frac{2}{(1+\sqrt{5})} = \frac{(\sqrt{5}-1)}{2}.$$

This time both masses perform S.H.M. with the same period, and moving in the same direction, with A's amplitude smaller than B's.

Returning to the general problem, since the equations (9.23) and (9.24) are linear, we can add solutions, namely (9.31) and (9.32) to give the general solution

$$\left.\begin{aligned} x_1 &= A_1^+ \cos(\omega_+ t + \epsilon_+) + A_1^- \cos(\omega_- t + \epsilon_-) \\ x_2 &= A_1^+(2-\omega_+^2/\beta_1)\cos(\omega_+ t + \epsilon_+) + A_1^-(2-\omega_-^2/\beta_1)\cos(\omega_- t + \epsilon_-). \end{aligned}\right\} \qquad (9.33)$$

Here A_1^+, A_1^-, ϵ_+ and ϵ_- are the arbitrary constants.

In our original problem we supposed that A was kept fixed while B was pulled down a distance, say b, then both A and B were released. Hence at $t = 0$

$$\left.\begin{aligned} x_1 &= 0, \quad \dot{x}_1 = 0 \\ x_2 &= b\,, \quad \dot{x}_2 = 0\,. \end{aligned}\right\} \qquad (9.34)$$

Let us also suppose that the masses are equal, so that

$$\left.\begin{aligned} x_1 &= A_1^+ \cos(\omega_+ t + \epsilon_+) + A_1^- \cos(\omega_- t + \epsilon_-) \\ x_2 &= -\frac{A_1^+}{2}(\sqrt{5}-1)\cos(\omega_+ t + \epsilon_+) + \frac{A_1^-}{2}(\sqrt{5}+1)\cos(\omega_- t + \epsilon_-) \end{aligned}\right\} \qquad (9.35)$$

The four conditions now give

$$0 = A_1^+ \cos\epsilon_+ + A_1^- \cos\epsilon_- , \tag{9.36}$$

$$0 = \omega_+ A_1^+ \sin\epsilon_+ + \omega_- A_1^- \sin\epsilon_- , \tag{9.37}$$

$$b = -\frac{A_1^+}{2}(\sqrt{5}-1)\cos\epsilon_+ + \frac{A_1^-}{2}(\sqrt{5}+1)\cos\epsilon_- , \tag{9.38}$$

$$0 = \omega_+ A_1^+ (\sqrt{5}-1)\sin\epsilon_+ - \omega_- A_1^- (\sqrt{5}+1)\sin\epsilon_- . \tag{9.39}$$

We must solve these equations for the constants A_1^+, A_1^-, ϵ_+ and ϵ_-. From (9.37) and (9.39), $A_1^+ \sin\epsilon_+ = A_1^- \sin\epsilon_- = 0$. If $A_1^+ = 0$, (9.36) gives $A_1^- \cos\epsilon_- = 0$. But since we also have $A_1^- \sin\epsilon_- = 0$, clearly $A_1^- = 0$ and so $x_1 = x_2 = 0$ for all t. This means that there is no motion, which is a contradiction and so we conclude that $A_1^+ \neq 0$. Hence $\epsilon_+ = 0$, and so $A_1^- \cos\epsilon_- = -A_1^+ (\neq 0)$ with $A_1^- \sin\epsilon_- = 0$. Clearly we must have $\epsilon_- = 0$, and it is a trivial matter to show that

$$A_1^+ = -b/\sqrt{5}, \quad A_1^- = b/\sqrt{5} .$$

Thus the complete solution is

$$x_1 = \frac{b}{\sqrt{5}}[\cos(\omega_- t) - \cos(\omega_+ t)] \tag{9.40}$$

$$x_2 = \frac{b}{2\sqrt{5}}[(\sqrt{5}-1)\cos(\omega_+ t) + (\sqrt{5}+1)\cos(\omega_- t)] , \tag{9.41}$$

where $\omega_+^2 = (3+\sqrt{5})\beta/2$, $\omega_-^2 = (3-\sqrt{5})\beta/2$, and $\beta = \lambda/am$. Fig. 9.5 illustrates the motion of the masses as time increases, in the case $b = a/5$.

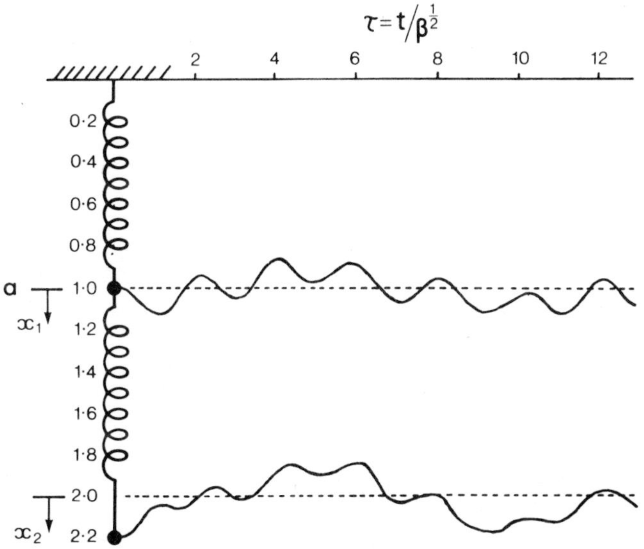

Fig. 9.5 Two mass-spring motion

9.5* Transverse Oscillations

Suppose a particle of mass m is suspended between two identical springs, as illustrated in Fig. 9.6. The whole system is assumed to lie on a smooth horizontal table, and the particle is displaced from its equilibrium position in a direction perpendicular to the initial line of the springs.

In equilibrium the springs, of natural length l and modulus λ, are extended a distance d, and in motion extended a further distance d'. If x is the distance of the particle from its equilibrium position, then

$$(l + d + d')^2 = (l + d)^2 + x^2$$

giving

$$d' = [(l + d)^2 + x^2]^{1/2} - (l + d) . \qquad (9.42)$$

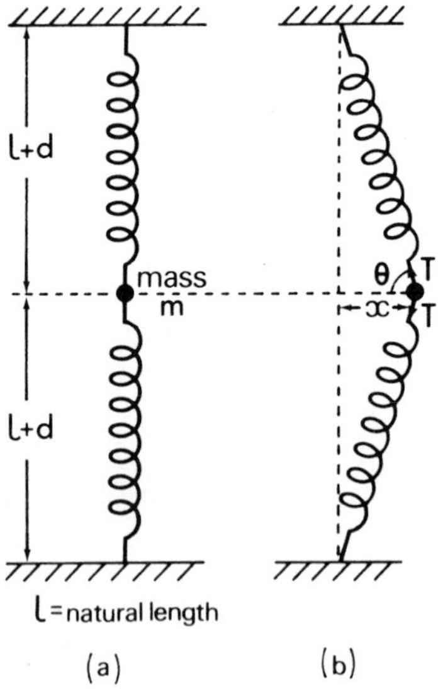

Fig. 9.6 Transverse oscillations: (a) equilibrium, (b) motion

The equation of motion for the particle in its direction of motion (the x direction) is

$$m \frac{d^2x}{dt^2} = - 2T \cos\theta \qquad (9.43)$$

where T is the tension in the spring, given by

$$T = - \frac{\lambda(d' + d)}{l} \qquad (9.44)$$

and the angle θ is defined by

$$\cos\theta = x/(l+d+d')\,. \tag{9.45}$$

Substituting (9.44) and (9.45) in (9.43) gives

$$m\,\frac{d^2x}{dt^2} = -\,\frac{2\lambda(d'+d)x}{l(l+d+d')}. \tag{9.46}$$

If we assume the motion to be only a small perturbation from the equilibrium position, x will be small, and from (9.42),

$$\begin{aligned} d' &= (l+d)\left[1+\left(\frac{x}{l+d}\right)^2\right]^{1/2} - (l+d) \\ &= (l+d)\left[1+\frac{x^2}{2(l+d)^2}+\mathrm{O}(x^4)\right] - (l+d) \\ &= x^2/2(l+d) + \mathrm{O}(x^4)\,, \end{aligned}$$

showing, as expected, that d' will also be small, and of order x^2. Substituting d' in (9.46) shows that

$$\begin{aligned} m\,\frac{d^2x}{dt^2} &= -\,\frac{2\lambda x}{l(l+d)}\left(d+\frac{x^2}{2(l+d)}\right)\left(1-\frac{x^2}{2(l+d)^2}\right) + \mathrm{O}(x^5) \\ &= -\,\frac{2\lambda x}{l(l+d)}\,(d+lx^2/2(l+d)^2) + \mathrm{O}(x^5) \\ &= -\left[\frac{2\lambda d}{l(l+d)}\right]x - \left[\frac{\lambda}{(l+d)^3}\right]x^3 + \mathrm{O}(x^5)\,. \end{aligned} \tag{9.47}$$

Hence, if x is very small and provided d is non-zero,

$$\frac{d^2x}{dt^2} \approx -\left[\frac{2\lambda d}{ml(l+d)}\right]x\,, \tag{9.48}$$

which shows that the particle performs S.H.M. about its equilibrium position. Since the direction of motion is perpendicular to the initial direction of the springs, the oscillations are called *transverse* oscillations. Oscillations in the direction of the springs are called *longitudinal* oscillations (see Problem 9.3).

One interesting aspect of this problem is that the character of the motion is completely altered if initially the springs are unextended (i.e. $d = 0$) or if d is so small so that $dx < x^3$. In this special case, (9.47) reduces to

$$\frac{d^2x}{dt^2} = -\left(\frac{\lambda}{ml^3}\right)x^3 \tag{9.49}$$

and this is clearly not the equation of S.H.M. In fact it is not possible to find the complete *analytic* solution to this non-linear differential equation.

9.6* Complex Substitution Techniques

If we are solving differential equations and suspect that the solution, $x = x(t)$, is oscillatory we can use the substitution

$$\boxed{x = \exp(i\omega t)} \tag{9.50}$$

to find the possible values of ω. Provided the differential equation is linear, both the real and imaginary parts of x are solutions of the differential equation. For instance, if x satisfies

$$\frac{d^2x}{dt^2} + 4x = 0\,,$$

try substituting $x = \exp(i\omega t)\,.$

Then

$$\frac{d^2x}{dt^2} = -\,\omega^2 \exp(i\omega t)$$

and on substitution in the differential equation,

$$-\,\omega^2 + 4 = 0$$

giving $\omega = \pm\,2$. Hence we see that both $\cos 2t$ and $\sin 2t$ are solutions of the differential equation, and the complete solution is

$$x = A\cos 2t + B\sin 2t\,,$$

A and B being arbitrary constants.

We use this technique in Ch. 9.7.

9.7* Vibrations in Crystal Lattices

Crystalline solids are characterized by a regular and symmetric arrangement of the atoms in the solid. A structure of this type is normally called a 'crystal lattice' and Fig. 9.7 shows the structure of the sodium chloride crystal, NaCl.

The atoms are bound together by forces, acting along the lines joining them. We may regard the system as equivalent to a set of particles joined by elastic strings. Clearly the system will vibrate in some manner when subjected to an external

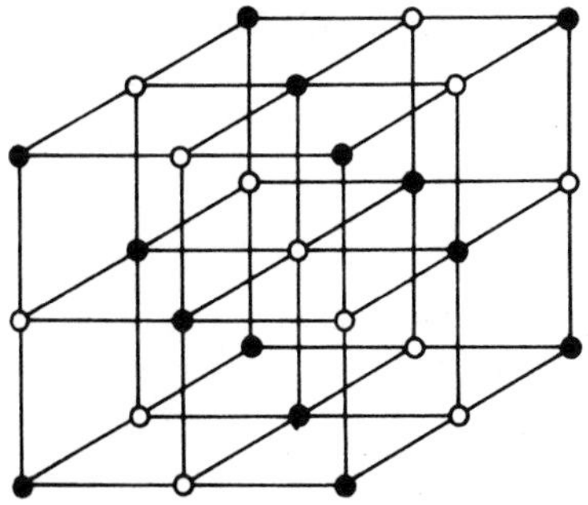

Fig. 9.7 Sodium chloride crystal lattice

force. In this section we will consider the vibrations of a particularly simple crystal lattice. The model consists of a chain of two different types of particles occupying alternate positions, as illustrated in Fig. 9.8.

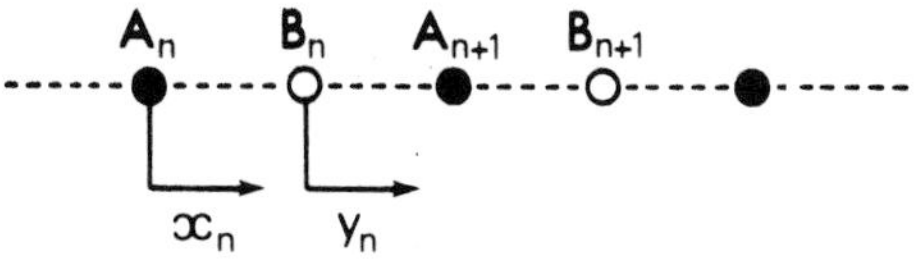

Fig. 9.8 Simple lattice model

In equilibrium, the particles will be spaced at equal intervals, say d, along the chain. We assume that the force in the chain can be represented by a tension that obeys Hooke's law and only acts between neighbouring particles. We will only consider *longitudinal* vibrations of this chain.

Suppose the particle A_n moves a longitudinal distance x_n along the chain, and B_n moves a distance $y_n (n = 0, 1, 2, \ldots)$. The equations of motion for the two types of particle are

$$A_n: \quad m_A \ddot{x}_n = T_{A_n B_n} - T_{A_n B_{n-1}},$$

$$B_n: \quad m_B \ddot{y}_n = T_{B_n A_{n+1}} - T_{B_n A_n},$$

where $T_{A_n B_n}$ is the tension in the chain $A_n B_n$, and m_A, m_B are the respective masses of the particles at A_n, B_n. Hence

$$m_A \ddot{x}_n = \lambda(y_n - x_n) - \lambda(x_n - y_{n-1}),$$

$$m_B \ddot{y}_n = \lambda(x_{n+1} - y_n) - \lambda(y_n - x_n),$$

where λ is the modulus of elasticity divided by the natural length. This gives

$$\left.\begin{aligned} m_A \ddot{x}_n &= \lambda(y_n - 2x_n + y_{n-1}) \\ m_B \ddot{y}_n &= \lambda(x_{n+1} - 2y_n + x_n) \quad (n = 0, 1, 2, \ldots) \end{aligned}\right\} \tag{9.51}$$

The above equations form a system of simultaneous differential equations, infinite in number. One class of solutions of these equations is of the form

$$\left.\begin{aligned} x_n &= A \exp i\,(2\pi\eta n - \omega t)\,, \\ y_n &= B \exp i\,(2\pi\eta n - \omega t)\,, \quad (n = 0, 1, 2, \ldots)\,, \end{aligned}\right\} \tag{9.52}$$

where ω and η are constants. As the differential equations are linear, both the real and imaginary parts of (9.52) are possible solutions. Choosing the real parts to represent the actual displacements, we have

$$\left.\begin{aligned} x_n &= Re\,\{A \exp i\,(2\pi\eta n - \omega t)\}\,, \\ x_{n+1} &= Re\,\{A \exp i(2\pi\eta n - \omega t + 2\pi\eta)\} \\ &= Re\,\{A \exp (2\pi i\eta) \exp i\,(2\pi\eta n - \omega t)\} \end{aligned}\right\} \tag{9.53}$$

We note that this solution represents simple harmonic motion of frequency ω, with the particle's amplitudes dependent. This type of vibration is termed a *progressive wave.* If, for example, $\eta = 0$,

$$x_{n+1} = Re\,\{A \exp(-i\omega t)\} = x_n,$$

showing that the amplitudes are equal.

Similarly, if $\eta = ½$

$$\begin{aligned} x_{n+1} &= Re\,\{A \exp(i\pi) \exp i(\pi n - \omega t)\} \\ &= Re\,\{-A \exp i(\pi n - \omega t)\} \\ &= -\,x_n, \end{aligned}$$

showing that the amplitudes are of equal magnitude but alternately in opposite directions. A similar result holds for the displacement y_n.

Substituting (9.52) in (9.51) gives

$$\begin{aligned} (2\lambda - m_A\omega^2)A \;-\; \lambda\,[1 + \exp(-2\pi\eta i)]\,B &= 0 \\ -\,\lambda\,[1 + \exp(2\pi\eta i)]\,A \;+\; (2\lambda - m_B\omega^2)B &= 0\,. \end{aligned} \tag{9.54}$$

These equations are consistent only if

$$\begin{vmatrix} 2\lambda - m_A\omega^2 & -\lambda[1+\exp(-2\pi\eta i)] \\ -\lambda[1+\exp(2\pi\eta i)] & 2\lambda - m_B\omega^2 \end{vmatrix} = 0 \qquad (9.55)$$

giving $m_A m_B \omega^4 - 2\lambda(m_A + m_B)\omega^2 + 2\lambda[1-\cos(2\pi\eta)] = 0.$

Hence $$\omega^2 = \frac{\lambda}{m_A m_B}\left\{(m_A+m_B) \pm [(m_A+m_B)^2 - 2m_A m_B(1-\cos(2\pi\eta))]^{1/2}\right\},$$

and for a given value of η there are two admissible values for the frequency, called the acoustic and optical frequencies, defined by

$$\omega^2 = \begin{cases} \dfrac{\lambda}{m_A m_B}\left\{(m_A+m_B) - [(m_A+m_B)^2 - 4m_A m_B \sin^2(\pi\eta)]^{1/2}\right\} & \text{acoustic} \\[2ex] \dfrac{\lambda}{m_A m_B}\left\{(m_A+m_B) + [(m_A+m_B)^2 - 4m_A m_B \sin^2(\pi\eta)]^{1/2}\right\} & \text{optical} \end{cases} \qquad (9.56)$$

Substituting these values in (9.54) gives the corresponding complex amplitude ratios:

$$A/B = \begin{cases} \dfrac{-m_B[1+\exp(2\pi\eta i)]}{m_A - m_B - [(m_A+m_B)^2 - 4m_A m_B \sin^2(\pi\eta)]^{1/2}} & \text{acoustic} \\[2ex] \dfrac{-m_B[1+\exp(2\pi\eta i)]}{m_A - m_B + [(m_A+m_B)^2 - 4m_A m_B \sin^2(\pi\eta)]^{1/2}} & \text{optical} \end{cases} \qquad (9.57)$$

We note that the actual amplitude ratio is the modulus of A/B. Thus both particles are performing simple harmonic motion with the same frequency.

The solutions for ω^2 and A/B are both periodic in η with period 1, and all distinct solutions are obtained if we restrict η to a unit range, say

$$-\tfrac{1}{2} \leqslant \eta \leqslant \tfrac{1}{2}. \qquad (9.58)$$

The two admissible frequencies are plotted against η in Fig. 9.9 for several different mass ratios $\sigma = m_B/m_A$. We now examine the two branches in more detail.

ACOUSTIC BRANCH:

On this branch the frequency is given by

$$\omega^2 = \frac{\lambda}{m_A m_B}\left\{m_A + m_B - [(m_A + m_B)^2 - 4m_A m_B \sin^2(\pi\eta)]^{\frac{1}{2}}\right\} \tag{9.59}$$

and so if η is small,

$$\omega^2 = \left(\frac{2\pi^2\lambda}{m_A + m_B}\right)\eta^2 + O(\eta^4)\,.$$

Thus as $\eta \to 0$, $\omega \to 0$, and for small η

$$\omega \approx \left[\frac{\lambda}{2(m_A + m_B)}\right]^{\frac{1}{2}}(2\pi\eta)\,,$$

showing that the frequency is linearly proportional to η, which is called the *wave number*. Also as $\eta \to 0$, $A/B \to 1$, showing that each pair of particles moves in unison as a rigid unit.

These waves of the acoustic branch are identical with longitudinal elastic vibrations, regarding the chain as an elastic string.

OPTICAL BRANCH:

Here the frequency is given by

$$\omega^2 = \frac{\lambda}{m_A m_B}\left\{m_A + m_B + [(m_A + m_B)^2 - 4m_A m_B \sin^2(\pi\eta)]^{\frac{1}{2}}\right\} \tag{9.60}$$

and as $\eta \to 0$, ω tends to a finite non-zero value, namely

$$\omega = \left[\frac{2(m_A + m_B)\lambda}{m_A m_B}\right]^{\frac{1}{2}}.$$

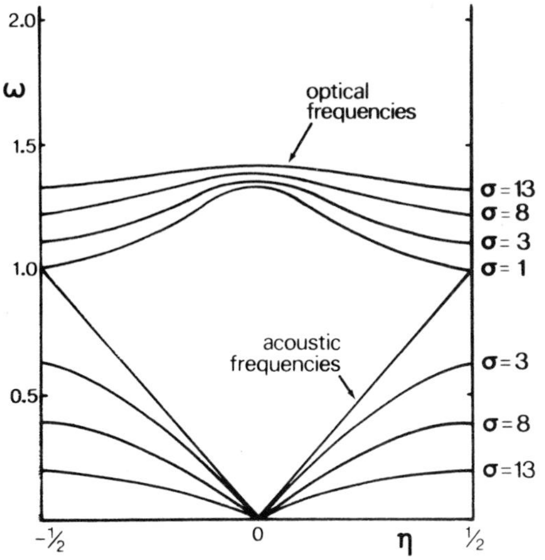

Fig. 9.9 Vibration frequencies

Also $A/B = -m_B/m_A$; i.e. $Am_A + Bm_B = 0$, showing that a pair of particles moves such that their motion opposes each other while keeping their centre of mass stationary. The optical vibrations for small η are of special importance in the problem of the interaction of crystals with light.

9.8 Damped and Forced Oscillations

In this section we are concerned with a special case of (9.2), namely

$$\boxed{\ddot{x} + 2a\dot{x} + b^2x = F(t)} \tag{9.61}$$

where a is a positive constant, and F is a given function of time t. This equation frequently occurs in both mechanical and electrical problems, and we look at its derivation in two important cases:

MECHANICAL SYSTEM:

A 'spring-mass' system is shown in Fig. 9.10. We assume the particle has unit mass, and the spring, of natural length a, has modulus of elasticity λ. We also include a damping or resistive force (due to such forces as air resistance) which is represented by a dashpot mechanism. A dashpot is a piston moving in a pot of oil, which provides a resistance proportional to the velocity of the piston. If x measures the displacement of the mass from equilibrium, the damping force will be $-c\dot{x}$, c being a positive constant. Suppose that there is also an applied external force, $F = F(t)$ say, then the equation of motion is

$$\boxed{\ddot{x} = -\lambda x - c\dot{x} + F(t)} \quad , \tag{9.62}$$

which is of the form (9.61) with $a = c/2$ and $b = \sqrt{\lambda}$.

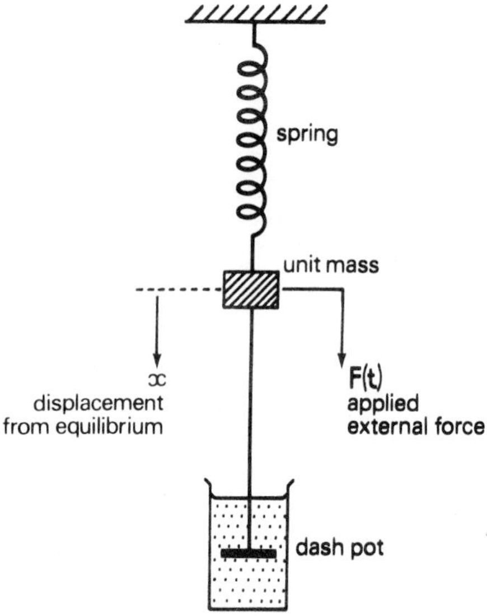

Fig. 9.10 Spring-mass system

ELECTRICAL NETWORK:

A simple closed electrical network is illustrated in Fig. 9.11. The current flowing round the circuit, I (amps), is a function of time t. The resistance R (ohms), capacitance C (farads) and inductance L (henrys) are all positive and may depend on t, but for simplicity we assume that they are all constants.

The applied voltage E (volts) is a known function of time and the total charge Q (coulombs) in the capacitor at time t is related to the current by

$$I = \frac{dQ}{dt} . \tag{9.63}$$

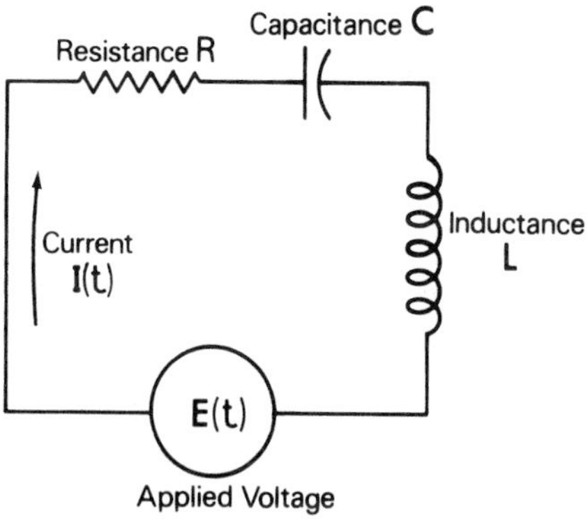

Fig. 9.11 Electrical network

Kirchoff's second law for electrical networks states that "In a closed circuit the impressed voltage is equal to the sum of the voltage drops in the rest of the circuit". The voltage drop across the resistance is IR, across the capacitor is $(1/C)\,Q$, and across the inductance is $L\,(dI/dt)$.

Hence

$$L\frac{dI}{dt} + RI + \frac{1}{C}Q = E(t)\,.$$

Using (9.63), we see that

$$\boxed{L\frac{d^2Q}{dt^2} + R\frac{dQ}{dt} + \frac{1}{C}Q = E(t)} \quad , \qquad (9.64)$$

which is of the form (9.61) with x replaced by Q, $F(t)$ by $E(t)/L$, and with $a = R/2L$, $b = 1/\sqrt{CL}$.

There are in fact a multitude of physical problems which lead to differential equations of the form (9.61), and this situation illustrates an important characteristic of mathematical physics; namely, that many different physical situations, when mathematically formulated, are identical. This means that the solution of a single mathematical problem will give solutions to many different physical problems.

We now return to the mathematical problem of solving (9.61) and first consider the case $F(t) = 0$. In this case the motion is termed 'damped free vibrations'.

9.8.1 DAMPED FREE VIBRATIONS:

The governing differential equation is

$$\boxed{\ddot{x} + 2a\dot{x} + b^2x = 0} \quad , \tag{9.65}$$

and the solution will clearly depend on the sign of $(a^2 - b^2)$.

(i) $\underline{(a^2 - b^2) > 0}$

In this case $$x = A\,e^{\alpha_1 t} + B\,e^{\alpha_2 t}, \tag{9.66}$$

where α_1 and α_2 are given by $\alpha = -a \pm (a^2 - b^2)^{1/2}$

and so both α_1 and α_2 are negative.

(ii) $\underline{(a^2 - b^2) = 0}$

In this case $$x = (A + Bt)e^{-\alpha t}. \tag{9.67}$$

(iii) $\underline{(a^2 - b^2) < 0}$

In this case $$x = e^{-\alpha t}\,[A\cos(ct) + B\sin(ct)]\ , \tag{9.68}$$

where $c = (b^2 - a^2)^{1/2}$, and in all the above cases A and B are arbitrary constants.

Since $a > 0$, in all three cases $x \to 0$ as $t \to \infty$, showing that the motion always dies out eventually. In cases (i) and (ii), the motion is damped without oscillation, and these cases are referred to as *overdamped* and *critically damped* motion respectively. Figures 9.12 and 9.13 illustrate typical solutions in these two cases. In case (iii), part of the solution is periodic whereas the other part is damped. Combining the two parts the motion will be oscillatory but with decreasing amplitude. A typical solution is illustrated in Fig. 9.14. This motion is called *underdamped* motion.

We have seen that the resistance term $(2a\dot{x})$ in all cases dampens the motion. If $a = 0$, the motion is pure oscillatory, but if $a > 0$ the amplitude of the oscillations decreases, and if $a^2 > b^2$, the oscillatory motion is completely eliminated by the resistance.

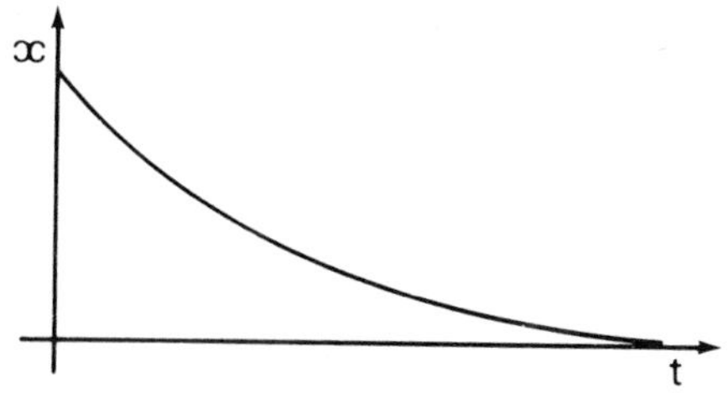

Fig. 9.12 Overdamped motion; $(a^2 - b^2) > 0$

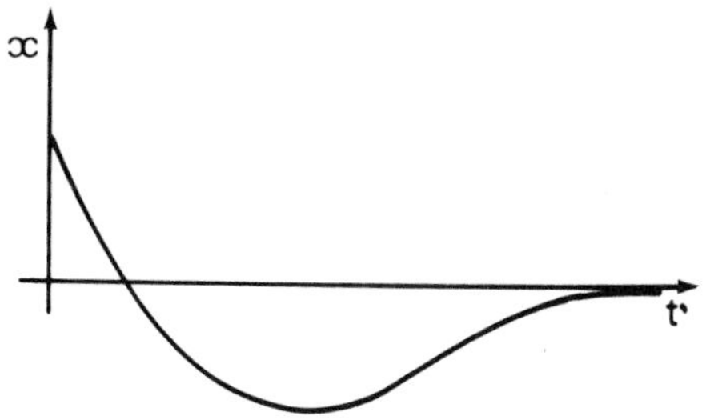

Fig. 9.13 Critically damped motion; $(a^2 - b^2) = 0$

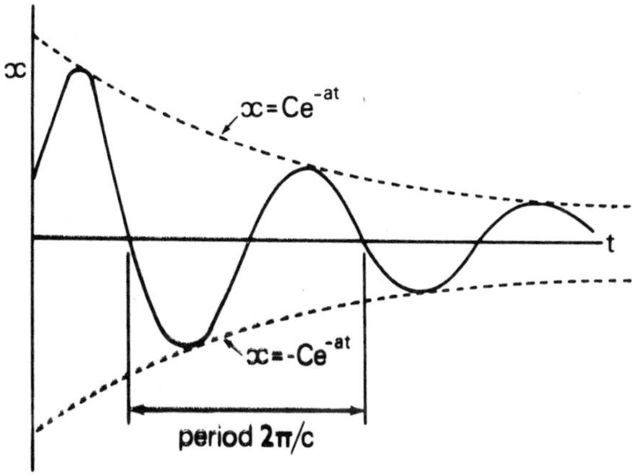

Fig. 9.14 Underdamped motion; $C^2 = A^2 + B^2$

9.8.2 FORCED OSCILLATIONS:

We now consider solutions of (9.61) when $F(t) \neq 0$. The motion is termed 'forced oscillations', and the function $F(t)$ is called the forcing function.

We study one particular case in some detail, namely when $F = F_o\cos(\omega t)$, where F_o and ω are positive constants. The governing differential equation is

$$\boxed{\ddot{x} + 2a\dot{x} + b^2x = F_o\cos(\omega t)} \quad . \qquad (9.69)$$

The solution of this differential equation is of the form

$$x = x_c + x_p \tag{9.70}$$

where x_c is the complementary function and x_p is a particular integral. The complementary function is the solution of

$$\ddot{x} + 2a\dot{x} + b^2 x = 0.$$

and we have already seen that, provided $a > 0$, all solutions of this equation tend to zero as t tends to infinity. This part of the solution is called the 'transient'. To find the particular integral, which will be the dominant term in (9.70) as $t \to \infty$, we assume

$$x_p = A\cos(\omega t) + B\sin(\omega t), \tag{9.71}$$

where A and B are constants to be found. Substituting into (9.69) gives

$$-\omega^2 [A\cos(\omega t) + B\sin(\omega t)] + 2a\omega [-A\sin(\omega t) + B\cos(\omega t)]$$

$$+ b [A\cos(\omega t) + B\sin(\omega t)] = F_o \cos(\omega t),$$

and equating coefficients of $\cos(\omega t)$ and $\sin(\omega t)$ gives

$$(b^2 - \omega^2)A + 2a\omega B = F_o,$$

$$2a\omega A - (b^2 - \omega^2)B = 0.$$

Solving for A and B gives

$$A = \frac{F_o(b^2 - \omega^2)}{[(b^2 - \omega^2)^2 + 4a^2\omega^2]},$$

$$B = \frac{2a\omega F_o}{[(b^2 - \omega^2)^2 + 4a^2\omega^2]},$$

and so $x = \text{'transient'} + \dfrac{F_o [(b^2 - \omega^2)\cos(\omega t) + 2a\omega\sin(\omega t)]}{[(b^2 - \omega^2)^2 + 4a^2\omega^2]}$;

i.e.

$$x = \text{'transient'} + \frac{F_o \cos(\omega t - \phi)}{[(b^2 - \omega^2)^2 + 4a^2\omega^2]^{1/2}} \tag{9.72}$$

where

$$\tan\phi = 2a\omega/(b^2 - \omega^2). \tag{9.73}$$

As $t \to \infty$,

$$x \to \frac{F_o \cos(\omega t - \phi)}{[(b^2 - \omega^2)^2 + 4a^2\omega^2]^{1/2}}$$

(since the transient $\to 0$ as $t \to \infty$), showing that eventually the solution for x oscillates with frequency ω, the same frequency as the applied force. This is why the motion is given the name 'forced oscillations'. We also note that the solution has a phase shift ϕ from the applied force, and its amplitude, say D, is given by

$$D = F_0/[(b^2 - \omega^2)^2 + 4a^2\omega^2]^{1/2} . \tag{9.75}$$

Writing $v = bD/F_0$, $u = \omega^2/b^2$, reduces (9.75) to

$$v = 1/[(1-u)^2 + 4cu]^{1/2}, \tag{9.76}$$

where $c = a^2/b^2$. The quantity v is a measure of the amplitude, and a function of u, the ratio of the square of the forced frequency to the square of the 'natural' frequency. [The *natural frequency* is defined as the frequency of the solution of

$$\ddot{x} + b^2 x = 0 ;$$

i.e. the natural frequency is b.]

Figure 9.15 shows the function v against the frequency square ratio ω^2/b^2 for varying values of $c = a^2/b^2$. It can easily be shown that v has a maximum or minimum at $u = 1 - 2a^2/b^2$, and the corresponding value of the amplitude is

$$F_0 b^2/2a^2(3b^2 - 4a^2) . \tag{9.77}$$

We look at two special features of forced oscillations:

(i) RESONANCE:

Fig. 9.15 shows one very significant feature, namely that as $a \to 0$, that is $c \to 0$, the maximum value of the amplitude, (9.77), tends to infinity. This implies that if there is no damping (i.e. $a = 0$), the amplitude increases indefinitely as $u \to 1$. Now $u = 1$ corresponds to the forcing and natural frequencies being equal. This phenomenon is called *resonance*.

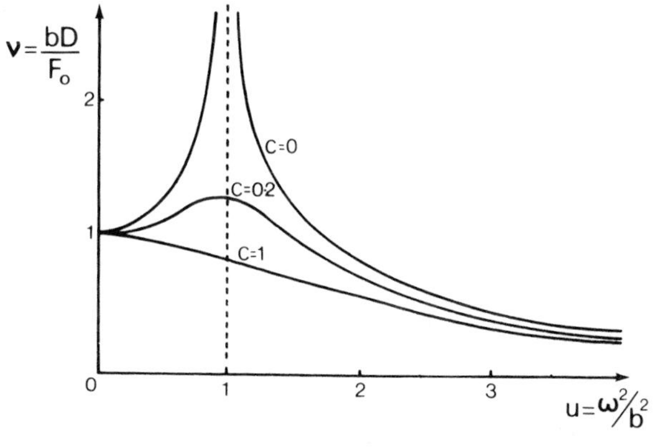

Fig. 9.15 Variation of amplitude with frequency ratio

Even when there is a small amount of resistance, that is a is small, Fig. 9.15 shows that the amplitude will still have a considerable maximum value.

Resonance is an extremely important phenomenon, and often one that we wish to guard against. For, at resonance, the amplitude increases indefinitely and this type of oscillation can be destructive if it occurs in machinery. For instance, the initial persistent trouble with Cunard's liner *Queen Elizabeth II* (Q.E.2) was due to resonance in the ship's turbines. Resonance occurs when the natural and forced frequencies are equal, and this phenomenon can often be observed, for instance, the opera singer's voice cracking a wine glass. Resonance is also the reason why a marching column of men breaks step while crossing a bridge because if it did not, the marching steps would apply a periodic force whose frequency might be equal to one of the natural frequencies of the bridge. If this occurred the bridge would begin to oscillate, with increasing amplitude, until it collapsed!

(ii) BEATS:

Another phenomenon which occurs when there is negligible resistance is called beats. We illustrate this by solving the differential equation

$$\ddot{x} + bx = F_o \cos(\omega t) \tag{9.78}$$

subject to the initial conditions $x = 0$, $\dot{x} = 0$ at $t = 0$. It can soon be shown that, provided $\omega^2 \neq b$, the general solution of (9.78) is

$$x = A \cos\sqrt{b}t + B \sin\sqrt{b}t + \frac{F_o \cos(\omega t)}{(b - \omega^2)}.$$

Applying the initial conditions gives

$$A = -F_o/(b - \omega^2), \quad B = 0,$$

and so
$$x = \frac{F_o}{(b - \omega^2)} [\cos(\omega t) - \cos(\sqrt{b}t)]$$

i.e.
$$x = \frac{2F_o}{(b - \omega^2)} \sin[(\sqrt{b} - \omega)t/2] \sin[(\sqrt{b} + \omega)t/2]. \tag{9.79}$$

If $(\sqrt{b} - \omega)$ is small, $\sin[(\sqrt{b} + \omega)t/2]$ will be a rapidly oscillating function compared with $\sin[(\sqrt{b} - \omega)t/2]$. Fig. 9.16 illustrates the motion. There is a periodic variation in the amplitude as well as a rapid periodic motion about the origin. This phenomenon can be observed in acoustics when two tuning forks of nearly equal frequency are sounded simultaneously. The periodic variation of the amplitude will be clear to the unaided ear.

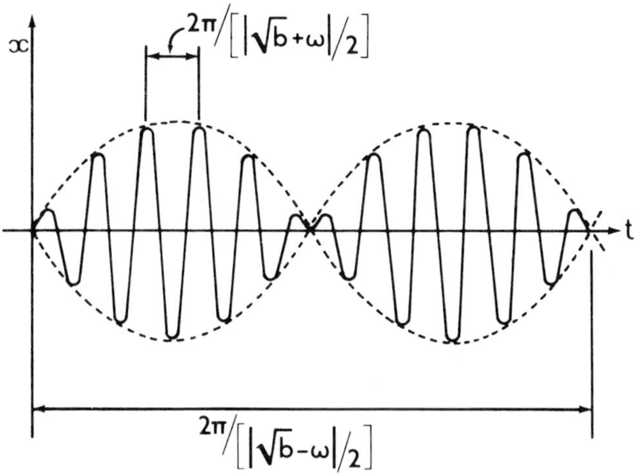

Fig. 9.16 Beats

9.9* The Seismograph

The seismograph is an instrument which measures and records the movements of the ground during an earthquake or tremor. The first seismograph worthy of mention is due to J.D. Forbes in 1841. In this section we describe and analyse a simple seismograph, similar to that due to Forbes. The model is illustrated in Fig. 9.17.

A mass M is suspended by both a spring and dashpot from a platform AB, fixed to the earth. If the ground oscillates the platform will transmit the oscillations to the mass. A printer, attached to the mass, records its displacement on a scale attached to the platform.

During a ground disturbance, let the displacement of the platform from its initial position be $y = y(t)$. If x denotes the displacement of the mass relative to the platform, the velocity of the mass (and therefore the dashpot) is

$$\frac{d}{dt}(x+y) = \dot{x} + \dot{y}$$

The plunger in the dashpot, though, will move with speed $\dot{y}$, and so the dashpot will be resisted by a force of the form $c\dot{x}\,(c > 0)$. If a is the natural length of the spring, the equation of motion of the mass M is

$$M(\ddot{x} + \ddot{y}) = Mg - c\dot{x} - k(x-a),$$

where k is the modulus of elasticity per unit length; and assuming, for simplicity, unit mass we have

$$\ddot{x} + c\dot{x} + kx = g + ka - \ddot{y}. \tag{9.80}$$

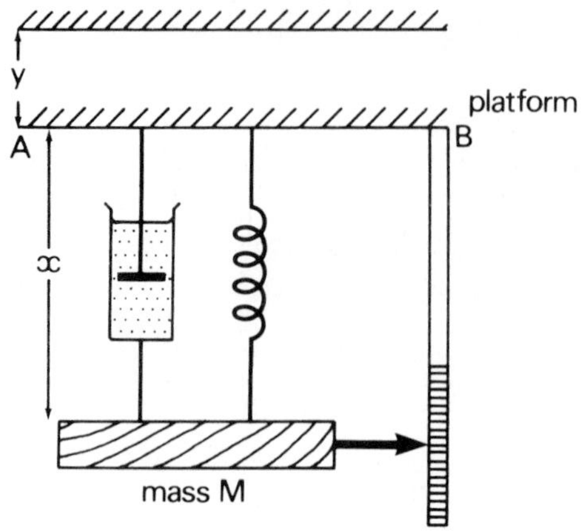

Fig. 9.17 Seismograph

If $\tilde{x}$ is the displacement of the mass from its equilibrium position, then

$$\tilde{x} = x - (g + ka)/k , \tag{9.81}$$

and $\tilde{x}$ is the recorded displacement on the scale. Hence

$$\ddot{\tilde{x}} + c\dot{\tilde{x}} + k\tilde{x} = -\ddot{y} . \tag{9.82}$$

The motion of the mass clearly depends on the impressed motion $y = y(t)$.

We assume $$y = A\cos(\omega t) ,$$

giving $$\ddot{\tilde{x}} + c\dot{\tilde{x}} + k\tilde{x} = A\omega^2 \cos(\omega t) ,$$

and using the results from Ch. 9.8,

$$\tilde{x} = \text{‘transient’} + D\sin(\omega t + \phi). \tag{9.83}$$

From (9.72), we see that

$$D = \frac{A(\omega^2/k)}{[(1-\omega^2/k)^2 + (c^2/k)(\omega^2/k)]^{1/2}} . \tag{9.84}$$

The constants c and k are assumed known, and so observations of D and ω enable one to determine the amplitude, A, of the ground's oscillation.

Worked Examples

EXAMPLE 9.1:

A circular cylinder of radius a and height h, composed of material of uniform density ρ_1, floats in a liquid of density $\rho_2 (> \rho_1)$. Find the period of vertical oscillations of the cylinder. (Archimedes Principle states that the upthrust on the cylinder equals the weight of liquid displaced.)

SOLUTION:

We first find the cylinder's position when in equilibrium (see Fig. 9.18).

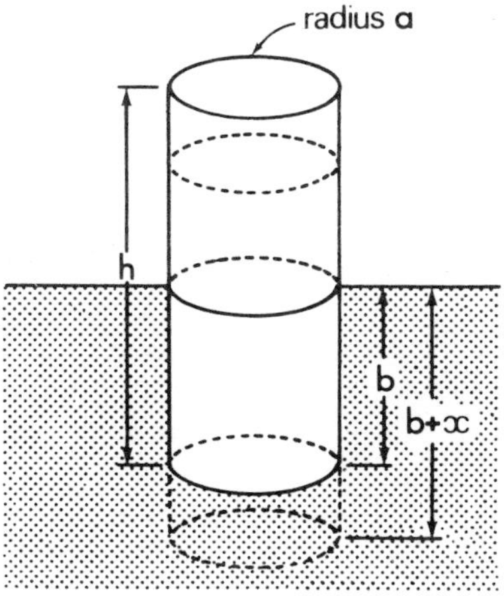

Fig. 9.18 Cylinder in liquid

Equating the gravitational force of the cylinder to the upthrust exerted on the cylinder gives

$$\rho_1 (\pi a^2 h) g = \rho_2 (\pi a^2 b) g ,$$

which gives the length of cylinder immersed as $b = (h\rho_1)/\rho_2$.

Suppose the cylinder is displaced downwards a distance x from the equilibrium position. Then the equation of motion for the cylinder is

$$\begin{aligned} \rho_1 (\pi a^2 h)\ddot{x} &= \rho_1 (\pi a^2 h) g - \rho_2 [\pi a^2 (b + x)] g \\ &= - \rho_2 \pi a^2 g x . \end{aligned}$$

Thus
$$\ddot{x} + \omega^2 x = 0\,,$$

where
$$\omega^2 = \frac{\rho_2 g}{\rho_1 h}.$$

This is the equation of S.H.M. and its solution has period

$$2\pi/\omega = 2\pi(\rho_1 h/\rho_2 g)^{1/2}.$$

EXAMPLE 9.2:

A spring-mass system, illustrated in Fig. 9.19 lies on a smooth horizontal table, the ends A and D being fixed. The particles B and C both have mass m, and the springs AB, BC and CD each have natural length $a/2$ and modulus λ.

If AD = $3a$, show that for oscillations along the line of the springs, there are two possible frequencies. Describe each mode of oscillation.

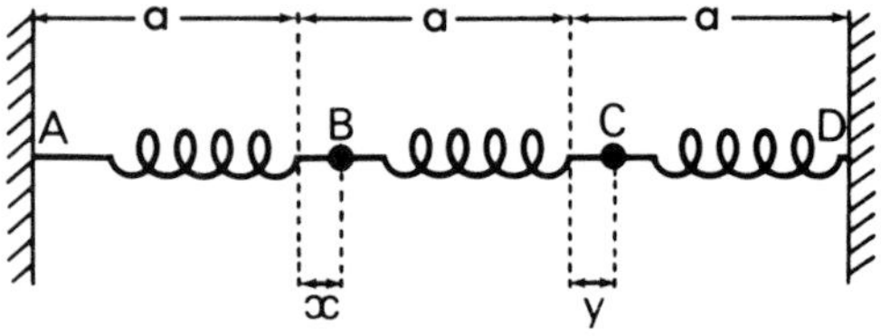

Fig. 9.19 Spring-mass system

SOLUTION:

In equilibrium AB = BC = CD = a. Suppose the masses B and C are displaced from equilibrium distances x and y respectively. The equations of motion for B and C are

$$m\ddot{x} = T_2 - T_1$$

$$m\ddot{y} = T_3 - T_2\,,$$

where T_1, T_2 and T_3 are the tensions in the springs AB, BC and CD respectively. Now

$$T_1 = \lambda(\tfrac{1}{2}a + x)/(\tfrac{1}{2}a), \quad T_2 = \lambda(\tfrac{1}{2}a + y - x)/(\tfrac{1}{2}a), \quad T_3 = \lambda(\tfrac{1}{2}a - y)/(\tfrac{1}{2}a).$$

Note that in equilibrium, the extension in each spring is $\frac{1}{2}a$. Substituting into the equations of motion

$$m\ddot{x} = \left(\frac{2\lambda}{a}\right)(y - 2x)$$

and $$m\ddot{y} = \left(\frac{2\lambda}{a}\right)(-2y + x)\,.$$

Writing $\Omega^2 = (2\lambda/am)$,

$$\ddot{x} + 2\Omega^2 x - \Omega^2 y = 0\,,$$

$$-\Omega^2 x + \ddot{y} + 2\Omega^2 y = 0\,.$$

Assuming a solution of the form

$$x = \alpha\cos(\omega t + \epsilon)$$

$$y = \beta\cos(\omega t + \epsilon)\,,$$

we must satisfy,

$$(-\omega^2 + 2\Omega^2)\alpha - \Omega^2\beta = 0\,,$$

$$-\Omega^2\alpha + (-\omega^2 + 2\Omega^2)\beta = 0\,.$$

Eliminating the ratio α/β from these two equations gives

$$(-\omega^2 + 2\Omega^2)^2 - \Omega^4 = 0\,,$$

which simplifies to $$\omega^4 - 4\Omega^2\omega^2 + 3\Omega^4 = 0\,.$$

Thus $\omega^2 = 3\Omega^2$ or Ω^2

i.e. $$\underline{\omega_1 = \sqrt{3}\Omega} \quad \text{and} \quad \underline{\omega_2 = \Omega}\,.$$

When $\omega_1 = \sqrt{3}\Omega$, the ratio α/β is given by

$$\frac{\alpha}{\beta} = \frac{-\omega_1^2 + 2\Omega^2}{\Omega^2} = -1; \qquad \text{i.e. } \alpha = -\beta\,.$$

When oscillating in this mode, $\alpha = -\beta$, so that the amplitudes of oscillations of B and C are equal in magnitude but opposite in direction. Similarly when $\omega_2 = \Omega$, $\alpha = \beta$. The oscillations are illustrated in Fig. 9.20.

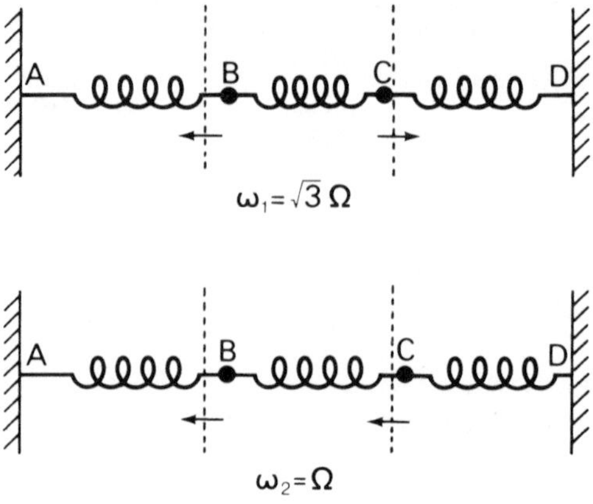

Fig. 9.20 Modes of oscillation

EXAMPLE 9.3:

Solve the equation $$\ddot{x} + \omega^2 x = F_o \cos(\omega t)$$

with $x = \dot{x} = 0$ when $t = 0$. Show that the oscillation has period $2\pi/\omega$, but with an amplitude increasing linearly with time.

SOLUTION:

The solution of the differential equation is of the form

$$x = x_c + x_p ,$$

where x_c is the complementary function, and x_p is a particular solution. Thus x_c satisfies

$$\ddot{x}_c + \omega^2 x_c = 0 ,$$

and solving $$x_c = A\cos(\omega t) + B\sin(\omega t)$$

where A and B are arbitrary constants.

In Ch. 9.8, we tried to find a particular solution of the form

$$x_p = A\cos(\omega t) + B\sin(\omega t) .$$

In this particular case, when the natural frequency and forced frequency are equal, this method will not work. This is because the suggested x_p is in fact the same as the complementary function. Thus if we substitute this x_p into the full differential equation, we would obtain $0 = F_o\cos(\omega t)$.

Consequently there is *no* particular solution of this form.

Let us try another form for x_p.

If
$$x_p = Ct\cos(\omega t) + Dt\sin(\omega t)\,,$$

where C and D are constants to be determined, then

$$\dot{x}_p = (C + \omega Dt)\cos(\omega t) + (D - \omega Ct)\sin(\omega t)$$

and
$$\ddot{x}_p = (2\omega D - \omega^2 Ct)\cos(\omega t) + (-2\omega C - \omega^2 Dt)\sin(\omega t)\,.$$

Substituting into the full differential equation,

$$\ddot{x}_p + \omega^2 x_p = 2\omega D\cos(\omega t) - 2\omega C\sin(\omega t) = F_o\cos(\omega t)\,.$$

This is satisfied if

$$D = F_o/2\omega, \qquad C = 0\,.$$

Hence a particular solution is

$$x_p = \frac{F_o t}{2\omega}\sin(\omega t)$$

and the complete solution is

$$x = A\cos(\omega t) + \frac{F_o t}{2\omega}\sin(\omega t)\,.$$

Thus if $x = \dot{x} = 0$ at $t = 0$, we must have

$$0 = A\,,$$

$$0 = B\omega\,,$$

and we have the solution

$$x = \frac{F_o t}{2\omega}\sin(\omega t)\,.$$

There is an oscillating term, with period $2\pi/\omega$, and the amplitude increases linearly with time.

EXAMPLE 9.4:

An electrical circuit, as illustrated in Fig. 9.11, contains a resistance R, capacitance C and inductance L where R, C and L are constants and $4L > CR^2$. Find the current, I, when there is zero initial current and charge, and when the impressed voltage is given by $E = t$ for $t > 0$.

SOLUTION:

From (9.64), the equation for the charge Q on the capacitor is given by

$$\frac{d^2Q}{dt^2} + \frac{R}{L}\frac{dQ}{dt} + \frac{1}{CL}Q = \frac{t}{L}.$$

This differential equation has solution

$$Q = Q_c + Q_p$$

where Q_c is the solution of

$$\frac{d^2Q_c}{dt^2} + \frac{R}{L}\frac{dQ_c}{dt} + \frac{1}{\mathrm{CL}}Q_c = 0$$

and Q_p is a particular solution of the full differential equation. Thus

$$Q_c = \mathrm{e}^{-Rt/2L}\,[A\sin(\omega t) + B\cos(\omega t)]\ ,$$

where $\omega^2 = (4L - CR^2)/(4CL^2)$. To find a particular solution, we try $Q_p = \alpha t + \beta$, where α and β are constants to be determined. Substituting into the full differential equation gives

$$\frac{R\alpha}{L} + \frac{1}{CL}(\alpha t + \beta) = \frac{1}{L}t\,.$$

Thus Q_p is a solution when $\alpha = C$ and $\beta = -RC^2$. Hence the general solution is

$$Q = \mathrm{e}^{-Rt/2L}\,[A\sin(\omega t) + B\cos(\omega t)] + Ct - RC^2.$$

The initial conditions are $Q = 0$ at $t = 0$ and $I = dQ/dt = 0$ at $t = 0$.

But

$$I = \left(-\frac{R}{2L}\right)\mathrm{e}^{-Rt/2L}\,[A\sin(\omega t) + B\cos(\omega t)]$$

$$+\ \mathrm{e}^{-Rt/2L}\,[A\omega\cos(\omega t) - B\omega\sin(\omega t)] + C$$

i.e. $$I = e^{-Rt/2L}\left\{-\left(\frac{RA}{2L} + \omega B\right)\sin(\omega t) + \left(-\frac{RB}{2L} + \omega A\right)\cos(\omega t)\right\} + C.$$

Applying the initial conditions gives

$$0 = B - RC^2$$

and $$0 = \left(-\frac{RB}{2L} + \omega A\right) + C.$$

Thus $B = RC^2$ and $A = C(R^2C - 2L)/(2L\omega)$, and substituting into the expression for I and using $\omega^2 = (4L - CR^2)/(4CL^2)$ results in

$$I = C - e^{-Rt/2L}\left\{\left(\frac{RC}{2L\omega}\right)\sin(\omega t) + C\cos(\omega t)\right\}.$$

Note that, although the applied voltage, E, and charge, Q, are increasing without limit, the current I is finite and in fact tends to C as $t \to \infty$.

EXAMPLE 9.5:

A particle of mass m is attached to the lower end, B, of an elastic spring AB, which is hanging vertically. The upper end, A, is forced to undergo a vertical oscillation, its downward displacement from its initial position being $a\sin(\Omega t)$. During its motion for $t > 0$, the mass is subject to a resistance equal in magnitude to mk times its speed. The spring has modulus λ and natural length l.

If the spring's length at time $t\ (> 0)$ is $l + mgl/\lambda + x$, show that x satisfies

$$\ddot{x} + k\dot{x} + \left(\frac{\lambda}{ml}\right)x = a\Omega\{\Omega\sin(\Omega t) - k\cos(\Omega t)\}.$$

Find the amplitude of the forced oscillation of the mass when $k = \Omega$ and $\lambda/ml = \Omega^2$, and find the phase angle of this oscillation relative to A's oscillation.

SOLUTION:

In order to write down an equation of motion, we must first define a *fixed* origin and axes. The point A is *not* fixed, but if we take its initial position, say O, as the origin, and choose one axis vertically downwards from O, then the particle's displacement in this direction from O is:

$$\mathbf{r} = \left\{a\sin(\Omega t) + \left(l + \frac{mgl}{\lambda} + x\right)\right\}\mathbf{i}$$

where **i** is a unit vector in the vertically downwards direction. The situation at some time $t\ (>0)$ is illustrated in Fig. 9.21.

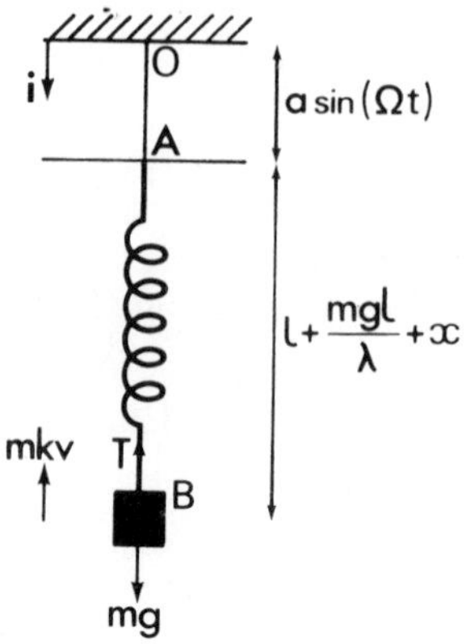

Fig. 9.21 Spring-mass system

The equation of motion for the particle, in this inertial reference frame, is

$$m\frac{d^2\mathbf{r}}{dt^2} = mg\mathbf{i} - mk\frac{d\mathbf{r}}{dt} - T\mathbf{i}$$

where T is the tension in the spring. Hence $T = (\lambda/l)\,(mgl/\lambda + x)$ and evaluating $d\mathbf{r}/dt$ and $d^2\mathbf{r}/dt^2$, we obtain

$$m\{-a\Omega^2\sin(\Omega t) + \ddot{x}\} = -mk\{a\Omega\cos(\Omega t) + \dot{x}\} - \frac{\lambda x}{l}.$$

Hence we have the differential equation

$$\ddot{x} + k\dot{x} + \left(\frac{\lambda}{ml}\right)x = a\Omega\,\{\Omega\sin(\Omega t) - k\cos(\Omega t)\}.$$

Solving this equation, as usual,

$$x = x_c + x_p\,,$$

where x_c is the complementary function ($\to 0$ as $t\to\infty$) and x_p is the particular solution, that is the forced oscillation. Thus we just require a knowledge of x_p. Let us assume that x_p is of the form

$$x_p = A\sin(\Omega t) + B\cos(\Omega t)$$

where A and B are constants to be determined. Then

$$\dot{x}_p = A\Omega\cos(\Omega t) - B\Omega\sin(\Omega t)$$

and
$$\ddot{x}_p = -A\Omega^2\sin(\Omega t) - B\Omega^2\cos(\Omega t)\,.$$

Substituting into the differential equation gives

$$\left(-A\Omega^2 - Bk\Omega + \frac{A\lambda}{ml}\right)\sin(\Omega t) + \left(-B\Omega^2 + Ak\Omega + \frac{B\lambda}{ml}\right)\cos(\Omega t)$$

$$= a\Omega^2\sin(\Omega t) - a\Omega k\cos(\Omega t)\,.$$

When $k = \Omega$ and $\lambda/ml = \Omega^2$, the equations reduce to

$$\Omega^2\,[(-B-a)\sin(\Omega t) + (A+a)\cos(\Omega t)] = 0\,.$$

Hence $A = B = -a$ for a solution, and

$$x_p = -a\{\sin(\Omega t) + \cos(\Omega t)\}$$

$$= a\sqrt{2}\left\{-\frac{1}{\sqrt{2}}\sin(\Omega t) - \frac{1}{\sqrt{2}}\cos(\Omega t)\right\}$$

$$= a\sqrt{2}\,\sin\left(\Omega t - \frac{3\pi}{4}\right).$$

We see that the amplitude of x_p is $a\sqrt{2}$, and the phase angle relative to A's oscillation, $a\sin(\Omega t)$, is $3\pi/4$.

Problems

9.1 A particle of mass m moves in a straight line under a force per unit mass, $\omega^2 x$, directed *towards* a fixed point O on the line, where x measures the distance from O and ω is a constant. If initially the particle is projected from O with a given speed u, show that in its subsequent motion $x = (u/\omega)\sin(\omega t)$.

9.2 For simple harmonic motion, prove that

$$v^2 = \omega^2(a^2 - x^2)$$

where ω, a and x are defined as in Ch. 9.2, and $v = \dot{x}$.

A particle is suspended from a fixed point O by an elastic string. In equilibrium the extension of the string is b. The particle is released initially from rest at a point A, where the extension of the string is $3b$. Find the particle's speed at the instant when the string first becomes slack, and show that this occurs at time $2\pi(b/g)^{1/2}/3$.

Find also the time taken for the particle to reach A again.

9.3 A light elastic string of natural length $2a$ is stretched between two pegs, distance $4a$ apart, on a smooth horizontal table, and a particle of mass m is attached to the mid-point of the string. Show that the particle can execute simple harmonic motion of small amplitude both along the line of the string (longitudinal oscillations) and at right angles to it (transverse oscillations). Show that the respective frequencies are in the ratio $\sqrt{2} : 1$.

9.4 AB is a light elastic string of natural length $2a$ and modulus λ. A particle of mass m is attached to the string at the midpoint C of AB and another particle of mass $\frac{2}{3}m$ is attached to the end B. The end A of the string is fixed and the particle hangs at rest in equilibrium. Show that the periods of normal modes of small vertical oscillations of the particles are

$$2\pi\sqrt{2am/\lambda}\,, \qquad 2\pi\sqrt{am/3\lambda}\,.$$

The heavier particle is held at its equilibrium position and the lighter particle is held at a distance d below its equilibrium position, and the system is then released. Find the positions of the particles at any subsequent time.

9.5* Beads, fastened at equal intervals along the whole length of an infinite elastic string under tension and resting on a smooth horizontal table, are alternately of masses m and $2m$. Show that, for small vibrations parallel to the length of the string, the frequencies corresponding to wave number η are in the ratio

$$\{3 + [1 + 8\cos^2(\pi\eta)]^{1/2}\}^{1/2} : \{3 - [1 + 8\cos^2(\pi\eta)]^{1/2}\}^{1/2}.$$

Describe the nature of the vibrations.

9.6 A simple pendulum performs small oscillations of period 2λ in the absence of resisting forces. The pendulum is placed in a medium which exerts on the pendulum bob a resistive force per unit mass equal to $2n$ times its speed, where n is a constant. The bob is released from rest at a small angle to the downward vertical. If $n < \pi/\lambda$ show that the new period of oscillation is

$$2\pi\lambda/[\pi^2 - n^2\lambda^2]^{1/2}.$$

What happens if $n > \pi/\lambda$?

9.7 A particle of mass m moves in a straight line along the x-axis under the action of a force $m\omega^2 x$ directed towards the origin, ω being a constant. It is also subjected to a damping force of magnitude mkv where v is the particle's speed and k is a positive constant such that $k^2 < 4\omega^2$. If the particle starts from rest at a distance a from the origin, show that it will momentarily come to rest again at a distance

$$a \exp\left[-\pi k/(4\omega^2 - k^2)^{1/2}\right]$$

on the other side of the origin.

9.8 A particle is performing simple harmonic motion of period $2\pi/b$. If the particle is now subjected to a resisting force of magnitude $2av$ per unit mass, where a is a positive constant and v is the particle's speed, show that, provided $b^2 - a^2 > 0$,

(i) the oscillations are damped;
(ii) the period of oscillations is increased in the ratio $b:(b^2 - a^2)^{1/2}$;
(iii) the ratio of successive positive amplitude peaks is given by $\exp\left[2\pi a/(b^2 - a^2)^{1/2}\right] : 1$.

9.9 A block of mass 5kg is attached to a fixed point by a horizontal spring which obeys Hooke's law. The block moves in a straight line on a horizontal surface which is coated with a viscous lubricant. The resulting viscous force acting on the block is assumed to be proportional to its speed. The block is set oscillating.

It is found that successive maximum displacements of the block from equilibrium (i.e. when the spring has maximum extension) diminish in the ratio 0·85, and that the period of oscillations is 0·8 s. Show that the resistance to motion is approximately $2v$ N, where v is the speed of the block in $\mathrm{m\,s^{-1}}$. Also find the modulus of elasticity per unit length of the spring.

9.10 A particle of mass m is tied to the mid-point of a light elastic string of modulus λ and natural length $2a$, whose end points are tied to two fixed points on a smooth horizontal table, distance $4a$ apart. The resistance to the motion is mkv, where v is the speed of the particle. The particle is displaced from its equilibrium position in the line of the string through a distance a, and released from rest. Show that, if $mak < 8\lambda$, damped oscillations will occur. Also show that in the first complete oscillation the particle will travel a total distance $a(1 + e^{-k\pi/\alpha})^2$, where $\alpha^2 = (8\lambda/ma) - k^2$.

9.11 A block of mass m lies at rest on a horizontal table and is attached to one end of a light spring which, when stretched, exerts a tension of magnitude $m\omega^2$ times its extension, where ω is a constant. The surface of the table offers a resistance to the motion of the block of magnitude mk times the relative speed between the block and the table, k being a constant.

The end of the spring not attached to the mass is given a uniform velocity u along the table away from the mass and along the line of the spring. Obtain a differential equation for the displacement x of the mass at time t and show that for the case $k = 2\omega$,

$$x = (u/\omega)[(2 + \omega t)\,e^{-\omega t} + (\omega t - 2)]\ .$$

9.12 An electrical circuit contains a resistance R, capacitance C, and inductance L, where R, C and L are constants. In the absence of any impressed voltage, show that

(i) when there is no resistance, the charge Q is oscillatory with frequency $1/\sqrt{LC}$;

(ii) if a resistance is present, the current I always tends to zero as t tends to infinity.

If the impressed voltage is given by

$$E = E_0 \cos(\omega t)\,,$$

where E_0 and ω are constants, show that

(i) in the absence of any resistance, the charge on the capacitor is unbounded as $t \to \infty$ if $\omega = 1/\sqrt{LC}$;

(ii) the charge will always remain finite, no matter what the value of ω, provided there is some resistance in the circuit.

Chapter 10

Variational Calculus

10.1 Introduction

The first problem in variational calculus was posed by a Swiss mathematician, JOHANN BERNOULLI (1667-1748). The problem was to find the curve joining two fixed points, A and B, in a vertical plane for which a body sliding down the curve (under constant gravity and neglecting friction) travels from A to B in a minimum time. The answer to this problem is not trivial and it will be shown in Ch. 10.4 that the required curve is in fact an arc of a cycloid. This curve of quickest descent is termed the 'brachistochrone'.

This and a number of similar problems led to the development of variational calculus although the topic attracted little attention for about two centuries. In the last forty years however interest in the calculus of variations has been greatly revived due to its application to problems of optimization and control, such as rocket trajectory paths and optimal economic growth. In this chapter the classical problems of variational calculus are solved, while in Ch. 11 we study applications to control problems.

10.2 Classical Problems

Three classical problems in variational calculus, to which we shall refer later, are stated below:

10.2.1 SHORTEST PATH BETWEEN TWO POINTS

The problem is to determine the shortest plane curve joining two fixed points A and B. Without loss of generality we can choose axes Ox, Oy, Oz such that A and B have coordinates (0,0,0) and $(a,b,0)$. If $\mathscr{C}$ is any plane curve joining A and B, as illustrated in Fig. 10.1, let s denote arc length along $\mathscr{C}$ from A. Now,

$$\left(\frac{ds}{dx}\right)^2 = 1 + \left(\frac{dy}{dx}\right)^2$$

so that, integrating from A to B, we obtain the arc length

$$s = \int_0^a \left[1 + \left(\frac{dy}{dx}\right)^2\right]^{1/2} dx \quad . \tag{10.1}$$

We wish to find the curve $\mathscr{C}$, i.e. $y = y(x)$ say, from A to B which minimizes s. Clearly this problem has an elementary solution, namely the straight line from A to B, but we use this example in Ch. 10.3 to develop the calculus of variations.

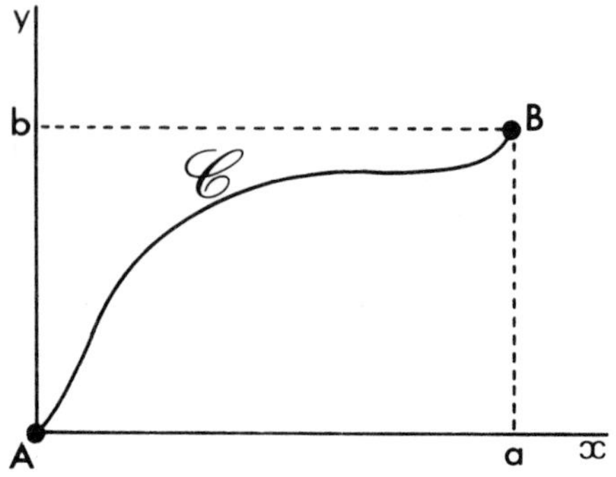

Fig. 10.1 Curve $\mathscr{C}$ from A to B

10.2.2 The Brachistochrone Problem

The problem is to determine the curve of quickest descent for a particle moving under constant gravity on a smooth path between two fixed points A and B.

Let A and B be the points (0,0) and (a,b) as illustrated in Fig. 10.2. Assuming that the particle is released from rest at A, conservation of energy gives

$$\tfrac{1}{2} mv^2 - mgx = 0,$$

where v is the particle's speed. But $v = |\mathbf{v}|$, where $\mathbf{v} = (dx/dt, dy/dt)$, giving

$$v = \left[\left(\frac{dx}{dt}\right)^2 + \left(\frac{dy}{dt}\right)^2\right]^{1/2} = \frac{ds}{dt} \; .$$

Thus $ds/dt = (2gx)^{1/2}$, where s is the arc length along the curve from A. Integrating this equation, the time of descent is given by

$$t = \int \frac{ds}{(2gx)^{1/2}} \; .$$

Since
$$\left(\frac{ds}{dx}\right)^2 = 1 + \left(\frac{dy}{dx}\right)^2 ,$$

the time of descent from A to B is given by

$$t = \frac{1}{(2g)^{1/2}} \int_0^a \left[\frac{1 + (dy/dx)^2}{x}\right]^{1/2} dx \quad . \tag{10.2}$$

We require the path $y = y(x)$ which minimizes this integral.

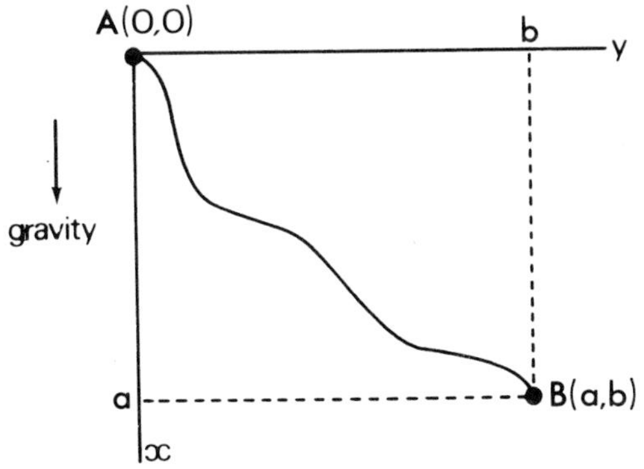

Fig. 10.2 The Brachistochrone problem

10.2.3 MINIMUM SURFACE AREA

A plane curve, $y = y(x)$, passes through the points $(0,b)$ and (a,c), $b \neq c$, in the xy-plane, where a,b and c are positive constants, and the curve lies above the x-axis as illustrated in Fig. 10.3. The problem is to find the curve $y = y(x)$ which, when rotated about the x-axis, yields a surface of revolution with minimum area.

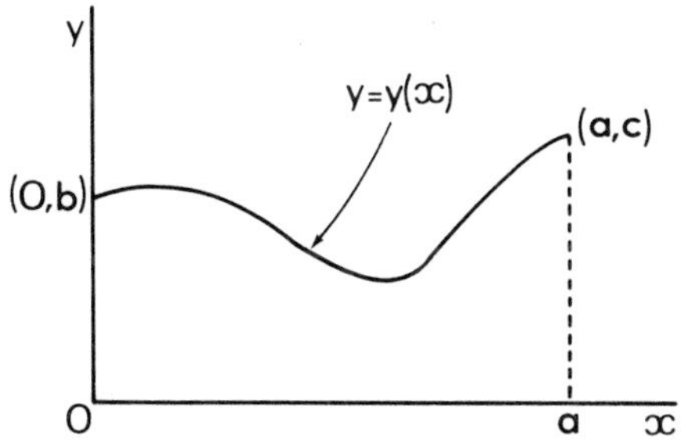

Fig. 10.3 Curve rotated about x-axis

It is easy to see that the area of surface of revolution is given by

$$A = \int 2\pi y \, ds \ ,$$

where s is the arc length along the curve. Thus

$$A = 2\pi \int_0^a y[1 + (dy/dx)^2]^{1/2} \, dx \qquad (10.3)$$

and we wish to find the curve $y = y(x)$ which minimizes this integral.

10.3 Functionals

The right-hand sides of equations (10.1), (10.2) and (10.3) are examples of functionals. A functional can be defined loosely as a rule which assigns a real number to each function belonging to some class of functions. For instance, equation (10.1) defines a real number s for any choice of the continuous function $y(x)$ which is such that $y(0) = 0$ and $y(a) = b$. The examples cited in Ch. 10.2 are all of the type

$$\boxed{J = \int_a^b F\left(y, \frac{dy}{dx}, x\right) dx} \ , \qquad (10.4)$$

where F is a given function of y, dy/dx and x, and $y(x)$ is some function of x, the independent variable. For a given function $y = y_1(x)$ we can evaluate J, resulting in a value $J = J_1$. For a second function $y = y_2(x)$ we can again evaluate J, obtaining $J = J_2$. In general $J_1 \neq J_2$, for the value of J depends on the path $y = y(x)$ taken between $x = a$ and $x = b$. We call J a *functional.*

A simple example is given by

$$J = \int_0^1 \left(y +, x\frac{dy}{dx}\right) dx \ .$$

Consider the following paths from $x = 0$ to $x = 1$:

(i) $y = x$;

(ii) $y = 2x^2$;

(iii) $y = e^x$.

For path (i) $J_1 = \int_0^1 (x + x) \, dx = 1$;

for path (ii) $J_2 = \int_0^1 (2x^2 + 4x^2) \, dx = 2$;

and for (iii) $J_3 = \int_0^1 (e^x + xe^x)\,dx = e.$

We note that the value of J does indeed depend on the function $y(x)$ used in the integrand.

Variational calculus attempts to find the function $y(x)$, if it exists, which yields a maximum or a minimum value of J. We introduce the method by considering the problem described in Ch. 10.2.1, in which we wish to find the shortest plane curve joining points A and B. In mathematical terms we wish to find the path $y = y_0(x)$ from A to B which minimizes the functional 10.1.

Note that there are some restrictions on the permissible curves $y = y(x)$ in this example; for A and B are fixed points and we must therefore have (see Fig. 10.1)

$$y(0) = 0, \; y(a) = b \; . \tag{10.5}$$

In particular, for the optimum curve $y = y_0(x)$,

$$y_0(0) = 0, \; y_0(a) = b \; . \tag{10.6}$$

Consider a neighbouring curve $\mathscr{C}_\epsilon$ given by

$$y = y_\epsilon(x) = y_0(x) + \epsilon\eta(x) \; , \tag{10.7}$$

where ϵ is a small constant and $\eta(x)$ is a differentiable function of x, arbitrary except for the restriction

$$\eta(0) = \eta(a) = 0 \; . \tag{10.8}$$

This ensures that y_ϵ satisfies condition (10.5). Thus for a given choice of $y(x)$, we can generate a whole class of neighbouring curves $\mathscr{C}_\epsilon$, depending on the value of ϵ. Note that for $\epsilon = 0$, we obtain the optimal curve $\mathscr{C}_0$. Fig. 10.4 illustrates the situation.

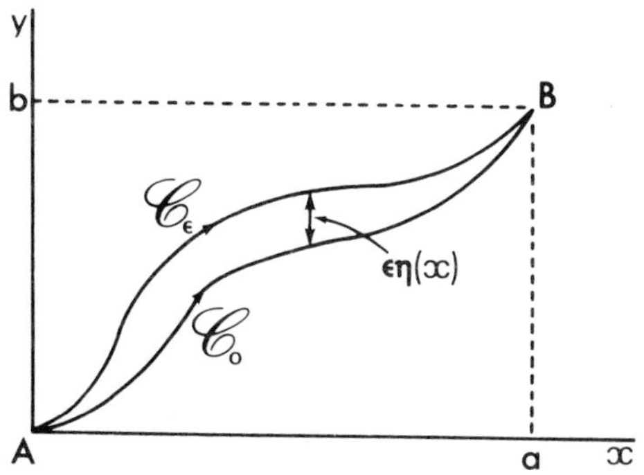

Fig. 10.4 Neighbouring curve $\mathscr{C}_\epsilon$

For a given $\eta(x)$, the arc length s along $\mathscr{C}_\epsilon$ depends on ϵ, and using (10.1),

$$s(\epsilon) = \int_0^a \left[1 + \left(\frac{dy_\epsilon}{dx}\right)^2\right]^{1/2} dx ,$$

$$= \int_0^a \left[1 + \left(\frac{dy_0}{dx} + \epsilon\frac{d\eta}{dx}\right)^2\right]^{1/2} dx .$$

Hence the minimum value of s occurs when

$$\frac{ds}{d\epsilon} = 0.$$

But by our choice of y_0, we know that this minimum occurs when $\epsilon = 0$, so that the condition for the minimum is

$$\boxed{\frac{ds}{d\epsilon} = 0 \text{ when } \epsilon = 0} \quad . \tag{10.9}$$

Now differentiating with respect to ϵ through the integral sign,

$$\frac{ds}{d\epsilon} = \int_0^a \frac{1}{2}\left[1 + \left(\frac{dy_0}{dx} + \epsilon\frac{d\eta}{dx}\right)^2\right]^{-1/2} 2\left(\frac{dy_0}{dx} + \epsilon\frac{d\eta}{dx}\right)\frac{d\eta}{dx}\, dx ,$$

and equating to zero at $\epsilon = 0$ gives

$$\int_0^a \left[1 + \left(\frac{dy_0}{dx}\right)^2\right]^{-1/2} \frac{dy_0}{dx}\frac{d\eta}{dx}\, dx = 0 .$$

Integrating by parts,

$$\left\{\left[1 + \left(\frac{dy_0}{dx}\right)^2\right]^{-1/2} \frac{dy_0}{dx}\,\eta\right\}_0^a - \int_0^a \frac{d}{dx}\left\{\left[1 + \left(\frac{dy_0}{dx}\right)^2\right]^{-1/2} \frac{dy_0}{dx}\right\}\eta\, dx = 0 ,$$

and the first term is zero by virtue of (10.8). Hence the optimal curve $y = y_0(x)$ must satisfy

$$\boxed{\int_0^a \frac{d}{dx}\left\{\left[1 + \left(\frac{dy_0}{dx}\right)^2\right]^{-1/2} \frac{dy_0}{dx}\right\}\eta\, dx = 0} \tag{10.10}$$

for *all* differentiable functions η such that $\eta(0) = \eta(a) = 0.$

Before continuing with this example, we state an important theorem, the proof of which is beyond the scope of this text.

THEOREM 10.1:

If $\int_\alpha^\beta \varphi(x)\,\eta(x)\,dx = 0$ for all differentiable functions $\eta(x)$ such that $\eta(\alpha) = \eta(\beta) = 0$, then $\varphi(x) = 0$ for $\alpha < x < \beta$.

We can apply this theorem directly to (10.10) which then shows that $y_0(x)$ must satisfy

$$\boxed{\frac{d}{dx}\left\{\left[1 + \left(\frac{dy_0}{dx}\right)^2\right]^{-1/2} \frac{dy_0}{dx}\right\} = 0} \; .$$

Integrating, $$\frac{dy_0}{dx} \Bigg/ \left[1 + \left(\frac{dy_0}{dx}\right)^2\right]^{1/2} = \text{constant} = A,$$

i.e. $$\left(\frac{dy_0}{dx}\right)^2 = A^2\left[1 + \left(\frac{dy_0}{dx}\right)^2\right];$$

which shows that dy_0/dx is a constant, say B. Hence, on integration,

$$y_0 = Bx + C \; ,$$

where C is a constant. Applying the conditions (10.6) shows that

$$\boxed{y_0 = (b/a)x} \; . \qquad (10.11)$$

As expected, we have shown that the shortest distance between A and B is a straight line. The result is certainly obvious, but we have developed a technique which can be applied to more general problems.

10.4 Euler's Equation and Transversality Conditions

We now return to the general problem, namely finding the optimal path $x(t)$, $a \leqslant t \leqslant b$, which minimizes or maximizes the functional

$$\boxed{J = \int_a^b F\left(x, \frac{dx}{dt}, t\right) dt} \; , \qquad (10.12)$$

where F is a given function of x, dx/dt and t. Equation (10.12) is the same as (10.4), except that we are now using t in (10.12) (instead of x) as the independent variable and x (instead of y) as the dependent variable. We assume that the end points, a and b, are known constants but in this general theory we assume that there is not necessarily any restriction on the value of x or dx/dt at the end points.

The value of J depends on the path $x = x(t)$ taken between $t = a$ and $t = b$. Suppose J has in fact an extreme value (i.e. maximum or minimum) when the path taken, $\mathscr{C}_0$, is given by

$$x = x_0(t), \quad a \leqslant t \leqslant b \; . \tag{10.13}$$

We wish to find this path and consequently the extremum value of J. Following the analysis in Ch. 10.3, consider the class of neighbouring curves, $\mathscr{C}_\epsilon$, given by

$$x = x_\epsilon(t) = x_0(t) + \epsilon\eta(t) \; , \tag{10.14}$$

where ϵ is a small constant, and $\eta(t)$ is an arbitrary differentiable function of t. Note that, unlike the example in Ch. 10.3 where η had given values at the end points, in this theory there are not necessarily any conditions on η. A typical path, $\mathscr{C}_\epsilon$, is shown in Fig. 10.5.

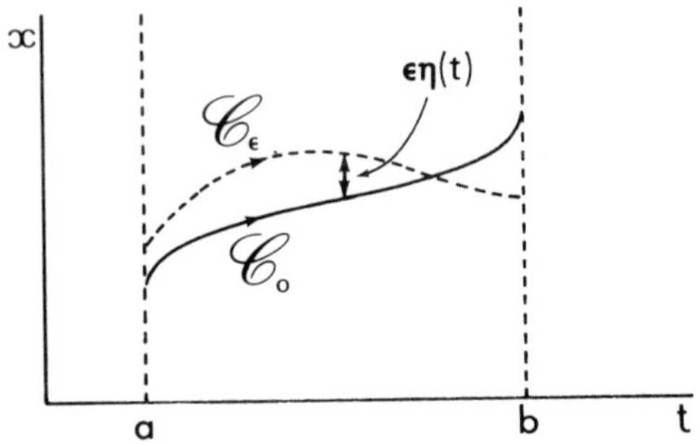

Fig. 10.5 Optimal curve $\mathscr{C}_0$ and neighbouring curve $\mathscr{C}_\epsilon$

Using (10.12) we see that the value of J for the path $\mathscr{C}_\epsilon$ is given by

$$J = \int_a^b F(x_0 + \epsilon\eta, \dot{x}_0 + \epsilon\dot{\eta}, t)dt \; ,$$

where dot denotes differentiation with respect to t. Hence, for a given $\eta(t)$, J is a function of ϵ, and we write

$$J(\epsilon) = \int_a^b F(x_0 + \epsilon\eta, \dot{x}_0 + \epsilon\dot{\eta}, t)dt \; . \tag{10.15}$$

For extremum values of J, we must have $dJ/d\epsilon = 0$. But by our choice of $\mathscr{C}_0$, we know that $J(0)$ is an extremum value. Thus the condition for maxima or minima of J is

$$\boxed{\frac{dJ}{d\epsilon} = 0 \text{ at } \epsilon = 0} \; . \tag{10.16}$$

Now

$$\frac{dJ}{d\epsilon} = \int_a^b \{\eta F_x(x_o + \epsilon\eta, \dot{x}_o + \epsilon\dot{\eta}, t) + \dot{\eta} F_{\dot{x}}(x_o + \epsilon\eta, \dot{x}_o + \epsilon\dot{\eta}, t)\} dt$$

where $F_x(x_o + \epsilon\eta, \dot{x}_o + \epsilon\dot{\eta}, t)$ means $\partial F/\partial x$ evaluated at $x = x_o + \epsilon\eta$, $\dot{x} = \dot{x}_o + \epsilon\dot{\eta}$, $t = t$, and similarly $F_{\dot{x}}(x_o + \epsilon\eta, \dot{x}_o + \epsilon\dot{\eta}, t)$ means $\partial F/\partial \dot{x}$ evaluated as above. Hence condition (10.16) gives

$$\int_a^b \{\eta F_x(x_o, \dot{x}_o, t) + \dot{\eta} F_{\dot{x}}(x_o, \dot{x}_o, t)\} \, dt = 0$$

and integrating the second term by parts,

$$\int_a^b \eta \frac{\partial F}{\partial x} dt + \left[\eta \frac{\partial F}{\partial \dot{x}}\right]_a^b - \int_a^b \eta \frac{d}{dt}\left(\frac{\partial F}{\partial \dot{x}}\right) dt = 0$$

i.e.

$$\boxed{\left[\eta \frac{\partial F}{\partial \dot{x}}\right]_a^b + \int_a^b \left\{\frac{\partial F}{\partial x} - \frac{d}{dt}\left(\frac{\partial F}{\partial \dot{x}}\right)\right\} \eta \, dt = 0} \quad . \tag{10.17}$$

Note that if x is fixed at the end points, we must have $\eta(a) = \eta(b) = 0$, and the first term in (10.17) is zero. In this case, a necessary and sufficient condition for the optimal path x_o is

$$\int_a^b \left\{\frac{\partial F}{\partial x} - \frac{d}{dt}\left(\frac{\partial F}{\partial \dot{x}}\right)\right\} \eta \, dt = 0 \tag{10.18}$$

where the functions are evalued at $x = x_o$, $\dot{x} = \dot{x}_o$, $t = t$. When the value of x is arbitrary at the end points, sufficient (but not necessary) conditions for the optimal path are

$$\left[\eta \frac{\partial F}{\partial \dot{x}}\right]_a^b = 0 \; ; \tag{10.19}$$

$$\int_a^b \left\{\frac{\partial F}{\partial x} - \frac{d}{dt}\left(\frac{\partial F}{\partial \dot{x}}\right)\right\} \eta \, dt = 0 \; . \tag{10.20}$$

Equation (10.20) holds for *all* differentiable functions $\eta(x)$, and this set of functions clearly includes the subset of differentiable functions $\eta(x)$ such that $\eta(a) = \eta(b) = 0$. Hence we can apply the theorem in Ch. 10.3 to equations (10.18) and (10.20), giving the condition for the optimal path

$$\boxed{\frac{\partial F}{\partial x} - \frac{d}{dt}\left(\frac{\partial F}{\partial \dot{x}}\right) = 0} \quad . \tag{10.21}$$

This is called *Euler's equation*†, and when x is given at the end points it is the condition to be satisfied by the optimum path. When x is not prescribed at the end points (including the case when x is only given at one point), Euler's equation together with

$$\boxed{\left[\eta \frac{\partial F}{\partial \dot{x}}\right]_a^b = 0} \quad , \tag{10.22}$$

are conditions for an extremum of J. When x is not given at either end point, η is completely arbitrary and its value at the end points is arbitrary. Thus (10.22) gives

$$\boxed{\frac{\partial F}{\partial \dot{x}} = 0 \text{ at } t = a \text{ and } t = b} \quad . \tag{10.23}$$

If, for instance, x is given at $t = a$, then we must have $\eta(a) = 0$, and (10.22) gives $\partial F/\partial \dot{x} = 0$ at $t = b$. These end point conditions, which are very important in the applications of variational calculus in control theory, are called the *transversality conditions.* They can be summarised by saying that

$$\boxed{\frac{\partial F}{\partial \dot{x}} = 0 \text{ at any end point where } x \text{ is not prescribed}} \quad . \tag{10.24}$$

We illustrate the use of these conditions in the following examples.

EXAMPLE 1:

Find the curve $x = x(t)$ which minimizes the functional

$$J = \int_0^1 (\dot{x}^2 + 1)dt$$

where $x(0) = 1$ and $x(1) = 2$.

SOLUTION:

There are no transversality conditions, for x is prescribed at both $t = 0$ and $t = 1$. The Euler equation (10.21) applies with

$$F = \dot{x}^2 + 1 \, .$$

† or Euler – Lagrange equation.

Thus
$$0 = \frac{\partial F}{\partial x} - \frac{d}{dt}\left(\frac{\partial F}{\partial \dot{x}}\right)$$
$$= 0 - \frac{d}{dt}(2\dot{x})$$
$$= -2\ddot{x} \; .$$

Hence, the optimal path must satisfy $\ddot{x} = 0$, which on integration gives

$$x = At + B \; ,$$

A and B being constants. Applying the conditions $x(0) = 1$, $x(1) = 2$ gives

$$x = 1 + t, \quad 0 \leqslant t \leqslant 1 \; ,$$

and this is the optimal path. The corresponding value of J is given by

$$J = \int_0^1 2\,dt = 2 \; ,$$

and this in fact can be shown to be a minimum for J.

EXAMPLE 2 (The Brachistochrone problem):

Determine the plane curve of quickest descent as a particle, moving on a smooth surface, falls under gravity from a fixed point A to another fixed point B.

SOLUTION:

In Ch. 10.2.2 it was shown that this problem requires the path $y = y(x)$ from A to B (Fig. 10.2) which minimizes the integral

$$t = \frac{1}{(2g)^{1/2}} \int_0^a \left[\frac{1 + (dy/dx)^2}{x}\right]^{1/2} dx \; .$$

There are no transversality conditions as $y(0) = 0$ and $y(a) = b$, but we can apply Euler's equation

$$\frac{\partial F}{\partial y} - \frac{d}{dx}\left(\frac{\partial F}{\partial \dot{y}}\right) = 0$$

where dot denotes d/dx, and

$$F(y, \dot{y}, x) = [(1 + \dot{y}^2)/x]^{1/2} \; .$$

Clearly $\partial F/\partial y = 0$ which means that

$$\frac{d}{dx}\left(\frac{\partial F}{\partial \dot{y}}\right) = 0 \ .$$

Integrating, we see that for the optimum path

$$\frac{\partial F}{\partial \dot{y}} = \text{constant} = c\text{, say.}$$

Hence
$$\frac{\dot{y}}{[x(1 + \dot{y}^2)]^{1/2}} = c \ ,$$

and squaring, $y^2 = c^2x(1 + \dot{y}^2)$.

Rearranging,
$$\frac{dy}{dx} = \left[\frac{c^2x}{1 - c^2x}\right]^{1/2} ,$$

and integrating,
$$y = c\int\frac{x\,dx}{(x - c^2x^2)^{1/2}} .$$

This integral can be evaluated, using the substitution

$$x = \frac{1}{2c^2}(1 - \cos\theta)$$

so that
$$\frac{dx}{d\theta} = \frac{1}{2c^2}\sin\theta \ ,$$

and
$$y = \frac{1}{2c^2}\int(1 - \cos\theta)d\theta \ .$$

Hence,
$$y = \frac{1}{2c^2}(\theta - \sin\theta) + \text{const.}$$

The constant must be zero in order that $y = 0$ when $x = 0$ (i.e. $\theta = 0$). Thus the curve of quickest descent is given by the parametric equations

$$\boxed{\begin{aligned} x &= \frac{1}{2c^2}(1 - \cos\theta) \\ y &= \frac{1}{2c^2}(\theta - \sin\theta) \end{aligned}} \ .$$

This is the equation of a cycloid, and the constant c must be chosen so that the curve passes through the point (a,b).

The above theory shows that the cycloid gives an extremum value for the time of descent from A to B. This extremum is in fact a minimum. A typical example is illustrated in Fig. 10.6.

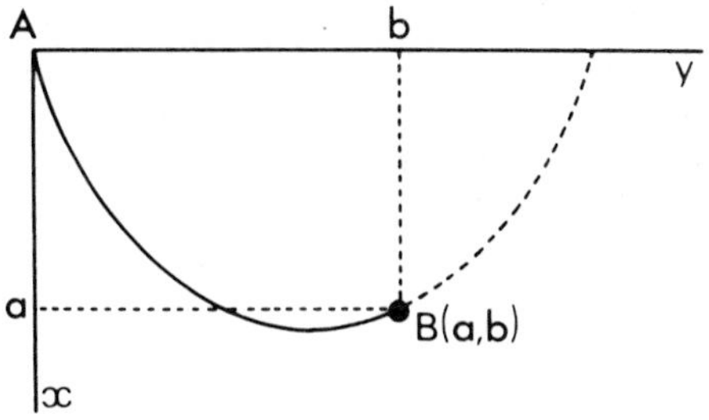

Fig. 10.6 The cycloid from A *to* B

EXAMPLE 3:

Find the curve $x = x(t)$ which minimizes the functional

$$J = \int_0^1 (\dot{x}^2 + 1)dt ,$$

where $x(0) = 1$ but x can take any value at $t = 1$.

SOLUTION:

Since x is not given at $t = 1$, we have the transversality condition $\partial F/\partial \dot{x} = 0$ at $t = 1$, where the integrand F is given by

$$F = \dot{x}^2 + 1 .$$

Thus at $t = 1$, $\dot{x} = 0$; i.e. $\dot{x}(1) = 0$.

Euler's equation gives

$$0 = \frac{\partial F}{\partial x} - \frac{d}{dt}\left(\frac{\partial F}{\partial \dot{x}}\right) = -2\ddot{x} .$$

Integrating gives $\dot{x}(t) =$ constant, and using the transversality condition,

$$\dot{x}(t) = 0 .$$

Integrating, $$x(t) = \text{constant},$$

and since $x(0) = 1$ we see that the optimum path is

$$x(t) = 1, \quad 0 \leqslant t \leqslant 1 .$$

The corresponding value of J is 1, which is clearly a minimum.

Notice that the minimum value of J in this example is less than the minimum obtained for the same functional in example 1, where both end points were fixed. This is to be expected, as the first example is more restricted than the example above.

The application of Euler's equation determines the path which makes the functional stationary; i.e. maximum, minimum or inflection. It is often clear from the physical context which type of stationary value is involved, but the theory can be developed further to give specific conditions for maxima or minima. We refer the interested reader to 'Calculus of Variations' by J. C. Clegg (Oliver and Boyd, 1968).

The functional J is often referred to as the *cost functional.* In many practical problems in control theory, J does in fact represent the cost of a particular operation. Also the integrand is often a function of more than one dependent variable. The results obtained above can soon be generalized. For instance, suppose we wish to find extemum of the cost functional

$$J = \int_a^b F(x_1, x_2, \ldots x_n; \dot{x}_1, \dot{x}_2, \ldots, \dot{x}_n; t)dt \ . \tag{10.25}$$

where $x_1, x_2, \ldots, x_n$ are functions of t.

It can be shown that the conditions to be satisfied on the optimum path are

$$\boxed{\frac{\partial F}{\partial x_i} - \frac{d}{dt}\left(\frac{\partial F}{\partial \dot{x}_i}\right) = 0 \quad (i = 1, 2, \ldots, n)} \ , \tag{10.26}$$

and the transversality conditions are

$$\boxed{\frac{\partial F}{\partial \dot{x}_i} = 0 \text{ at any end point where } x_i \ (i = 1, 2, \ldots, n) \text{ is not prescribed.}} \tag{10.27}$$

Thus we have n Euler equations, (10.26), and a maximum of $2n$ transversality conditions, (10.27).

10.5* Hamilton's Principle

In earlier chapters we have applied Newton's equation of motion $\mathbf{F} = m\ddot{\mathbf{r}}$ to the motion of a single particle. This is not the only way in which the motion of a particle can be described. An alternative method is to employ Lagrange's equations which follow from Hamilton's Principle and are described below.

This approach does not tell us anything that could not be derived from Newton's equations. In fact we can derive Newton's equations from Hamilton's Principle and it is also possible to derive Hamilton's Principle from Newton's equations. On the other hand, Hamilton's Principle can be applied to a wider range of physical phenomena, and in this sense can be regarded as more versatile than Newton's equations. The complete theory which can be developed from Hamilton's Principle is known as analytical mechanics and is beyond the scope of this book. We give only a short introduction to the subject.

HAMILTON'S PRINCIPLE may be stated as follows: 'Of all possible paths along which a dynamical system may move from one point to another, within a specified time interval and consistent with any constraints, the actual path followed is that which minimizes the integral with respect to time of the difference between the kinetic and potential energies.'

In mathematical terms the principle states that the path followed by the system minimizes

$$\int_{t_1}^{t_2} (T - V)dt \ , \tag{10.28}$$

where T is the kinetic energy and V the potential energy.

Let us apply the principle to the one-dimensional motion of a particle of mass m along the x-axis. In this case we have

$$T = \tfrac{1}{2}m\dot{x}^2$$

and, using (6.23),

$$V = -\int_{x_0}^{x} f(x)dx \ ,$$

where $f(x)$ is the force acting on the particle. Thus we require to minimize

$$\int_{t_0}^{t_1} L(x, \dot{x})dt \ , \tag{10.29}$$

where

$$L = T - V = \tfrac{1}{2}m\dot{x}^2 + \int_{x_0}^{x} f(x)dx \ , \tag{10.30}$$

is called the *Lagrangian*.

Euler's equation (10.21) requires the optimum path to satisfy

$$\boxed{\frac{\partial L}{\partial x} - \frac{d}{dt}\left(\frac{\partial L}{\partial \dot{x}}\right) = 0} \tag{10.31}$$

i.e.
$$f(x) - \frac{d}{dt}(m\dot{x}) = 0$$

which gives
$$m\ddot{x} = f(x) , \qquad (10.32)$$

the usual Newtonian equation of motion. Equation (10.31) is called the *Lagrangian equation of motion.*

As an example, consider motion under gravity with the x-axis vertically upwards. Then $V = mgx$, and

$$L = \tfrac{1}{2}m\dot{x}^2 - mgx .$$

Applying (10.31),
$$-mg - \frac{d}{dt}(m\dot{x}) = 0 ,$$

which gives, as expected,
$$\ddot{x} = -g .$$

For motion in three dimensions we again define

$$L = T - V ,$$

where $T = T(\dot{x}, \dot{y}, \dot{z})$ and $V = V(x, y, z)$. Applying Hamilton's Principle now results in a Lagrangian equation of motion of the form (10.31) for each co-ordinate x, y and z.

At first sight it would appear that the Lagrangian method is a complicated way of obtaining the simple results of the Newtonian method. However, this is far from being the case in general. Let us now consider the problem of a plane pendulum as illustrated in Fig. 10.7.

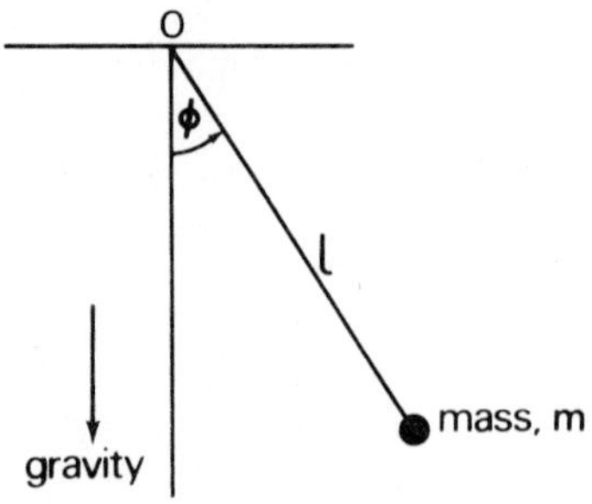

Fig. 10.7 Plane pendulum

We know that

$$T = \tfrac{1}{2}m(l\dot{\phi})^2$$

and $$V = mgl(1 - \cos\phi),$$

so that $$L = \tfrac{1}{2}m\,l^2\dot{\phi}^2 - mg\,l(1 - \cos\phi)\,. \tag{10.33}$$

Now L depends on ϕ and $\dot{\phi}$ so that the Lagrangian equation (10.31) with x replaced by ϕ becomes

$$\frac{\partial L}{\partial \phi} - \frac{d}{dt}\left(\frac{\partial L}{\partial \dot{\phi}}\right) = 0\,, \tag{10.34}$$

whence we obtain

$$-\,mgl\sin\phi - \frac{d}{dt}(ml^2\dot{\phi}^2) = 0$$

i.e. $$\ddot{\phi} + \frac{g}{l}\sin\phi = 0\,. \tag{10.35}$$

This is the equation found in Ch. 9.3 after applying Newton's equation of motion in the transverse direction. [Notice that a Lagrangian equation can be formulated in any co-ordinate system and is not restricted to Cartesians.] The Lagrangian method has some advantage in the example above for no knowledge of acceleration components in polar co-ordinates is needed. We do not pursue the method further in this book, but interested readers are referred to *Lagrangian Dynamics* by C. W. Kilmister (Lagos Press, 1967).

10.6 Optimization with constraints

In many problems the variables x_i might not be completely independent. For instance one of the x_i, say x_j, might be required to satisfy a differential equation, or some other relationship. We say that x_j satisfies a *constraint.* Also the constraint might involve both x_j and x_k, say. Suppose we wish to find the path $x_i = x_i(t)$, $a \leqslant t \leqslant b$, $(i = 1, 2, \ldots, n)$ which minimizes or maximizes the cost functional

$$\boxed{J = \int_a^b F(x_1, x_2, \ldots, x_n;\ \dot{x}_1, \dot{x}_2, \ldots, \dot{x}_n;\ t)dt} \tag{10.36}$$

where the x_i satisfy the constraints

$$\begin{aligned} g(x_1, x_2, \ldots, x_n;\ \dot{x}_1, \dot{x}_2, \ldots, \dot{x}_n;\ t) &= 0 \\ h(x_1, x_2, \ldots, x_n;\ \dot{x}_1, \dot{x}_2, \ldots, \dot{x}_n;\ t) &= 0\,, \end{aligned} \tag{10.37}$$

g and h being functions of x_i, $\dot{x}_i$, t $(i = 1, 2, \ldots, n)$. We treat this problem in the same way as we treat an ordinary extremum of a function subject to a constraint, by introducing two Lagrange multipliers, λ and μ, and forming the *augmented cost functional*

$$J^* = \int_a^b (F + \lambda g + \mu h)dt \ . \tag{10.38}$$

We now find extremum of this functional as in Ch. 10.4 and this will determine extremum of J (10.36) subject to the constraints (10.37). The Lagrange multipliers are found from Euler's equations for (10.38) and the constraints (10.37). In principle, the Lagrange multipliers can be eliminated and the optimum path is found. This is illustrated in the following examples.

EXAMPLE 1:

Minimize the cost functional

$$t = \frac{1}{\sqrt{2g}} \int_0^a \frac{(1 + \dot{y}\dot{z})^{1/2}}{x^{1/2}} dx$$

where the variables $y = y(x)$, $z = z(x)$ are subject to the constraint

$$y = z + 1$$

and where $y(0) = 0$, $y(a) = b$.

Note:

The constraint is a very simple one, and in this easy example it is possible to eliminate one of the variables, say z, to give

$$t = \frac{1}{\sqrt{2g}} \int_0^a \frac{(1 + \dot{y}^2)^{1/2}}{x^{1/2}} dx \ .$$

This is just the cost functional in the brachistochrone problem and we have already seen that the equation for the optimum path is

$$\frac{\dot{y}}{[x(1 + \dot{y}^2)]^{1/2}} = c \ .$$

We check this result using the constraint with a Lagrange multiplier.

SOLUTION:

Introducing the Lagrange multiplier λ, the augmented cost functional is

$$t^* = \int_0^a \left[\frac{1}{\sqrt{2g}} \left(\frac{1 + \dot{y}\dot{z}}{x} \right)^{1/2} + \lambda(y - z - 1) \right] dx \;,$$

since the constraint is $y - z - 1 = 0$.

The transversality conditions are automatically satisfied, since y and hence z are fixed at the end points. Euler's equations, (10.26), with $n = 2$ and x_1, x_2 replaced by y, z respectively (and $t = x$) give

$$y\colon\ \lambda - \frac{d}{dx}\left\{ \frac{\dot{z}}{2\sqrt{2g}\,[x(1 + \dot{y}\dot{z})]^{1/2}} \right\} = 0$$

$$z\colon\ -\lambda - \frac{d}{dx}\left\{ \frac{\dot{y}}{2\sqrt{2g}\,[x(1 + \dot{y}\dot{z})]^{1/2}} \right\} = 0 \;.$$

Eliminating λ, we obtain

$$\frac{d}{dx}\left\{ \frac{\dot{y} + \dot{z}}{2\sqrt{2g}\,[x(1 + \dot{y}\dot{z})]^{1/2}} \right\} = 0 \;.$$

Integrating,

$$\frac{\dot{y} + \dot{z}}{[x(1 + \dot{y}\dot{z})]^{1/2}} = A \;;$$

and using the constraint, $z = y - 1$, we obtain for the optimum path

$$\frac{\dot{y}}{[x(1 + \dot{y}^2)]^{1/2}} = \frac{A}{2} \;,$$

as expected. In this example, the Lagrange multiplier method has been shown to yield the expected result.

EXAMPLE 2:

Minimize the cost functional $J = \frac{1}{2}\int_0^2 (\ddot{x})^2\, dt$, (10.39) where $x = x(t)$ satisfies

$$\begin{aligned} x(0) &= 1, \quad \dot{x}(0) = 1; \\ x(2) &= 0, \quad \dot{x}(2) = 0. \end{aligned} \tag{10.40}$$

SOLUTION:

The integrand contains the second derivative $\ddot{x}$, and our theory only allows the integrand to be a function of x, $\dot{x}$ and t. We overcome this by introducing two variables, x_1 and x_2 say, where

$$x_1 = x, \quad x_2 = \dot{x}_1 .$$

The cost functional is now

$$J = \frac{1}{2}\int_0^2 \dot{x}_2{}^2\, dt \tag{10.41}$$

and we have the constraint

$$x_2 - \dot{x}_1 = 0 .$$

The augmented cost functional is

$$J^* = \int_0^2 \left[\frac{1}{2}\dot{x}_2{}^2 + \lambda(x_2 - \dot{x}_1)\right] dt . \tag{10.42}$$

Forming Euler's equations (10.26) for x_1 and x_2 gives

$$x_1:\ 0 - \frac{d}{dt}(-\lambda) = 0 \quad \text{i.e. } \dot{\lambda} = 0 ; \tag{10.43}$$

$$x_2:\ \lambda - \frac{d}{dt}(\dot{x}_2) = 0 \quad \text{i.e. } \lambda - \ddot{x}_2 = 0 . \tag{10.44}$$

From (10.43) and (10.44), $$\frac{d^3 x_2}{dt^3} = 0 ,$$

and integrating, $$x_2 = At^2 + Bt + C ,$$

where A, B and C are constants of integration. Since $\dot{x}_1 = x_2$, we obtain

$$x_1 = \frac{At^3}{3} + \frac{Bt^2}{2} + Ct + D ,$$

D being a further constant of integration. Applying the end conditions (10.40) gives

$$x_1 = \frac{1}{2}t^3 - \frac{7}{4}t^2 + t + 1 ,$$

$$x_2 = \frac{3}{2}t^2 - \frac{7}{2}t + 1 .$$

We have determined the optimal path and substitution into (10.41) gives

$$J = \frac{1}{2}\int_0^2 \left(3t - \frac{7}{2}\right)^2 dt = 13/4 .$$

[This value of J is in fact a minimum.]

Worked Examples

EXAMPLE 10.1:

Find the path $y = y(x)$ which minimizes the functional

$$J = \int_0^1 (y'^2 + 1)^2 \, dx$$

when (i) $y(0) = 1$, $y(1) = 2$;

(ii) $y(0) = 1$ and $y(1)$ is not specified.

Here dash denotes differentiation with respect to x.

SOLUTION:

In each case the optimal path must satisfy the Euler equation

$$\frac{\partial F}{\partial y} - \frac{d}{dx}\left(\frac{\partial F}{\partial y'}\right) = 0 ,$$

where $F = y'^2 + 1$. Since $\partial F/\partial y = 0$, $\partial F/\partial y'$ is a constant; i.e.

$$2(y'^2 + 1)2y' = \text{constant} .$$

This is a cubic in y' with real constant coefficients. Clearly each root will be of the form $y' = A$, a constant. Thus, integrating, optimal paths are given by

$$y = Ax + B .$$

(i) $y(0) = 1$, $y(1) = 2$.

Since y is specified at both end points there are no transversality conditions and application of the end point values gives

$$B = 1 , \; A + B = 2 .$$

Thus $A = B = 1$, and the optimal path is

$$y = 1 + x , \; 0 \leqslant x \leqslant 1 .$$

The corresponding value of J is given by

$$J = \int_0^1 (1 + 1)^2 \, dx = 4 ,$$

and this can in fact be shown to be a minimum.

(ii) $y(0) = 1$, $y(1)$ not specified.

At $x = 1$ we have the transversality condition $\partial F/\partial y' = 0$; i.e. $y'(y'^2 + 1) = 0$ at $x = 1$. This only satisfied by taking $y' = 0$ at $x = 1$. Thus, since y is of the form $y = Ax + B$, we have $A = 0$. Since $y(0) = 1$, we must have $B = 1$, and the optimal path is

$$y = 1 , \; 0 \leqslant x \leqslant 1 .$$

The corresponding value of J is given by

$$J = \int_0^1 1 \, dx = 1 ,$$

and this is clearly a minimum.

EXAMPLE 10.2:

Show that the optimal trajectory $y = y(x)$ for the functional

$$J = \int_a^b F(y,y') \, dx ,$$

where F is a function of y and y' ($= dy/dx$), satisfies

$$y' \frac{\partial F}{\partial y'} - F = \text{constant} .$$

SOLUTION:

We know that $y = y(x)$ must satisfy the Euler equation

$$\frac{\partial F}{\partial y} - \frac{d}{dx}\left(\frac{\partial F}{\partial y'}\right) = 0$$

i.e.
$$y'\frac{\partial F}{\partial y} - y'\frac{d}{dx}\left(\frac{\partial F}{\partial y'}\right) = 0 \ .$$

We can re-write this as

$$y'\frac{\partial F}{\partial y} + y''\frac{\partial F}{\partial y'} - \frac{d}{dx}\left(y'\frac{\partial F}{\partial y'}\right) = 0 \ .$$

But, since F depends on x only through y and y', we know that

$$\frac{dF}{dx} = \frac{\partial F}{\partial y}\frac{dy}{dx} + \frac{\partial F}{\partial y'}\frac{dy'}{dx} = y'\frac{\partial F}{\partial y} + y''\frac{\partial F}{\partial y'} \ .$$

Hence $$\frac{dF}{dx} - \frac{d}{dx}\left(y'\frac{\partial F}{\partial y'}\right) = 0 \text{ ; or } \frac{d}{dx}\left(F - y'\frac{\partial F}{\partial y'}\right) = 0 \ .$$

Integrating gives the required result.

EXAMPLE 10.3:

Find the plane curve passing through the points $(0,b)$ and (a,c) and lying above the x-axis for which the surface area of revolution, when rotated about the x-axis, is a minimum. (See Ch. 10.2.3.)

SOLUTION:

As seen in Ch. 10.2.3, this problem requires the curve $y = y(x)$ which minimizes the integral

$$A = 2\pi\int_0^a y(1 + y'^2)^{\frac{1}{2}}\,dx$$

and is such that $y(0) = b$, $y(a) = c$ $(b \neq c)$. In this example the integrand,

$$F = y(1 + y'^2)^{\frac{1}{2}}$$

does not depend on x explicitly and we can apply the equation derived in

Worked Example 10.2. Thus the optimal path satisfies

$$y'\frac{\partial F}{\partial y'} - F = \text{constant};$$

i.e.
$$\frac{yy'^2}{(1+y'^2)^{1/2}} - y(1+y'^2)^{1/2} = \text{constant} = B.$$

Simplifying,
$$y^2 = B^2(1+y'^2)$$

or
$$\frac{dy}{dx} = \frac{1}{B}(y^2 - B^2)^{1/2}$$

i.e.
$$\int \frac{dy}{(y^2 - B^2)^{1/2}} = \int \frac{1}{B}\,dx\,.$$

Integrating,
$$\cosh^{-1}(y/B) = (x + C)/B$$

i.e.
$$y = B\cosh\left(\frac{x+C}{B}\right),$$

where C is a constant. The constants B and C are chosen so that the curve passes through $(0,b)$ and (a,c); i.e.

$$b = B\cosh(C/B) \text{ and } c = B\cosh\left(\frac{a+C}{B}\right),$$

from which the constants B and C are determined.

The trajectory has now been determined. This curve is in fact known as the catenary and represents the shape assumed by a heavy chain of uniform density when suspended freely from its ends.

EXAMPLE 10.4*:

A particle of mass m is moving under an attractive force k/R^2 per unit mass, directed towards the origin of plane polar coordinates (R,ϕ). Write down the Lagrangian for the particle, evaluate Lagrange's equation of motion for R and ϕ, and show that

$$\ddot{R} - H^2/m^2R^3 = -k/R^2,$$

where H is a constant.

SOLUTION:

We have seen in Ch. 7.4 that the potential energy for an inverse square law of force is given by

$$V = -mk/R \ .$$

The kinetic energy is given, in terms of plane polar coordinates, by

$$T = \tfrac{1}{2}m(\dot{R}^2 + R^2\dot{\phi}^2) \ ,$$

since $\mathbf{v} = \dot{R}\mathbf{e}_R + R\dot{\phi}\mathbf{e}_\phi$, in the usual notation. Hence the Lagrangian for the particle is

$$L = T - V = \tfrac{1}{2}m(\dot{R}^2 + R^2\dot{\phi}^2) + mk/R \ .$$

Evaluating Lagrange's equations of motion for R and ϕ gives

$$R: \quad \frac{\partial L}{\partial R} - \frac{d}{dt}\left(\frac{\partial L}{\partial \dot{R}}\right) = 0 \ ; \ \text{i.e.} \ -\frac{mk}{R^2} + mR\dot{\phi}^2 - \frac{d}{dt}(m\dot{R}) = 0 \ ,$$

$$\phi: \quad \frac{\partial L}{\partial \phi} - \frac{d}{dt}\left(\frac{\partial L}{\partial \dot{\phi}}\right) = 0 \ ; \ \text{i.e.} \ \frac{\partial L}{\partial \dot{\phi}} = \text{constant or } mR^2\dot{\phi} = \text{constant} = H.$$

Eliminating $\dot{\phi}$ from the two equations gives

$$\ddot{R} - H^2/m^2R^3 = -k/R^2 \ ,$$

which is equation (7.17).

EXAMPLE 10.5:

Find stationary paths $x = x(t)$ for the functional

$$J = \int_0^1 \left[1 + \left(\frac{d^2x}{dt^2}\right)\right]dt$$

subject to the boundary conditions

$$x(0) = 0 \ , \ \frac{dx}{dt}(0) = 1 \ , \ x(1) = 1 \ , \ \frac{dx}{dt}(1) = 1 \ .$$

SOLUTION:

To eliminate the second derivative in the integrand we introduce two variables

$$x_1 = x\,,\ x_2 = \frac{dx}{dt}\,.$$

Then the functional is $$J = \int_0^1 (1 + \dot{x}_2^{\,2})\,dt$$

with boundary conditions

$$x_1(0) = 0\,,\ x_2(0) = 1\,,\ x_1(1) = 1\,,\ x_2(1) = 1\,,$$

and *constraint* $x_2 - \dot{x}_1 = 0$.

The constraint is incorporated into the functional by using a Lagrange multiplier λ and considering optimal paths for the augmented functional

$$J^* = \int_0^1 [1 + \dot{x}_2^{\,2} + \lambda(x_2 - \dot{x}_1)]\,dt\,.$$

Euler's equations for x_1 and x_2 are

$$x_1:\ \frac{\partial F}{\partial x_1} - \frac{d}{dt}\left(\frac{\partial F}{\partial \dot{x}_1}\right) = 0\,;\ \text{i.e.}\ \frac{\partial F}{\partial \dot{x}_1} = -\lambda = \text{constant}\,;$$

$$x_2:\ \frac{\partial F}{\partial x_2} - \frac{d}{dt}\left(\frac{\partial F}{\partial \dot{x}_2}\right) = 0\,;\ \text{i.e.}\ \lambda - 2\ddot{x}_2 = 0\,;$$

where $F = 1 + \dot{x}_2^{\,2} + \lambda(x_2 - \dot{x}_1)$.

Hence $\ddot{x}_2 = \lambda/2 = \text{constant} = A$ and, integrating,

$$x_2 = \tfrac{1}{2}At^2 + Bt + C\,,$$

$$x_1 = \tfrac{1}{6}At^3 + \tfrac{1}{2}Bt^2 + Ct + D\,,$$

where B, C and D are constants of integration. Applying the conditions at $t = 0$ gives

$$C = 1 \text{ and } D = 0\,;$$

and, at $t = 1$,

$$1 = \tfrac{1}{6}A + \tfrac{1}{2}B + 1 \text{ and } 1 = \tfrac{1}{2}A + B + 1\,.$$

Hence $A = B = 0$ and the optimal path is

$$x = t, \quad 0 \leqslant t \leqslant 1 .$$

The corresponding value of J is

$$J = \int_0^1 1\, dt = 1 ,$$

which is clearly a minimum value.

Problems

10.1 Evaluate the functional

$$J = \int_0^1 \left[y^2 + \left(\frac{dy}{dx} \right)^2 \right] dx$$

along the following paths from (0,0) to (1,1):

(i) $y = x$;

(ii) $y = x^2$;

(iii) $y = (e^x - 1)/(e - 1)$.

10.2 Find the optimal trajectory for

$$J = \int_0^1 \left\{ [1 + (dy/dx)^2]^{1/2}/x \right\} dx$$

where $y(0) = 0$ and $y(1) = 1$.

10.3 Find the curve $x = x(t)$ passing through $x(0) = x_0$, $x(T) = 0$ which minimizes the cost functional

$$J = \int_0^T (x^2 + \alpha^2 \dot{x}^2)\, dt$$

where α is a constant and T is fixed.

Minimize the same functional but without the restriction $x(T) = 0$. Determine the minimum cost in each case and compare the results obtained.

10.4 Minimize the functional

$$J = \int_0^2 \left(\tfrac{1}{2}\dot{x}^2 + x\dot{x} + \dot{x} + x \right) dt$$

[Note that x and $\dot{x}$ are not fixed at $t = 0$ and 2.]

10.5* A particle of mass m moves, under gravity, in two-dimensional space. Introducing rectangular axes Oxy, y being vertically upwards, write down the Lagrangian for the particle.

Evaluate Lagrange's equations for x and y, and show that the usual Newtonian equations of motion are obtained.

10.6* What is the potential energy of a stretched spring, modulus of elasticity λ, natural length a and extension y?

One end of a spring, modulus λ and natural length a, is fixed and a mass m is attached to the other end. The system is held vertically from the fixed point. When in equilibrium the mass is given a small push in the downward vertical direction. If x denotes the displacement of the mass from equilibrium, write down the Lagrangian and evaluate Lagrange's equation for x.

10.7* A particle of mass m is projected horizontally with velocity u along the inside smooth surface of a sphere of radius a at the level of the centre. Write down the Lagrangian for the motion (using θ and ϕ as variables) and evaluate Lagrange's equations for θ and ϕ. Hence show that

$$a^2\dot{\theta}^2 + u^2 \cot^2 \theta + 2ga \cos\theta = 0 .$$

Here (r, θ, ϕ) are spherical polar coordinates.

10.8 The cost of a chemical process is described by the functional

$$J = \int_0^1 \left(\tfrac{1}{2}y'^2 + y\right) dx ,$$

where $y(0) = 1$. If also $y(1) = 0$, find the optimal trajectory for J, and evaluate the corresponding minimum value of J.

If $y(1)$ is not specified, use the transversality condition at $x = 1$ to determine the optimal trajectory. Evaluate the corresponding value of J and compare with the value found above.

10.9 Find the curve joining the points (0,0) and (1,0) for which the integral

$$\int_0^1 \left(\frac{d^2y}{dx^2}\right)^2 dx$$

is a minimum if

(i) $\dfrac{dy}{dx}(0) = a , \quad \dfrac{dy}{dx}(1) = b ,$

(ii) no other conditions are prescribed.

Chapter 11

Optimal Control

11.1 Introduction

One of the most recent and interesting developments in applied mathematics is in the field of optimization and control. The theory has application not only in the dynamics and construction of machinery and work plants, but also in the economic stability of a community.

To illustrate, as simply as possible, one of the principal ideas in control theory, we suggest trying this simple experiment. Place one end of a long thick rod on the palm of your right hand and with your left hand hold the rod vertical. When you release the hold of your left hand the rod will soon fall, due to small vibrations in the palm of your right hand. But it should be possible to observe the commencement of the rod's fall, and, with a suitable movement of your right hand, stabilize the vertical position of the rod. This is *control*, since you are observing the unstable motion of the rod, and then attempting to stabilize it by an appropriate action.

In this chapter it will only be possible to look at one branch of control theory, namely the application of variational calculus to optimal control problems.

11.2 The Basic Problem

Most mathematical models for systems, whether physical, mechanical or economic, etc, can be characterized by a set of functions $x_1(t), x_2(t), \ldots,$ $x_n(t)$ which satisfy a set of first order differential equations of the form

$$\frac{dx_i}{dt} = f_i(x_1, x_2, \ldots, x_n; u_1, u_2, \ldots, u_m, t) \qquad (11.1)$$

$(i = 1, 2, \ldots, n)$ where f_i is a function of $x_1, x_2, \ldots x_n, u_1, u_2, \ldots u_m$ and t. The functions $u_1, u_2, \ldots, u_m$ are called *control variables.* In order to solve (11.1) for $x_1, x_2, \ldots, x_n,$ called the *state variables* it is necessary to prescribe the control variables.

For a consistent set of equations there will also be some boundary conditions. If $t_0 \leqslant t \leqslant t_1$ is the time interval under consideration, the x_i are normally specified at $t = t_0$; i.e. $x_i(t_0)$ specified for $i = 1, 2, \ldots, n$. Thus for given $u_j (j = 1, 2, \ldots, m)$ equation (11.1) will determine the x_i uniquely. The normal control problem is to choose the controls $u_1, \ldots, u_m$ so that the system moves from its given initial state: $(x_1(t_0), x_2(t_0), \ldots, x_n(t_0))$ to some prescribed state at $t = t_1$. If this can be achieved we say the system is *controllable*. If there exists more than one set of controls which control the system, the *optimal control* problem is to find the set of controls which controls the system in some 'best' way. For example we might wish to minimize (or maximize) a functional of the form

$$J = \int_{t_0}^{t_1} F(x_1, x_2, \ldots, x_n; u_1, u_2, \ldots, u_m; t)\, dt \,. \tag{11.2}$$

At first sight this formulation might appear to be rather complicated. This is of course to be expected if we are attempting to represent widely differing types of system. To assist understanding of the basic ideas we first consider a simple physical problem.

A particle, attached to the lower end of a vertical spring whose other end is fixed, is oscillating about its equilibrium position. If x denotes the particle's displacement from its equilibrium position the governing differential equation for the motion is

$$\ddot{x} = -\omega^2 x \,. \tag{11.3}$$

Suppose the particle is at its maximum displacement $x = a$ at time $t = 0$. At this instant of time we wish to apply a force u per unit mass to the particle in order to bring the particle to rest when its displacement is zero: i.e. when $x = 0$. Is it possible to find such a force?

The governing differential equation is now

$$\ddot{x} = -\omega^2 x + u \,. \tag{11.4}$$

This system can be expressed in the form (11.1) by defining

$$x_1 = x \,, \quad x_2 = \dot{x} \,.$$

The governing differential equations then become

$$\dot{x}_1 = x_2 \,; \quad \dot{x}_2 = -\omega^2 x_1 + u \,. \tag{11.5}$$

The control problem is to choose u so that the system (x_1, x_2) moves from $(a,0)$ at $t = 0$ to $(0,0)$ at some subsequent time. As a first guess, try taking u to be a constant, say C, for all time $t \geqslant 0$. Then

$$\ddot{x}_1 + \omega^2 x_1 = C \,,$$

which has solutions

$$x_1 = A \sin(\omega t) + B \cos(\omega t) + C/\omega^2$$

and

$$x_2 = A\,\omega \cos(\omega t) - B\,\omega \sin(\omega t)\ .$$

Applying the initial conditions $x_1 = a$, $x_2 = 0$ gives $A = 0$ and $a = B + C/\omega^2$. Thus

$$x_1 = (a - C/\omega^2)\cos(\omega t) + C/\omega^2$$

and

$$x = -(a - C/\omega^2)\omega \sin(\omega t)\ .$$

The value of x_2 is again zero when $t = \pi/\omega$, and at this time

$$x_1 = -(a - C/\omega^2) + C/\omega^2 = -a + 2C/\omega^2\ .$$

We also want $x_1 = 0$ at this time. Hence we take $C = a\omega^2/2$.

Thus we have seen that the control

$$u = a\omega^2/2 \tag{11.6}$$

takes the system from $(a,0)$ at $t = 0$ to $(0,0)$ at $t = \pi/\omega$ and the system is controllable.

It is not difficult to find a second control (Problem 11.1).

A problem in *optimal control* is to find the control which takes the system from $(a,0)$ to $(0,0)$ in minimum time. This is not a trivial problem. It can be formulated in terms of equation (11.2) by taking $F \equiv 1$ and $t_0 = 0$. We then require the control which minimizes

$$J = \int_0^{t_1} 1\,dt = t_1\ . \tag{11.7}$$

11.3 An Electrochemical Process

Consider an electrochemical system, which is modelled by the differential equation

$$\boxed{\ddot{x} = -\dot{x} + u} \tag{11.8}$$

where x and u are functions of t. We wish to minimize the cost functional

$$J = \tfrac{1}{2}\int_0^{\infty} (x^2 + \alpha u^2)\,dt \tag{11.9}$$

where α is a disposable constant, by choosing the control variable u appropriately.

Following the technique evolved in Ch. 10.6, we introduce two state variables, x_1 and x_2, by defining

$$x_1 = x \,,$$

$$x_2 = \dot{x}_1 \; (= \dot{x}) \,.$$

The constraints that x_1 and x_2 must satisfy are

$$\left.\begin{aligned} \dot{x}_1 - x_2 &= 0 \,, \\ \dot{x}_2 + x_2 - u &= 0 \,. \end{aligned}\right\} \tag{11.10}$$

Typical end conditions for an electrochemical system are x and $\dot{x}$ given at $t = 0$, and both tend to zero as $t \to \infty$. In terms of the state variables, x_1 and x_2, the end conditions are

$$\left.\begin{aligned} &x_1(0) = a, \;\; x_2(0) = b \,; \\ &x_1 \to 0, \;\; x_2 \to 0 \;\text{ as }\; t \to \infty \end{aligned}\right\} \tag{11.11}$$

where a and b are given constants.

The augmented cost functional is

$$J^* = \int_0^\infty \left[\tfrac{1}{2}(x_1{}^2 + \alpha u^2) + \lambda(\dot{x}_1 - x_2) + \mu(\dot{x}_2 + x_2 - u)\right] dt \,, \tag{11.12}$$

λ and μ being Lagrange multipliers. Applying Euler's equation with $x = x_1$, x_2 and u respectively gives

$$[x_1]: \qquad x_1 - \frac{d}{dt}(\lambda) = 0 \qquad \text{i.e. } x_1 = \dot{\lambda}; \tag{11.13}$$

$$[x_2]: \qquad -\lambda + \mu - \frac{d}{dt}(\mu) = 0 \qquad \text{i.e. } \lambda = \mu - \dot{\mu}; \tag{11.14}$$

$$[u]: \qquad \alpha u - \mu - \frac{d}{dt}(0) = 0 \qquad \text{i.e. } \mu = \alpha u \,. \tag{11.15}$$

From equations (11.13) – (11.15) and with the constraints (11.10), i.e. five equations, we must eliminate λ and μ and solve for x_1, x_2 and u. Now from (11.10)

$$\frac{d^4x_1}{dt^4} = \frac{d^3x_2}{dt^3} = -\ddot{x}_2 + \ddot{u}$$

$$= -(-\dot{x}_2 + \dot{u}) + \ddot{u}$$

$$= \ddot{x}_1 + \frac{1}{\alpha}(-\dot{\mu} + \ddot{\mu}), \text{ using (11.15)} \quad ,$$

$$= \ddot{x}_1 + \frac{1}{\alpha}(-\dot{\lambda}), \text{ from (11.14)} \ ,$$

$$= \ddot{x}_1 - \frac{x_1}{\alpha}, \text{ using (11.13)} \ .$$

Writing $\mathscr{D} \equiv \dfrac{d}{dt}$,

$$\left(\mathscr{D}^4 - \mathscr{D}^2 + \frac{1}{\alpha}\right)x_1 = 0 \ . \tag{11.16}$$

Clearly the solution for x_1, and hence the control variable u, depends on the choice of the disposable constant α. For simplicity we will choose $\alpha = 4$, which implies that u has a high 'weighting' in the cost functional, compared with x. When $\alpha = 4$, (11.16) reduces to

$$\left(\mathscr{D} - \frac{1}{\sqrt{2}}\right)^2 \left(\mathscr{D} + \frac{1}{\sqrt{2}}\right)^2 x_1 = 0$$

and solving, $\quad x_1 = (At + B)e^{-t/\sqrt{2}} + (Ct + D)e^{t/\sqrt{2}}$.

Since $x_1 \to 0$ as $t \to \infty$, we see immediately that $C = D = 0$.

Also $\quad x_2 = \dot{x}_1 = \left(A - \dfrac{1}{\sqrt{2}}B - \dfrac{1}{\sqrt{2}}At\right)e^{-t/\sqrt{2}}$.

Applying the conditions at the end $t = 0$ gives $B = a$ and $A - \dfrac{1}{\sqrt{2}}B = b$.

Thus

$$x_1 = \left[\left(b + \frac{1}{\sqrt{2}}a\right)t + a\right]e^{-t/\sqrt{2}}$$

$$x_2 = \left[-\frac{1}{\sqrt{2}}\left(b + \frac{1}{\sqrt{2}}a\right)t + b\right]e^{-t/\sqrt{2}} \ ;$$

and since $u = \dot{x}_2 + x_2$, we have

$$u = \left[\frac{1}{2}(1 - \sqrt{2})\left(b + \frac{1}{\sqrt{2}}a\right)t - (\sqrt{2} - 1)b - \frac{1}{2}a\right]e^{-t/\sqrt{2}} . \tag{11.17}$$

We have determined the control variable $u = u(t)$ which minimizes the cost functional. Substitution in (11.9) will in fact give the minimum value of J.

The solution of this control problem can be illustrated in a block diagram by noting that u is of the form

$$u = a_1 x_1 + a_2 x_2 \tag{11.18}$$

where a_1 and a_2 are constants that can be determined. For substituting into (11.18), using (11.17), gives

$$\frac{1}{2}(1 - \sqrt{2})\left(b + \frac{1}{\sqrt{2}}a\right)t - (\sqrt{2} - 1)b - \frac{1}{2}a$$

$$= a_1\left[\left(b + \frac{1}{\sqrt{2}}a\right)t + a\right] + a_2\left[-\frac{1}{\sqrt{2}}\left(b + \frac{1}{\sqrt{2}}a\right)t + b\right].$$

Comparing coefficients of t and the constant term:

$$\frac{1}{2}(1 - \sqrt{2}) = a_1 - \frac{1}{\sqrt{2}}a_2$$

$$-(\sqrt{2} - 1)b - \tfrac{1}{2}a = a\,a_1 + b\,a_2 .$$

Solving, it can soon be shown that

$$a_1 = -\tfrac{1}{2}$$

$$a_2 = -(\sqrt{2} - 1) ,$$

giving

$$\boxed{u(t) = -\tfrac{1}{2}x_1(t) - (\sqrt{2} - 1)x_2(t)}$$

or, $$u(t) \approx -(0{\cdot}5)x_1(t) - (0{\cdot}414)x_2(t) . \tag{11.19}$$

Given the values of x_1 and x_2 at any time t determines from (11.19) the value of $u(t)$ at that time. This is illustrated in a *block diagram*, Fig. 11.1.

The large blocks represent the differential relationships between the variables, the small blocks represent multiplication and the circles represent combination, according to the signs outside.

Starting with the upper left hand circle, the *input* consists of a value of u (chosen initially, but later produced from feedback). The first operation, $1/(1 + \mathscr{D})$, operates on u to produce a value of x_2 (i.e. according to the differential equation $\dot{x}_2 + x_2 = u$). Continuing along the top of the diagram the operation $(1/\mathscr{D})x_2$ produces a value of x_1. The combination of the appropriate multiples of x_1 and x_2 is represented by the lower circle, producing a new value of u which is fed back into the input. The cycle of events is repeated continuously.

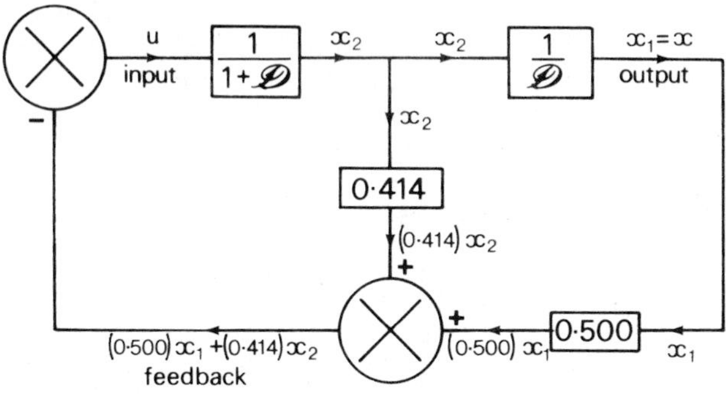

Fig. 11.1 Block diagram for control of electrochemical system

This type of system is called a *closed loop feedback* system. The control variable u is a continuous function of x_1 and x_2, and the problem is adequately solved using variational calculus. Note that there were no restrictions on the control function. In practice there are usually some bounds on any control; for example $|u(t)| \leqslant K$ for all t. The problem is then far more complicated, the control function often jumping from one extreme value to another; e.g. $+K$ to $-K$ in the example above. This type of problem is illustrated in Ch. 11.4.

11.4 Bang – Bang Control

The concept of bang-bang control is best illustrated by a simple physical example. Suppose we wish to drive a car from a stationary position on a horizontal driveway into a stationary position in a garage, moving a total distance a. The available controls for the driver are the accelerator and the brake (for simplicity we assume no gear change!) and the equation of motion for the car is

$$\frac{d^2x}{dt^2} = f \tag{11.20}$$

where $f = f(t)$ represents the applied acceleration or deceleration (braking).

Clearly f will be subject to both lower and upper bounds so that

$$-\alpha \leqslant f(t) \leqslant \beta \tag{11.21}$$

where β is the maximum acceleration possible and α the maximum deceleration. We wish to solve (11.20) subject to the constraint (11.21) and such that

$$x(0) = 0\,, \quad \frac{dx}{dt}(0) = 0\,; \quad x(T) = a\,, \quad \frac{dx}{dt}(T) = 0\,; \tag{11.22}$$

where T is the (unspecified) time of travel. Is the system controllable; i.e. can it be done? Quite clearly, common sense tells us that it can and that the result can be achieved in a variety of ways. An optimal control problem would be to ask what control f accomplishes the operation in a minimum time? Again the answer to this is probably fairly obvious, but it is instructive to see how we can solve the problem by mathematical reasoning.

The time of travel can be expressed as

$$T = \int_0^T dt = \int_0^a \frac{1}{v}\,dx\,,$$

where $v = dx/dt$. Regarding v as a function of x, $v = v(x)$, it must satisfy $v(0) = v(a) = 0$. Now

$$f = \frac{d^2x}{dt^2} = v\frac{dv}{dx} = \frac{d}{dx}\left(\tfrac{1}{2}v^2\right);$$

so that, defining

$$g(x) = \tfrac{1}{2}v^2\,,$$

we have

$$f = \frac{dg}{dx}$$

and

$$T = \int_0^a (2g)^{-\frac{1}{2}}\,dx\,. \tag{11.23}$$

Our problem is to find the control f which minimizes T subject to the inequality constraint (11.21) where $dg/dx = f$ and such that $g(0) = g(a) = 0$. We can deal with the equality constraint $dg/dx - f = 0$ by using the Lagrange multiplier method of Ch. 10.6. But this method does not immediately apply to the inequality constraints (11.21). However these can be changed to an equality constraint by introducing a second control variable z where

$$z^2 - (f + \alpha)(\beta - f) = 0 \tag{11.24}$$

and z is a real variable. We note that if (11.21) is satisfied then $z^2 \geqslant 0$ and z

is in fact real. We thus form the augmented functional

$$T^* = \int_0^a \left\{ (2g)^{-\frac{1}{2}} + \lambda\left(\frac{dg}{dx} - f\right) + \mu[z^2 - (f + \alpha)(\beta - f)] \right\} dx \,. \tag{11.25}$$

Here g is the state variable, f and z are control variables, x is the independent variable and λ and μ are Lagrange multipliers. The optimum path will satisfy Euler's equation for each of these variables, i.e.

$$[g]: \quad \frac{\partial F}{\partial g} - \frac{d}{dx}\left(\frac{\partial F}{\partial g'}\right) = 0 : \; -(2g)^{-\frac{3}{2}} - \frac{d\lambda}{dx} = 0 \tag{11.26}$$

$$[f]: \quad \frac{\partial F}{\partial f} - \frac{d}{dx}\left(\frac{\partial F}{\partial f'}\right) = 0 : \; -\lambda + \mu(2f + \alpha - \beta) = 0 \tag{11.27}$$

$$[z]: \quad \frac{\partial F}{\partial z} - \frac{d}{dx}\left(\frac{\partial F}{\partial z'}\right) = 0 : \; 2z\mu = 0 \tag{11.28}$$

where F is the integrand of (11.25) and dashes differentiation with respect to x.

Equation (11.28) gives either $z = 0$ or $\mu = 0$. If $\mu = 0$ then (11.27) gives $\lambda = 0$ and (11.26) will result in $(2g)^{-\frac{3}{2}} = 0$. This is clearly not possible so we conclude that $z = 0$. But when $z = 0$ we have, from (11.24),

$$(f + a)(\beta - f) = 0 \,, \tag{11.29}$$

so that either $f = -\alpha$ or $f = \beta$. Thus on the optimum path the acceleration takes either its maximum or its minimum value, but not intermediate values (or (11.28) would not be satisfied). Clearly initially we must have $f = \beta$ and finally $f = -\alpha$, so that, assuming there is only one switch in the control f, we have

$$\frac{d^2x}{dt^2} = f = \begin{cases} \beta \text{ for } 0 \leqslant t < \tau \\ -\alpha \text{ for } \tau < t \leqslant T \end{cases} \tag{11.30}$$

the switch taking place at time τ, to be determined.

Integrating,

$$\frac{dx}{dt} = \begin{cases} \beta t & \text{for } 0 \leqslant t < \tau \\ -\alpha(t - T) & \text{for } \tau < t \leqslant T \end{cases} \tag{11.31}$$

since $dx/dt = 0$ at $t = 0$ and $t = T$. Integrating again, using $x = 0$ at

$t = 0$ and $x = a$ at $t = T$, gives

$$x = \begin{cases} \frac{1}{2}\beta t^2 & \text{for } 0 \leqslant t < \tau \\ -\frac{1}{2}\alpha(t - T)^2 + a & \text{for } \tau < t \leqslant T \end{cases} \tag{11.32}$$

Now x and dx/dt must be continuous at $t = \tau$, so that, using (11.31) and (11.32) we must have

$$\beta\tau = -\alpha(t - T) \text{ and } \frac{1}{2}\beta\tau^2 = -\frac{1}{2}\alpha(\tau - T)^2 + a .$$

Eliminating T gives the switch time τ as

$$\tau = \left[\frac{2a\alpha}{\beta(\alpha + \beta)}\right]^{1/2}, \tag{11.33}$$

and substituting back gives the minimum time as

$$T = \left[\frac{2a(\alpha + \beta)}{\alpha\beta}\right]^{1/2} . \tag{11.34}$$

The problem has now been completely solved. The optimal control is given by

$$f = \begin{cases} \beta & \text{for } 0 \leqslant t < \tau \\ -\alpha & \text{for } \tau < t \leqslant T \end{cases} \tag{11.35}$$

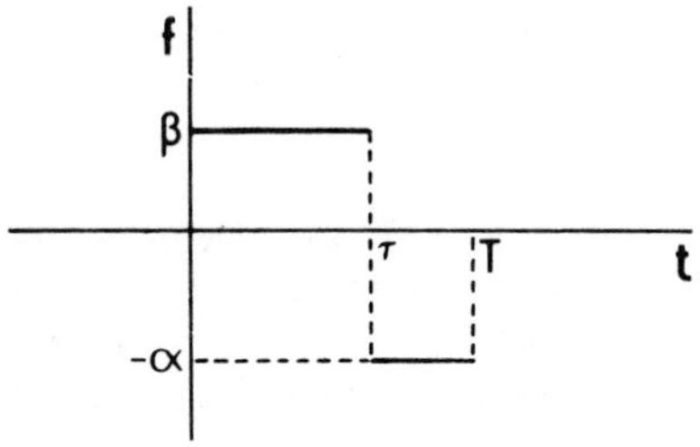

Fig. 11.2 The optimal control function

where τ and T are defined by (11.33) and (11.34). A typical example is illustrated in Fig. 11.2. This graph brings out clearly the fact that at $t = \tau$ there is a discontinuity in f, when it switches over from its maximum value to its minimum value. For obvious reasons this type of control is referred to as 'bang-bang control'. In the above example the nature of the optimum control was fairly obvious from intuition, but in other situations, not necessarily physical, it may not be so evident. Furthermore, we assumed that there was only one switch, and again this was a reasonable assumption on intuitive grounds, but in other optimal control problems it may not be at all obvious how many switches will occur on the optimal path. More advanced theory, first developed by Pontryagin and his Russian colleagues in the 1960's, lead to a 'maximum principle' which is suitable for such problems, but this is beyond the scope of this book.

11.5* Flight Trajectory Optimization

In this section we consider the problem of finding an optimum flight path for a rocket which maximizes its final velocity when placing a satellite into earth orbit, at perigee or apogee, at a specified height above the earth's surface.

Assuming the distance travelled by the rocket to be small compared with the radius of the earth, we can, to a good approximation, assume a constant gravitational field (Ch. 7.1), neglect air resistance and assume a flat surface for the earth. The flight path is illustrated in Fig. 11.3.

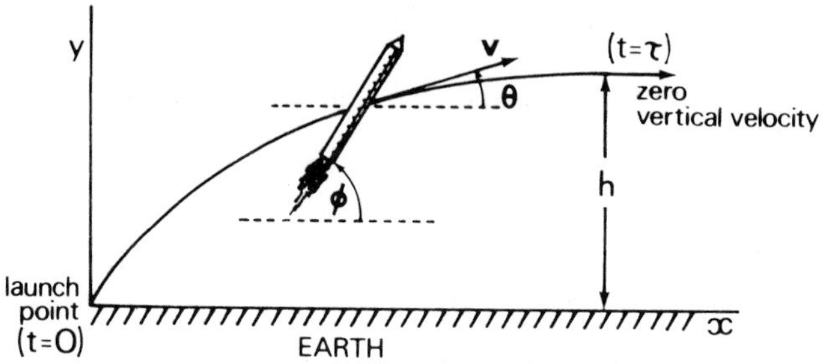

Fig. 11.3 Rocket flight path

Suppose F is the thrust produced by the rocket jet per unit mass. In Ch. 4.3, for a rocket moving in a straight line, we showed that the thrust per unit mass is given by

$$F = -\frac{c}{m}\frac{dm}{dt} , \tag{11.36}$$

m being the rocket mass at time t, and c being the relative exhaust velocity. Let ϕ be the angle between the rocket axis and the horizontal. We call this angle the *thrust attitude* angle, and we assume that it is possible for the rocket control (by using side retro-rockets) to alter ϕ throughout its motion. F is a given function of time, and ϕ is a function of time t which we will determine in order to maximize the final horizontal speed on reaching a specified height, h, with zero vertical velocity.

We define

$$\mathbf{r} = (x, y, 0), \quad \mathbf{v} = \dot{\mathbf{r}} = (v_1, v_2, 0) , \tag{11.37}$$

so that the equation of motion of the rocket is

$$m\dot{\mathbf{v}} = m(F \cos\phi, F \sin\phi, 0) + m(0, -g, 0) . \tag{11.38}$$

Equating components,

$$\dot{v}_1 = F\cos\phi \qquad \text{where } \dot{x} = v_1 ; \tag{11.39}$$

$$\dot{v}_2 = F\sin\phi - g \quad \text{where } \dot{y} = v_2 . \tag{11.40}$$

Assuming zero initial velocity, integrating (11.39) gives

$$v_1(t) = \int_0^t F\cos\phi\,dt\,.$$

Let $t = \tau$ be the time when the rocket reaches height h, assuming zero initial velocity. At this instant, its horizontal velocity is given by

$$\boxed{v_1(\tau) = \int_0^\tau F\cos\phi\,dt}\,. \tag{11.41}$$

This is the functional which we wish to maximize, ϕ being a control variable, subject to the constraints

$$\left.\begin{aligned} \dot{y} - v_2 &= 0 \\ \dot{v}_2 - F\sin\phi + g &= 0\,. \end{aligned}\right\} \tag{11.42}$$

The end conditions satisfied by the state variables, y and v_2, are

$$\left.\begin{aligned} y(0) &= 0\,,\ y(\tau) = h\,; \\ v_2(0) &= 0\,,\ v_2(\tau) = 0\,. \end{aligned}\right\} \tag{11.43}$$

Note that the equation $\dot{x} - v_1 = 0$ is not a constraint for the problem, since there is only one end condition on x, namely $x(0) = 0$. Thus for a given v_1 the equation $\dot{x} - v_1 = 0$ determines the function x.

Following the method evolved in previous sections, we maximize the augmented functional

$$\boxed{v_1^* = \int_0^\tau [F\cos\phi + \lambda(\dot{y} - v_2) + \mu(\dot{v}_2 - F\sin\phi + g)]\,dt}\,. \tag{11.44}$$

Here y and v_2 are state variables, ϕ is a control variable and λ and μ are Lagrange multipliers. Writing (11.44) in the form

$$v_1^* = \int_0^\tau f(\phi, y, \dot{y}, v_2, \dot{v}_2)\,dt$$

where

$$f(\phi, y, \dot{y}, v_2, \dot{v}_2) = F\cos\phi + \lambda(\dot{y} - v_2) + \mu(\dot{v}_2 - F\sin\phi + g)\,,$$

we see that the transversality conditions (10.24) are satisfied, since y and v_2 are fixed at $t = 0$ and $t = \tau$, and $\partial f/\partial\dot{\phi} = 0$ for all t.

Euler's equations for y, v_2 and ϕ are

$$[y]: \qquad \frac{\partial f}{\partial y} - \frac{d}{dt}\left(\frac{\partial f}{\partial \dot{y}}\right) = 0 \quad \text{i.e.} \quad \dot{\lambda} = 0 \,; \tag{11.45}$$

$$[v_2]: \qquad \frac{\partial f}{\partial v_2} - \frac{d}{dt}\left(\frac{\partial f}{\partial \dot{v}_2}\right) = 0 \quad \text{i.e.} \quad \lambda = -\dot{\mu} \,; \tag{11.46}$$

$$[\phi]: \qquad \frac{\partial f}{\partial \phi} - \frac{d}{dt}\left(\frac{\partial f}{\partial \dot{\phi}}\right) = 0 \quad \text{i.e.} \quad \mu = -\tan\phi \,. \tag{11.47}$$

From (11.46) and (11.45), we see that

$$\ddot{\mu} = 0$$

giving

$$\mu = at + b \,,$$

where a and b are constants.

But from (11.47),

$$\tan\phi = -(at + b) \,.$$

In order to put the satellite into orbit at perigee or apogee it is reasonable to assume that $\phi = 0$ at $t = \tau$, giving

$$0 = -(a\tau + b)$$

and so

$$\tan\phi = -b(1 - t/\tau) \,.$$

If $\phi = \phi_0$ at $t = 0$, we obtain

$$\boxed{\tan\phi = (1 - t/\tau)\tan\phi_0} \,. \tag{11.48}$$

Thus, for a given thrust program, $F = F(t)$, the optimum choice of the thrust attitude angle for maximum final horizontal velocity at a specified height is given by (11.48). But this equation only determines ϕ in terms of t and the *unknown* constants τ and ϕ_0. These constants are found by returning to the constraints (11.42) and solving for v_2 and y subject to the end conditions (11.43). Since there are *four* end conditions but only *two* first order differential equations, solving these equations will lead to the necessary values of the constants τ, and ϕ_0.

For any given function, $F = F(t)$, it would be a formidable task to solve (11.42) and in general we would resort to computational methods. In order to illustrate the way in which τ and ϕ_0 are determined, we look at the special case when F is a constant. In this case, we see that

$$v_2 = F\int \sin\phi \, dt - gt + A \,,$$

where A is a constant. But from (11.48)

$$\frac{d\phi}{dt} = -\frac{\tan\phi_0}{\tau\sec^2\phi}, \tag{11.49}$$

so that
$$\nu_2 = -\frac{F\tau}{\tan\phi_0}\int\tan\phi\sec\phi\,d\phi - gt + A$$

$$= -\frac{F\tau\sec\phi}{\tan\phi_0} - gt + A\,.$$

Since $\nu_2(0) = 0$, we see that $A = F\tau/\sin\phi_0$ giving

$$\nu_2 = \frac{F\tau}{\sin\phi_0}(1 - \cos\phi_0\sec\phi) - gt\,. \tag{11.50}$$

We are now in a position to determine one of the unknown constants: for at $t = \tau$, $\nu_2(\tau) = 0$ and $\phi = 0$ giving

$$0 = \frac{F\tau}{\sin\phi_0}(1 - \cos\phi_0) - g\tau\,.$$

Solving for ϕ_0 and writing $\alpha = g/F$, it can soon be shown that

$$\boxed{\begin{aligned}\sin\phi_0 &= \frac{2\alpha}{1+\alpha^2},\\ \cos\phi_0 &= \frac{1-\alpha^2}{1+\alpha^2}.\end{aligned}} \tag{11.51}$$

Substituting for ϕ_0 in (11.50) gives

$$\nu_2 = \frac{F\tau}{2\alpha}[(1+\alpha^2) - (1-\alpha^2)\sec\phi] - gt\,. \tag{11.52}$$

Integrating the other constraint, $\dot{y} - \nu_2 = 0$, gives

$$y = \frac{F\tau}{2\alpha}\left[(1+\alpha^2)t - (1-\alpha^2)\int\sec\phi\,dt\right] - \frac{1}{2}gt^2 + B\,,$$

where B is a constant. But from (11.49) and (11.51)

$$\int\sec\phi\,dt = -\frac{\tau(1-\alpha^2)}{2\alpha}\int\sec^3\phi\,d\phi$$

$$= -\frac{\tau(1-\alpha^2)}{4\alpha}[\sec\phi\tan\phi + \log(\sec\phi + \tan\phi)]\,.$$

Thus

$$y = \frac{F\tau}{8\alpha^2}\left\{4\alpha(1 + \alpha^2)t + (1 - \alpha^2)^2 t\,[\sec\phi\tan\phi\right.$$

$$\left. + \log(\sec\phi + \tan\phi)]\right\} - \frac{1}{2}gt^2 + B\,. \qquad (11.53)$$

Since $y(0) = 0$, we have

$$B = -\frac{F\tau^2(1 - \alpha^2)^2}{8\alpha^2}[\sec\phi_0\tan\phi_0 + \log(\sec\phi_0 + \tan\phi_0)]\,,$$

and using (11.51),

$$B = -\frac{F\tau^2}{8\alpha^2}\left[2\alpha(1 + \alpha^2) + (1 - \alpha^2)^2\log\left\{\frac{(1 + \alpha)}{(1 - \alpha)}\right\}\right]. \qquad (11.54)$$

We can now determine our second constant τ; for $y(\tau) = h$ and after some algebra, from (11.53) and (11.54) it can be shown that

$$\boxed{\tau^2 = 8\alpha^2 h\Big/F(1 - \alpha^2)\left[2\alpha - (1 - \alpha^2)\log\left\{\frac{1 + \alpha}{1 - \alpha}\right\}\right]}. \qquad (11.55)$$

Consequently the two constants, τ and ϕ_0, have been determined and the optimum thrust attitude program is now known. The value of the maximum horizontal velocity and the horizontal distance travelled can be determined from (11.39). These calculations are set in Problem 11.7.

It is of interest to see how in fact ϕ varies, and in particular the value of ϕ_0. Now from (11.51), we see that ϕ_0 depends only on $\alpha = g/F$. If for instance $\alpha = 0{\cdot}5$, $\phi_0 = \sin^{-1}(4/5) \approx 53^\circ 8'$; and if α is very small (i.e. $F \gg g$) the angle ϕ_0 is very small. We have of course omitted air resistance in our theory, and probably an improved path for the rocket would be first to move vertically upwards until above the earth's atmosphere. At this instant the rocket would be at height, say, y_0 and travelling with velocity $\mathbf{v}_0$ and we could then apply similar optimization techniques to find the optimum thrust attitude angle. The resulting calculations show that

$$\boxed{\tan\phi = (1 - t/\tau)\tan\theta_0} \qquad (11.56)$$

where θ_0 is the initial angle made by the velocity vector with the horizontal.

11.6* Optimal Economic Growth

In this section we introduce a model for an 'economy' which produces a single product. The formulation of this model might appear to be rather imprecise, but it should be noted that it is the underlying trends which are important in economic theory rather than precise results. Suppose we let $Y(t)$ be the output (per unit time) at time t, and assume that the output depends only on two factors, the labour force $L(t)$ and the capital $K(t)$. We then introduce a production function

$$Y = F(K, L)\,. \tag{11.57}$$

We assume that F has a 'return to scale' property which can be expressed as

$$F(\alpha K, \alpha L) = \alpha F(K, L) \tag{11.58}$$

for any positive constant α. For instance, using $\alpha = 1/L$ in (11.58), we can express the output per worker as

$$y = Y/L = \frac{F(K, L)}{L} = F(K/L, 1) = F(k, 1) = f(k), \text{ say}, \tag{11.59}$$

where $k(t) = K/L$ is the capital per worker. A typical production function $f(k)$ takes the form of the curve shown in Fig. 11.4.

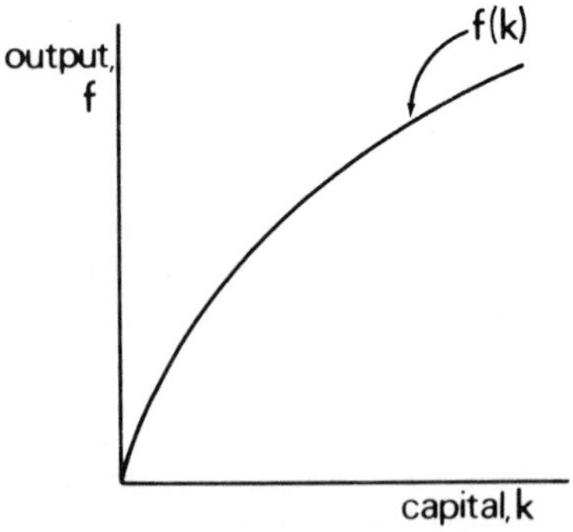

Fig. 11.4 Typical production function

an example of such a function is $f = k^{1/2}$, which is a special case of the Cobb-Douglas production function.

$$f(k) = k^{\alpha} \qquad (0 < \alpha < 1) \tag{11.60}$$

In this economic model we assume that the revenue derived from the output is either consumed (wages for the labour force) or invested (new machinery etc.) so that

$$Y(t) = C(t) + I(t) \tag{11.61}$$

where C is the consumption and I the investment per unit time at time t. Investment is used to increase the capital stock (for example extra plant or machinery), so that

$$I = dK/dt\,. \tag{11.62}$$

Thus, using (11.59) with (11.61) and (11.62), gives

$$f(k) = c(t) + \frac{1}{L}\frac{dK}{dt},$$

where $c = C/L$ is the consumption per worker. Writing $k = K/L$ we have

$$\frac{dk}{dt} = \frac{1}{L}\frac{dK}{dt} - \frac{K}{L^2}\frac{dL}{dt}\,.$$

Hence if the labour force is assumed to grow at a constant exponential rate so that $L = L_0 e^{\lambda t}$, we obtain

$$\frac{1}{L}\frac{dK}{dt} = \frac{dk}{dt} + \lambda k\,.$$

This gives

$$\boxed{f(k) = c + \lambda k + dk/dt}\quad, \tag{11.63}$$

which is the fundamental equation of *neoclassical economic growth.* We can re-write this equation as

$$f(k) - \lambda k = c + dk/dt\,.$$

Certain important characteristics of the differential equation can be investigated without actually specifying the output function. For instance, there are two special values of k:

(i) $\underline{k = \hat{k}}$ where $[f(k) - \lambda k]$ has a maximum, so that

$$df/dk = \lambda \quad \text{at} \quad k = \hat{k}; \tag{11.64}$$

(ii) $\underline{k = \tilde{k}}$ where $[f(k) - \lambda k]$ has a zero, so that

$$f(\tilde{k}) = \lambda\tilde{k}\,. \tag{11.65}$$

These values are illustrated in Fig. 11.5.

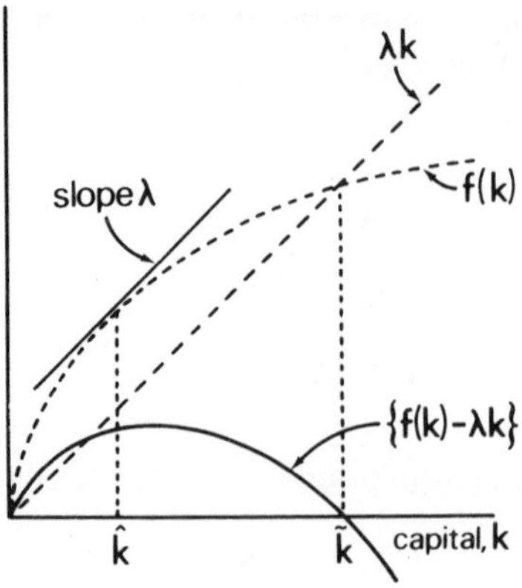

Fig. 11.5 Properties of $[f(k) - \lambda k]$

Clearly the solution of (11.63) depends on the form of the consumption rate. We consider two special cases of constant consumption rate.

(i) $c = \hat{c}$, which is the maximum value of c which can be sustained at equilibrium (i.e. when $dk/dt = 0$). Thus, at equilibrium, from (11.63),

$$c = f(k) - \lambda k \, , \tag{11.66}$$

and, from (11.64), this has a maximum value when $k = \hat{k}$. Hence (11.64) defines $\hat{k}$ and (11.66) gives the corresponding value of c; i.e. $\hat{c} = f(\hat{k}) - \lambda\hat{k}$. The quantities $\hat{c}$ and $\hat{k}$ are called the *golden rule level* of consumption and capital per worker respectively. The situation is illustrated in Fig. 11.6. To determine whether or not this equilibrium position is stable, we consider a position near $\hat{k}$. Thus we put

$$k = \hat{k} + \xi$$

in (11.63) when $c = \hat{c}$; i.e.

$$dk/dt = f(k) - k - \hat{c} \, ,$$

giving
$$\begin{aligned} \frac{d\xi}{dt} &= f(\hat{k} + \xi) - \lambda(\hat{k} + \xi) - \hat{c} \, , \\ &= \left[f(\hat{k}) + \xi f'(\hat{k}) + \xi^2 \frac{f''(\hat{k})}{2} + \ldots \right] - \lambda\hat{k} - \lambda\xi - \hat{c} \\ &= [f(\hat{k}) - \lambda\hat{k} - \hat{c}] + \xi\,[f'(\hat{k}) - \lambda] + \xi^2 \frac{f''(\hat{k})}{2} + \ldots \\ &\approx \xi^2 \frac{f''(\hat{k})}{2} \end{aligned}$$

for small ξ, having used (11.66) and (11.64). Integrating this equation gives

$$-\frac{1}{\xi} = t\frac{f''(\hat{k})}{2} + A\,,$$

so that

$$\xi = -\left[t\frac{f''(\hat{k})}{2} + A\right]^{-1}.$$

At $t = 0$, $\xi = a$, say, giving $a = -1/A$. Thus,

$$\xi = -2/[t\,f''(\hat{k}) - 2/a]\,,$$

and, writing $b = -f''(\hat{k})$ (>0 for most production functions), we have

$$\xi = 2/(bt + 2/a)\,.$$

For $a > 0$, ξ remains finite for all time, whereas if $a < 0$, $\xi \to \infty$ as $t \to -2/(ab)$. Thus a slight fall in the value of k below $\hat{k}$ will precipitate a fall in the capital, whereas an increase in k above $\hat{k}$ can be sustained. The equilibrium position is unstable.

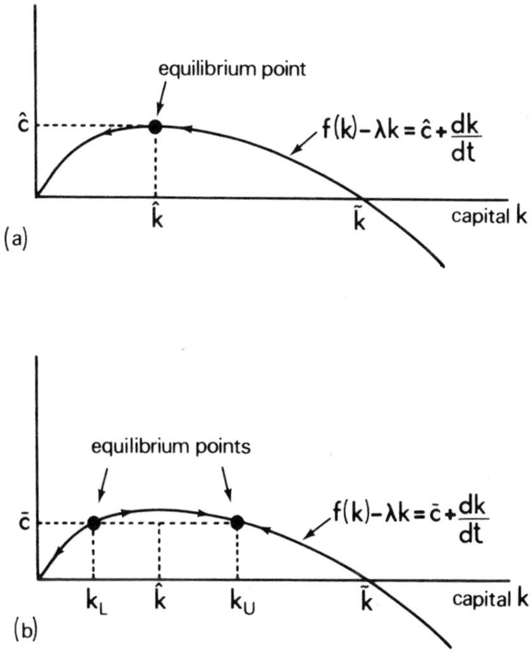

Fig. 11.6 Equilibrium points for given consumption, (a) Golden rule level of consumption, $\hat{c}$; (b) Constant consumption, $\bar{c}$

(ii) $c = \bar{c}$, where $\bar{c}$ is a constant such that $0 < \bar{c} < \hat{c}$. Fig. 11.6(b) shows that two equilibrium points (where $dk/dt = 0$) exist; namely $k = k_L$ and $k = k_U$. By means of a stability analysis as above it can be shown that $k = k_L$ is unstable, whereas $k = k_U$ is stable. This implies that, for fixed consumption $\bar{c}$, a big boost is needed to get beyond $k = k_L$. Once past this value there will be a tendency to gravitate up to k_U. This illustrates the concept of the "big push" needed to get an economy off the ground.

In the two special cases above, consumption per worker has been taken as a constant. In many situations it is possible to vary the consumption and to control this variation. Suppose there is a central planner (a chancellor or a manager) who is in a position to control the consumption. Consumption per worker, $c(t)$, is now a control variable and the central planner's job is to choose the time path

$$c = c(t), \qquad 0 \leqslant t \leqslant T$$

in some best way. We recall that $f = f(k(t))$ is a prescribed function of the capital per worker, $k(t)$, which is a state variable satisfying (11.63);

i.e.
$$dk/dt = f(k) - \lambda k - c, \tag{11.67}$$

where initially $k = k_0$, say. How can we identify the optimum consumption path? This clearly depends on the economic objectives. One possible objective is to maximize some function of consumption over a particular interval of time, so that, for instance, consumption is more important at earlier times than in the future. As an example, consider the function

$$e^{-\beta t} [c(t)]^{½}.$$

Since it is not possible to maximize at each instant of time, we define

$$W = \int_0^T e^{-\beta t} [c(t)]^{½} \, dt. \tag{11.68}$$

This functional, known as a *welfare integral*, is a measure of the usefulness of consumption over the time interval $0 \leqslant t \leqslant T$. Our aim is now to choose the consumption path so that W is maximized. The constant β is a measure of the importance attached to consumption at different times. If it is relatively important near $t = 0$, (which is often the case for governments and firms), then β is not small, whereas if consumption has equal importance at all future times we take $\beta = 0$.

As usual we introduce a Lagrange multiplier μ and form the augmented functional

$$W^* = \int_{t_0}^{t_1} \left\{ e^{-\beta t} c^{½} + \mu [\dot{k} - f(k) + \lambda k + c] \right. dt, \tag{11.69}$$

where $\dot{k} = dk/dt$. To maximize W^* there will be Euler equations for the state variable k and control variable c; i.e.

$$[k]: \quad \frac{\partial F}{\partial k} - \frac{d}{dt}\left(\frac{\partial F}{\partial \dot{k}}\right) = 0\,; \qquad \text{i.e. } \mu\left(\lambda - \frac{df}{dk}\right) - \frac{d\mu}{dt} = 0; \quad (11.70)$$

$$[c]: \quad \frac{\partial F}{\partial c} = 0\,; \qquad \text{i.e. } \tfrac{1}{2}\, c^{-\frac{1}{2}}\, e^{-\beta t} + \mu = 0\,; \quad (11.71)$$

where F is the integrand in (11.69). Substituting for μ from (11.71) into (11.70) gives

$$\dot{c} = 2(df/dk - \beta - \lambda)\, c \qquad (11.72)$$

and this equation, together with (11.67), determines the optimal paths for k and c. One further boundary condition is required; for example the final capital $k(T)$ might be specified.

For a given production function f it is usually a formidable task to solve for c and k and a numerical approach would have to be used. One simple example has been left to the reader (Problem 11.9).

Optimal control is now playing a leading part in the development of economic theories. One problem which is the subject of current research concerns the optimum depletion of natural resources. How quickly, for example, should Britain deplete its stocks of North Sea oil, and how much capital should be spent on developing a new energy resource to replace oil when all the oil stocks have been depleted?

Worked Examples

EXAMPLE 11.1:

The governing differential equation of a mechanical system is

$$\frac{d^2x}{dt^2} = u\,,$$

where x is the state variable and u the control variable. Find the optimal control u which takes the system from $x = 1$, $dx/dt = 1$ at $t = 0$ to $x = 0$, $dx/dt = 0$ at $t = 1$ and which minimizes

$$J = \int_0^1 u^2\, dt\,.$$

SOLUTION:

As usual, we reduce the second order differential equation to two first order equations by defining

$$v = dx/dt\,,$$

so that $$dv/dt = u\,.$$

Thus u is the control variable, and x and v are state variables subject to the constraints above and the boundary conditions

(i) $x = v = 1$ at $t = 0$;

(ii) $x = v = 0$ at $t = 1$.

We form the augmented functional

$$J^* = \int_0^1 [u^2 + \lambda(v - dx/dt) + \mu(dv/dt - u)]\,dt\,.$$

For optimum paths we have Euler equations for x, v and u; i.e.

$$[x]: \quad \frac{\partial F}{\partial x} - \frac{d}{dt}\left(\frac{\partial F}{\partial \dot{x}}\right) = 0\,; \qquad \text{i.e.} \quad \frac{d\lambda}{dt} = 0\,;$$

$$[v]: \quad \frac{\partial F}{\partial v} - \frac{d}{dt}\left(\frac{\partial F}{\partial \dot{v}}\right) = 0\,; \qquad \text{i.e.} \quad \lambda - \frac{d\mu}{dt} = 0\,;$$

$$[u]: \quad \frac{\partial F}{\partial u} - \frac{d}{dt}\left(\frac{\partial F}{\partial \dot{u}}\right) = 0\,; \qquad \text{i.e.} \quad 2u - \mu = 0\,;$$

where F is the integrand and a dot denotes time differentiation. Thus λ is a constant, say A, and

$$\frac{d\mu}{dt} = A,$$

giving $\mu = At + B$, where B is a constant. Hence the optimal control u is of the form $u = \frac{1}{2}At + \frac{1}{2}B$.

Using the constraints, we obtain

$$v = \tfrac{1}{4}At^2 + \tfrac{1}{2}Bt + C$$

and

$$x = \frac{At^3}{12} + \frac{Bt^2}{4} + Ct + D\,,$$

C and D being constants.

Applying the initial conditions gives $C = D = 1$, and applying conditions at $t = 1$ gives $A = 36$, $B = -20$. The optimal control u is thus given by

$$u = 2(9t - 5)\,, \qquad 0 \leqslant t \leqslant 1\,.$$

This control does in fact give a minimum for J.

EXAMPLE:11.2:

The angular motion of a ship is described in terms of the variable θ by the equation

$$\frac{d^2\theta}{dt^2} + \frac{d\theta}{dt} = p\,,$$

where p is the rudder setting which is subject to the constraint $|p| \leqslant 1$. A change in course from $\theta = \alpha$, $d\theta/dt = 0$, to $\theta = d\theta/dt = 0$ is required. Find p to minimize the time taken for the correction.

SOLUTION:

The time taken is given by

$$T = \int_0^T dt = \int_\alpha^0 \frac{dt}{d\theta}\, d\theta = \int_\alpha^0 \frac{1}{\omega}\, d\theta$$

where $\omega = d\theta/dt$. Regarding ω as a function of θ, then ω is a state variable, θ the independent variable and

$$p = \omega + \frac{d\omega}{dt} = \omega + \omega\,\frac{d\omega}{d\theta}\,,$$

p being the control variable.

Thus we wish to choose ω to minimize T subject to the constraints

$$p = \omega\left(1 + \frac{d\omega}{d\theta}\right)$$

and

$$-1 \leqslant p \leqslant 1\,.$$

The inequality constraint can be replaced by

$$z^2 - (p+1)(1-p) = 0\,,$$

where z is a second control variable. Thus we form the augmented functional

$$T^* = \int_\alpha^0 \left\{\frac{1}{\omega} + \lambda\left[p - \omega\left(1 + \frac{d\omega}{d\theta}\right)\right] + \mu\,[z^2 - (p+1)(1-p)]\right\} d\theta\,,$$

where λ and μ are Lagrange multipliers. For the optimum path we have Euler equations for ω, p and z; i.e.

$$[\omega] : \frac{\partial F}{\partial \omega} - \frac{d}{d\theta}\left(\frac{\partial F}{\partial \omega'}\right) = 0\,; \qquad \text{i.e.} \quad -\frac{1}{\omega^2} - \lambda + \omega\frac{d\lambda}{d\theta} = 0\,;$$

$$[p] : \frac{\partial F}{\partial p} - \frac{d}{d\theta}\left(\frac{\partial F}{\partial p'}\right) = 0\,; \qquad \text{i.e.} \quad \lambda + 2\mu p = 0\,;$$

$$[z] : \frac{\partial F}{\partial z} - \frac{d}{d\theta}\left(\frac{\partial F}{\partial z'}\right) = 0\,; \qquad \text{i.e.} \quad 2\mu z = 0\,;$$

where F is the integrand and a dash denotes differentiation with respect to θ. From the third equation, either $z = 0$ or $\mu = 0$. If $\mu = 0$, then the other two equations give $1/\omega^2 = 0$, which is not possible. Thus $\mu \neq 0$, and we have $z = 0$, which implies $p = \pm 1$, and we have a case of bang-bang control.

Since θ must decrease from $\theta = \alpha$ to $\theta = 0$, and at $t = 0$, $d\theta/dt = 0$, then initially $d^2\theta/dt^2 = d\omega/dt$ must be negative. Thus the optimal path must start with $p = -1$, since $p = \omega + d\omega/dt$. Similarly the path must finish on $p = 1$. Assuming that there is only one switch from $p = -1$ to $p = 1$, we must determine when this switch takes place.

On $p = -1$, the original differential equation can be written as

$$\frac{d\theta}{d\omega} = -1 + \frac{1}{(1+\omega)}\,,$$

so that, integrating,

$$\theta = -\omega + \log(1+\omega) + \alpha\,,$$

having applied the initial conditions, $\theta = \alpha$ and $\omega = 0$. Similarly on $p = 1$, we obtain

$$\theta = -\omega - \log(1-\omega)\,,$$

since finally $\theta = \omega = 0$. The switch occurs when the values of θ above are equal, i.e.

$$-\omega + \log(1+\omega) + \alpha = -\omega - \log(1-\omega)\,,$$

giving

$$\omega = (1 - e^{-\alpha})^{1/2}$$

and

$$\theta = -(1 - e^{-\alpha})^{1/2} - \log[1 - (1 - e^{-\alpha})^{1/2}]\,.$$

The time when this occurs can be determined by integrating the original differential equation.

EXAMPLE 11.3:

In an inventory-control production-scheduling problem the governing equation is

$$\frac{dI}{dt} = P,$$

where $I = I(t)$ is the inventory level and $P = P(t)$ is the production rate remaining after known sales demand has been met. It is planned that over a fixed time interval, $0 \leqslant t \leqslant T$, I should be increased from its value I_0 at $t = 0$ by increasing P in such a way that the cost functional

$$J = \int_0^T [(Q-I)^2 + \alpha^2 P^2]\, dt$$

is minimized. Here Q and α^2 are positive constants and $Q > I_0$. Determine the optimal production rate and inventory level.

SOLUTION:

We form the augmented functional

$$J^* = \int_0^T [(Q-I)^2 + \alpha^2 P^2 + \lambda(\dot{I} - P)]\, dt,$$

where λ is a Lagrange multiplier and a dot denotes differentiation with respect to time. The Euler equations for a state variable I and control variable P are

$$[I] : \frac{\partial F}{\partial I} - \frac{d}{dt}\left(\frac{\partial F}{\partial \dot{I}}\right) = 0; \qquad \text{i.e.} \quad -2(Q-I) - \dot{\lambda} = 0;$$

$$[P] : \frac{\partial F}{\partial P} - \frac{d}{dt}\left(\frac{\partial F}{\partial \dot{P}}\right) = 0; \qquad \text{i.e.} \quad 2\alpha^2 P - \lambda = 0;$$

where F is the integrand. Eliminating λ,

$$\frac{dP}{dt} = \frac{1}{\alpha^2}(I-Q).$$

Thus
$$\frac{d^2I}{dt^2} = \frac{dP}{dt} = \frac{1}{\alpha^2}(I-Q),$$

giving
$$\frac{d^2I}{dt^2} - \frac{1}{\alpha^2} I = -\frac{Q}{\alpha^2}.$$

This has general solution of the form

$$I = A \sinh(t/\alpha) + B \cosh(t/\alpha) + Q,$$

where A and B are constants. At $t = 0$, $I = I_0$, so that $I_0 = B + Q$, giving $B = I_0 - Q$. Now at $t = T$, I is not specified so that we must apply the transversality condition

$$\partial F/\partial \dot{I} = 0 \text{ at } t = T.$$

Thus $\lambda = 0$ at $t = T$, showing that $P = 0$ at $t = T$. Now $P = dI/dt$, so that we have, at $t = T$,

$$0 = \frac{A}{\alpha}\cosh(T/\alpha) + \frac{B}{\alpha}\sinh(T/\alpha).$$

This gives $$A = (Q - I_0)\tanh(T/\alpha)$$

and hence the optimum path is given by

$$I = (I_0 - Q)\frac{\cosh[(T-t)/\alpha]}{\cosh(T/\alpha)} + Q$$

and P is given by $$P = dI/dt.$$

EXAMPLE 11.4*

The production function in the neoclassical growth equation is given by

$$f(k) = k^\alpha, \qquad 0 < \alpha < 1.$$

What are the golden rule levels of c and k? Find k_L and k_U when $c = \bar{c}$, a constant, and $\alpha = \frac{1}{2}$.

SOLUTION:

The equations for golden rule levels [(11.64) and (11.66)] are

$$\hat{c} = f(\hat{k}) - \lambda\hat{k}$$

and $$\frac{df}{dk}(\hat{k}) - \lambda = 0.$$

The second equation gives $$\alpha\hat{k}^{\alpha-1} = \lambda;$$

i.e. $$\hat{k} = (\lambda/\alpha)^{1/(\alpha-1)}.$$

The first equation then gives

$$\begin{aligned}\hat{c} = \hat{k}^\alpha - \lambda\hat{k} &= \hat{k}(\hat{k}^{\alpha-1} - \lambda)\\ &= \hat{k}(\lambda/\alpha - \lambda)\\ &= \lambda\frac{(1-\alpha)}{\alpha}\left(\frac{\lambda}{\alpha}\right)^{1/(\alpha-1)}\end{aligned}$$

i.e. $$\hat{c} = (1-\alpha)\left(\frac{\lambda}{\alpha}\right)^{\alpha/(\alpha-1)}.$$

The values of k_L and k_U are given by

$$-\bar{c} + f(k) - \lambda k = 0,$$

so that

$$-\bar{c} + k^{1/2} - \lambda k = 0.$$

Hence,

$$k^{1/2} = -\frac{1}{2\lambda} \pm \frac{1}{2\lambda}(1-4\lambda\tau)^{1/2}.$$

Thus

$$k_U = [1+(1-4\lambda\tau)^{1/2}]^2/4\lambda^2$$

and

$$k_L = [1-(1-4\lambda\tau)^{1/2}]^2/4\lambda^2.$$

Problems

11.1 Show that the system (x_1, x_2), where

$$\dot{x}_1 = x_2,$$

$$\dot{x}_2 = -\omega^2 x_1 + u,$$

is controllable from $(0, 0)$ at $t = 0$ to $(a, 0)$ at some future time when the control u is of the form kt.

11.2 An electrochemical system is characterized by the differential equations

$$\frac{dx_1}{dt} = -x_1 + u$$

$$\frac{dx_2}{dt} = x_1,$$

where u is a control variable, chosen so as to minimize the cost functional,

$$\int_0^\infty (x_2{}^2 + 16u^2/3)\,dt.$$

Show that if the boundary conditions satisfied by the state variables are

$$x_1(0) = a, \qquad x_1(\infty) = 0$$

$$x_2(0) = b, \qquad x_2(\infty) = 0$$

then the optimum choice for u is

$$u(t) = -0{\cdot}366\,x_1(t) - 0{\cdot}433\,x_2(t).$$

Illustrate the feedback control in a block diagram.

11.3 A linear second order differential equation is described by

$$\dot{x}_1 = x_2, \qquad x_1(0) = 1.$$

$$\dot{x}_2 = u, \qquad x_2(0) = 1.$$

By using the Euler equations, find the optimal control $u(t)$ which minimizes:

$$\text{(i)} \quad J = \int_0^1 u^2 dt, \qquad x_1(1) = x_2(1) = 0;$$

$$\text{(ii)} \quad J = \int_0^1 u^2 dt, \qquad x_1(1) = 0.$$

11.4 A control system is described by the equation

$$\dot{x} + x = u,$$

where $x = x(t)$ is the state variable and $u = u(t)$ is the control variable. If $x = x_0$ at $t = 0$ and $x = Q$ at $t = T$, determine the state variable path when u is chosen to minimize

$$I = \int_0^T [(Q-x)^2 + u^2]\, dt.$$

11.5* A ballistic missile, launched with zero velocity from (0, 0) in the xy-plane, has a predetermined fuel propellant expenditure programme, but with a thrust direction $\phi(t)$ to be chosen. The motor is operative from $t = 0$ until $t = T$, at which instant the motor is shut down and the missile moves under gravity alone.

Denoting its velocity components by (u, v) and assuming constant gravity, show that the total range in the horizontal plane is given by

$$R = x_T + \frac{u_T}{g}[v_T + (v_T{}^2 + 2gy_T)^{1/2}]$$

where the suffix T indicates the function is to be evaluated at $t = T$.

If F is the rocket thrust per unit mass, show that the augmented cost functional to be considered in order to maximize the range in the horizontal plane is

$$R^* = \int_0^T \left[\frac{d}{dt}\left\{ x + \frac{u}{g}[v + (v^2 + 2gy)^{1/2}] \right\} + \lambda_1(\dot{u} - F\cos\phi) + \lambda_2(\dot{v} - F\sin\phi + g) + \lambda_3(\dot{x} - u) + \lambda_4(\dot{y} - v) \right] \mathrm{d}t,$$

where $\lambda_1, \lambda_2, \lambda_3, \lambda_4$, are Lagrange multipliers. Using Euler's equations show that

$$\dot{\lambda}_1 = -\lambda_3, \quad \dot{\lambda}_2 = -\lambda_4, \quad \dot{\lambda}_3 = \dot{\lambda}_4 = 0, \quad \tan\phi = \lambda_2/\lambda_1;$$

and using the transversality conditions show that at $t = T$,

$$\lambda_1 = -\frac{1}{g}(v_T + r), \quad \lambda_2 = -u_T(v_T + r), \quad \lambda_3 = -1, \quad \lambda_4 = -u_T/r$$

where $r = (v_T^2 + 2gy_T)^{1/2}$. Hence deduce that the control function ϕ is given by

$$\tan\phi = u_T/(v_T^2 + 2gy_T)^{1/2}$$

for $0 \leqslant t \leqslant T$.

If F is a constant, integrate the equations for u, v, x and y and show that ϕ satisfies

$$g\sin^3\phi + F\cos 2\phi = 0\,.$$

Show that this equation possesses one solution in the first quadrant if $F > g$, and show that the maximum range is given by

$$R = FT^2\left(\frac{F}{g}\cot\phi - \frac{1}{2}\cos\phi\right).$$

11.6* For the problem described in Ch. 11.5, evaulate ϕ at times $n\tau/10$ ($n = 0, 1, \ldots, 10$) when

(i) $\alpha = 0{\cdot}75$

(ii) $\alpha = 0{\cdot}5$

(iii) $\alpha = 0{\cdot}25$

Plot the graph of ϕ against τ for each value of α.

11.7* Continuing the problem in Ch. 11.5, when the thrust F is constant, integrate (11.39) and show that:

(i) the maximum final horizontal velocity is

$$v_1 = \frac{F\tau(1-\alpha^2)}{2\alpha}\log\left(\frac{1+\alpha}{1-\alpha}\right);$$

(ii) the horizontal distance travelled by the rocket at time t ($0 \leqslant t \leqslant \tau$) is given by

$$x = \frac{F\tau(1-\alpha^2)}{4\alpha^2}\left\{2\alpha(t-\tau)\log\left(\frac{1+\alpha}{1-\alpha}\right) + \tau(1+\alpha^2) + \tau(1-\alpha^2)\left[\tan\phi\log(\sec\phi + \tan\phi) - \sec\phi\right]\right\}.$$

11.8* For the problem in Problem 11.7, write a computer program to evaluate x, y, and ϕ at times $n\tau/m$ $(n = 0, 1, \ldots, m)$ for a given value of α.

Run this program with $m = 100$ and $\alpha = 0{\cdot}5$, plotting the path of the rocket. At times $0, 0{\cdot}1\tau, 0{\cdot}2\tau, \ldots, \tau$ indicate on the path the direction of the thrust attitude angle.

11.9* Develop the equations governing the optimal consumption and capital per worker trajectories as described in Ch. 11.6, when $\lambda = 0$ and

$$f(k) = k^{\alpha}, \qquad (0 < \alpha \leqslant 1),$$

using the welfare integral (11.68).

When $\alpha = 1$, solve for c and k with $\beta < 1$ over the time interval $[0, T]$, given that $k(0) = k_o$ and $k(T) = k_T$.

Appendix 1

Line Integrals

Let $Oxyz$ be rectangular Cartesian axes. If a scalar function $\Omega = \Omega(x, y, z)$ is a single-valued function of x, y and z, defined at all points within a region R, we say that Ω is a *scalar field* in R. Similarly if the components of a vector $\mathbf{a} = (a_1, a_2, a_3)$ are all scalar fields in R, we say that $\mathbf{a}$ is a *vector field* in R. Frequently R is all space and we drop the term 'in R'.

If $\Omega = \Omega(x, y, z)$ is a scalar field, and $\mathscr{C}$ is a curve from A to B defined by the parametric equation

$$\mathbf{r} = \mathbf{r}(t), \qquad t_0 \leqslant t \leqslant t_1, \tag{A1.1}$$

we define and denote the line integral of Ω along $\mathscr{C}$ from A to B by

$$\int_{\mathscr{C}} \Omega \, ds = \int_{t_0}^{t_1} \Omega(t) \frac{ds}{dt} \, dt, \tag{A1.2}$$

where $\Omega(t) = \Omega\{x(t), y(t), z(t)\}$ and s is the arc length along $\mathscr{C}$.

EXAMPLE:

Evaluate $\int_{\mathscr{C}} (x^2 - y) ds$, where $\mathscr{C}$ is the boundary of the circle $x^2 + y^2 = a^2$ in the positive quadrant from $(a, 0)$ to $(0, a)$ and in the xy-plane.

SOLUTION:

The curve $\mathscr{C}$ is illustrated in Fig. A1.1.

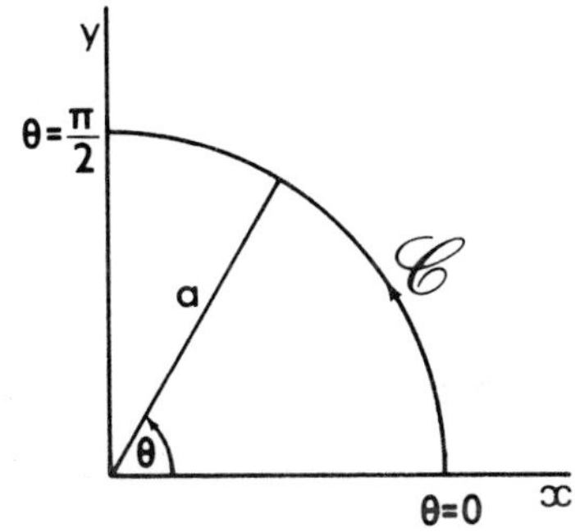

Fig. A1.1 Curve $\mathscr{C}$

A parametric equation for $\mathscr{C}$ is

$$\mathbf{r} = (a\cos\theta, a\sin\theta, 0), \qquad 0 \leqslant \theta \leqslant \pi/2\,,$$

so that
$$\frac{ds}{d\theta} = \left|\frac{d\mathbf{r}}{d\theta}\right| = [(-a\sin\theta)^2 + (a\cos\theta)^2]^{1/2} = a\,.$$

Hence
$$\int_{\mathscr{C}} (x^2 - y)\,ds = \int_0^{\pi/2} (a^2\cos^2\theta - a\sin\theta)\,a d\theta$$
$$= a^3\int_0^{\pi/2} \cos^2\theta\, d\theta - a^2\int_0^{\pi/2} \sin\theta\, d\theta$$
$$= \frac{a^3\pi}{4} - a^2\,.$$

In (A1.2) we defined the line integral of a scalar field. The definition of the line integral of a vector field along $\mathscr{C}$ from A to B is denoted and given by

$$\int_{\mathscr{C}} \mathbf{F}.d\mathbf{r} = \int_{t_0}^{t_1} \mathbf{F}.\frac{d\mathbf{r}}{dt}\,dt. \tag{A1.3}$$

Note that the right hand side can be written as

$$\int_{t_0}^{t_1} \mathbf{F}.\frac{d\mathbf{r}}{dt}dt = \int_{t_0}^{t_1} \mathbf{F}.\frac{d\mathbf{r}}{ds}\frac{ds}{dt}\,dt$$
$$= \int_{t_0}^{t_1} (\mathbf{F}.\hat{\mathbf{T}})\frac{ds}{dt}\,dt, \qquad \text{since } \frac{d\mathbf{r}}{ds} = \hat{\mathbf{T}} \text{ from (2.23),}$$
$$= \int_{\mathscr{C}} (\mathbf{F}.\hat{\mathbf{T}})\,ds, \quad \text{according to (A1.2);}$$

so that we can interpret the vector element $d\mathbf{r}$ as $d\mathbf{r} = \hat{\mathbf{T}}\,ds$ where $\hat{\mathbf{T}}$ is the unit tangent vector to $\mathscr{C}$. Definition (A1.3) is then compatible with the definition of a line integral of a scalar field.

EXAMPLE:

Evaluate $\int_{\mathscr{C}} \mathbf{F}.d\mathbf{r}$ where $\mathbf{F} = (y^2, xy, z^3)$ and $\mathscr{C}$ is the straight line from (0, 0, 0) to (1, 1, 1).

SOLUTION:

A parametric equation for $\mathscr{C}$ is

$$\mathbf{r} = (t, t, t), \qquad 0 \leqslant t \leqslant 1$$

Hence

$$\int_{\mathscr{C}} \mathbf{F}.d\mathbf{r} = \int_0^1 (t^2, t^2, t^3) \,.\, (1, 1, 1)\,dt$$

$$= \int_0^1 (2t^2 + t^3)dt$$

$$= \left[\frac{2}{3}t^3 + \frac{t^4}{4}\right]_0^1$$

$$= \frac{11}{12}.$$

Appendix 2

Grad, Div. and Curl

The partial differential operator *nabla* is defined and denoted by

$$\begin{aligned}\nabla &= \mathbf{i}\frac{\partial}{\partial x}+\mathbf{j}\frac{\partial}{\partial y}+\mathbf{k}\frac{\partial}{\partial z}\\ &= \left(\frac{\partial}{\partial x},\frac{\partial}{\partial y},\frac{\partial}{\partial z}\right)\end{aligned}\tag{A2.1}$$

and we define its operation on scalar and vector fields in the following way:

(i) *Gradient of a scalar field, Ω;*

$$\begin{aligned}\text{grad}\,\Omega &= \nabla\Omega\\ &= \left(\mathbf{i}\frac{\partial}{\partial x}+\mathbf{j}\frac{\partial}{\partial y}+\mathbf{k}\frac{\partial}{\partial z}\right)\Omega\\ &= \mathbf{i}\frac{\partial\Omega}{\partial x}+\mathbf{j}\frac{\partial\Omega}{\partial y}+\mathbf{k}\frac{\partial\Omega}{\partial z}\\ &= \left(\frac{\partial\Omega}{\partial x},\frac{\partial\Omega}{\partial y},\frac{\partial\Omega}{\partial z}\right),\end{aligned}\tag{A2.2}$$

which we note is a vector field.

(ii) *Divergence of a vector field,* $\mathbf{a} = (a_1, a_2, a_3)$;

$$\begin{aligned}\text{div}\,\mathbf{a} &= \nabla.\mathbf{a}\\ &= \left(\frac{\partial}{\partial x},\frac{\partial}{\partial y},\frac{\partial}{\partial z}\right).\left(a_1,a_2,a_3\right)\\ &= \frac{\partial a_1}{\partial x}+\frac{\partial a_2}{\partial y}+\frac{\partial a_3}{\partial z}\end{aligned}\tag{A2.3}$$

which we note is a scalar field.

(iii) <u>*Curl of a vector field,*</u> $\mathbf{a} = (a_1, a_2, a_3)$;

$$\operatorname{curl}\mathbf{a} = \nabla \times \mathbf{a}$$

$$= \begin{vmatrix} \mathbf{i} & \mathbf{j} & \mathbf{k} \\ \frac{\partial}{\partial x} & \frac{\partial}{\partial y} & \frac{\partial}{\partial z} \\ a_1 & a_2 & a_3 \end{vmatrix}$$

$$= \left(\frac{\partial a_3}{\partial y} - \frac{\partial a_2}{\partial z}, \frac{\partial a_1}{\partial z} - \frac{\partial a_3}{\partial x}, \frac{\partial a_2}{\partial x} - \frac{\partial a_1}{\partial y}\right) \tag{A2.4}$$

which we note is a vector field.

EXAMPLES

(i) If $\Omega = xyz$, $\operatorname{grad}\Omega = (yz, xz, xy)$.

(ii) If $\Omega = x^2 + y^2 + z^2$, $\operatorname{grad}\Omega = (2x, 2y, 2z) = 2\mathbf{r}$.

(iii) If $\mathbf{a} = (xy, yz, zx)$,

$$\operatorname{div}\mathbf{a} = y + z + x,$$

$$\operatorname{curl}\mathbf{a} = (-y, -z, -x).$$

(iv) If $\mathbf{a} = (x, y, z) = \mathbf{r}$,

$$\operatorname{div}\mathbf{a} = 3;$$

$$\operatorname{curl}\mathbf{a} = \mathbf{0}.$$

We also have the following identities

$$\operatorname{div}(\operatorname{curl}\mathbf{a}) = 0, \tag{A2.5}$$

$$\operatorname{curl}(\operatorname{grad}\Omega) = \mathbf{0}, \tag{A2.6}$$

which can easily be proved. For instance, if $\mathbf{a} = (a_1, a_2, a_3)$

$$\operatorname{div}(\operatorname{curl}\mathbf{a}) = \operatorname{div}\left(\frac{\partial a_3}{\partial y} - \frac{\partial a_2}{\partial z}, \frac{\partial a_1}{\partial z} - \frac{\partial a_3}{\partial x}, \frac{\partial a_2}{\partial x} - \frac{\partial a_1}{\partial y}\right)$$

$$= \frac{\partial^2 a_3}{\partial x \partial y} - \frac{\partial^2 a_2}{\partial x \partial z} + \frac{\partial^2 a_1}{\partial y \partial z} - \frac{\partial^2 a_3}{\partial y \partial x} + \frac{\partial^2 a_2}{\partial z \partial x} - \frac{\partial^2 a_1}{\partial z \partial y}$$

$$= \left(\frac{\partial^2 a_1}{\partial y \partial z} - \frac{\partial^2 a_1}{\partial z \partial y}\right) + \left(\frac{\partial^2 a_2}{\partial z \partial x} - \frac{\partial^2 a_2}{\partial x \partial z}\right) + \left(\frac{\partial^2 a_3}{\partial x \partial y} - \frac{\partial^2 a_3}{\partial y \partial x}\right)$$

$$= 0, \text{ since } \frac{\partial^2 a_1}{\partial y \partial z} = \frac{\partial^2 a_1}{\partial z \partial y} \text{ etc.}$$

Appendix 3

Curvilinear Coordinates

Suppose the Cartesian coordinates of a point $P(x, y, z)$ can be expressed as functions of u, v, and w; i.e.

$$x = x(u, v, w), \quad y = y(u, v, w), \quad z = z(u, v, w). \qquad \text{(A3.1)}$$

Also suppose that (A3.1) can be solved for u, v, w, to give

$$u = u(x, y, z), \quad v = v(x, y, z), \quad w = w(x, y, z). \qquad \text{(A3.2)}$$

If (A3.1) and (A3.2) both give unique solutions, we call (u, v, w) curvilinear coordinates of the point P.

Consider a point P which has curvilinear coordinates (u, v, w). If v and w are kept constant, the locus of P defines a curve for varying u. We call this the *u-coordinate line.* Similarly for $v-$ and $w-$ coordinate lines. If the tangents to the $u-$, $v-$ and $w-$ coordinate lines at P are mutually perpendicular, we say that the curvilinear system is *orthogonal.* We denote unit tangents to the coordinate lines by $\mathbf{e}_u, \mathbf{e}_v, \mathbf{e}_w$, and using (2.22) we see that

$$\left.\begin{aligned} \mathbf{e}_u &= \frac{1}{h_1}\frac{\partial \mathbf{r}}{\partial u}, & h_1 &= \left|\frac{\partial \mathbf{r}}{\partial u}\right| \\ \mathbf{e}_v &= \frac{1}{h_2}\frac{\partial \mathbf{r}}{\partial v}, & h_2 &= \left|\frac{\partial \mathbf{r}}{\partial v}\right| \\ \mathbf{e}_w &= \frac{1}{h_3}\frac{\partial \mathbf{r}}{\partial w}, & h_3 &= \left|\frac{\partial \mathbf{r}}{\partial w}\right| \end{aligned}\right\} \qquad \text{(A3.3)}$$

EXAMPLES:

(i) *Cylindrical Polar Coordinates* (R, ϕ, z)

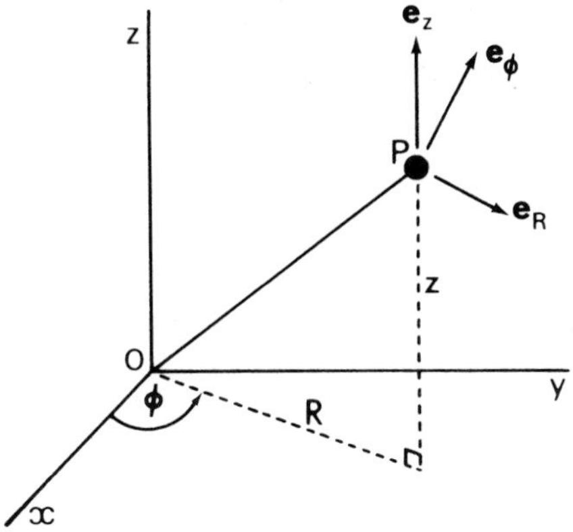

Fig. A3.1 Cylindrical polar coordinates

We see that $\mathbf{r} = (R\cos\phi, R\sin\phi, z)$

so that $h_1 = 1, \quad h_2 = R, \quad h_3 = 1\,.$

Thus
$$\mathbf{e}_R = (\cos\phi, \sin\phi, 0)$$
$$\mathbf{e}_\phi = (-\sin\phi, \cos\phi, 0)$$
$$\mathbf{e}_z = (0, 0, 1).$$

This system is orthogonal (for $\mathbf{e}_R . \mathbf{e}_\phi = \mathbf{e}_\phi . \mathbf{e}_z = \mathbf{e}_z . \mathbf{e}_R = 0$),.

(ii) *Spherical Polar Coordinates* (r, θ, ϕ)

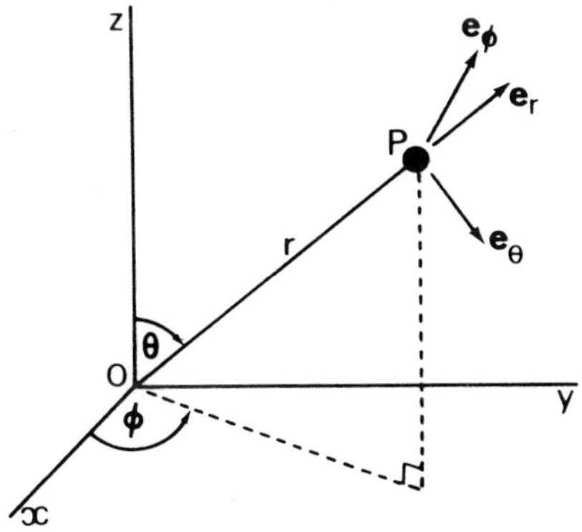

Fig. A3.2 Spherical polar coordinates

We see that $\mathbf{r} = (r\sin\theta\cos\phi,\ r\sin\theta\sin\phi, r\cos\theta)$

so that $h_1 = 1, \quad h_2 = r, \quad h_3 = r\sin\theta\,.$

Thus

$$\mathbf{e}_r = (\sin\theta\cos\phi, \sin\theta\sin\phi, \cos\theta)$$

$$\mathbf{e}_\theta = (\cos\theta\cos\phi, \cos\theta\sin\phi, -\sin\theta)$$

$$\mathbf{e}_\phi = (-\sin\phi, \cos\phi, 0)$$

This system is orthogonal.

Note that $\mathbf{i}, \mathbf{j}, \mathbf{k}$, the unit tangents to the $x-$, $y-$, $z-$ coordinate lines, are fixed in direction, whereas in general $\mathbf{e}_u, \mathbf{e}_v, \mathbf{e}_w$ will change in direction as P moves in space. Thus, for instance

$$\frac{\partial}{\partial u}(\mathbf{e}_u) \neq \mathbf{0}\,,$$

in general. For this reason, the expressions for grad, div and curl are modified in an orthogonal curvilinear system. If Ω is a scalar field, then it can be shown that

$$\text{grad}\ \Omega = \frac{1}{h_1}\frac{\partial\Omega}{\partial u}\mathbf{e}_u + \frac{1}{h_2}\frac{\partial\Omega}{\partial v}\mathbf{e}_v + \frac{1}{h_3}\frac{\partial\Omega}{\partial w}\mathbf{e}_w\,. \tag{A3.4}$$

If $\mathbf{a}$ is a vector field, which is expressed in terms of $\mathbf{e}_u, \mathbf{e}_v, \mathbf{e}_w$ as

$$\mathbf{a} = a_u\mathbf{e}_u + a_v\mathbf{e}_v + a_w\mathbf{e}_w\,,$$

then:

$$\text{div}\ \mathbf{a} = \frac{1}{h_1h_2h_3}\left\{\frac{\partial}{\partial u}(h_2h_3a_u) + \frac{\partial}{\partial v}(h_3h_1a_v) + \frac{\partial}{\partial w}(h_1h_2a_w)\right\} \tag{A3.5}$$

$$\text{curl}\ \mathbf{a} = \frac{1}{h_1h_2h_3}\begin{vmatrix} h_1\mathbf{e}_u & h_2\mathbf{e}_v & h_3\mathbf{e}_w \\ \dfrac{\partial}{\partial u} & \dfrac{\partial}{\partial v} & \dfrac{\partial}{\partial w} \\ h_1a_u & h_2a_v & h_3a_w \end{vmatrix}$$

$$= \frac{1}{h_2h_3}\left[\frac{\partial}{\partial v}(h_3a_w) - \frac{\partial}{\partial w}(h_2a_v)\right]\mathbf{e}_u + \frac{1}{h_1h_3}\left[\frac{\partial}{\partial w}(h_1a_u) - \frac{\partial}{\partial u}(h_3a_w)\right]\mathbf{e}_v$$

$$+ \frac{1}{h_1h_2}\left[\frac{\partial}{\partial u}(h_2a_v) - \frac{\partial}{\partial v}(h_1a_u)\right]\mathbf{e}_w\,. \tag{A3.6}$$

EXAMPLE:

Find the divergence and curl of the vector field

$$\mathbf{a} = R^2 \cos\phi \, \mathbf{e}_R + R^2 \sin\phi \, \mathbf{e}_\phi + R^2 \mathbf{e}_z$$

where (R, ϕ, z) are cylindrical polar coordinates.

SOLUTION:

Using (A3.5), with $h_1 = 1, \; h_2 = R, \; h_3 = 1$, we have

$$\operatorname{div} \mathbf{a} = \frac{1}{R}\left\{\frac{\partial}{\partial R}(Ra_R) + \frac{\partial a_\phi}{\partial \phi} + R\frac{\partial a_z}{\partial z}\right\}$$

$$= \frac{1}{R}\left\{\frac{\partial}{\partial R}(R^3\cos\phi) + \frac{\partial}{\partial\phi}(R^2\sin\phi) + R\frac{\partial}{\partial z}(R^2)\right\}$$

$$= \frac{1}{R}\left\{3R^2\cos\phi + R^2\cos\phi + 0\right\}$$

$$= 4R\cos\phi.$$

Using (A3.6),

$$\operatorname{curl} \mathbf{a} = \frac{1}{R}\begin{vmatrix} \mathbf{e}_R & R\mathbf{e}_\phi & \mathbf{e}_z \\ \dfrac{\partial}{\partial R} & \dfrac{\partial}{\partial \phi} & \dfrac{\partial}{\partial z} \\ R^2\cos\phi & R^3\sin\phi & R^2 \end{vmatrix}$$

$$= \frac{1}{R}\{0\mathbf{e}_R - 2R^2\mathbf{e}_\phi + (3R^2\sin\phi + R^2\sin\phi)\mathbf{e}_z\}$$

$$= -2R\mathbf{e}_\phi + 4R\sin\phi\,\mathbf{e}_z\,.$$

Appendix 4

Lagrange Multiplier Techniques

Suppose we wish to find extremum values of a function $f(x, y, z)$ subject to a constraint of the form

$$g(x, y, z) = 0 . \tag{A4.1}$$

Assuming that it is possible to solve (A4.1) for z, say, so that

$$z = h(x, y) , \tag{A4.2}$$

we can substitute z in (A4.1) to give

$$f\{x, y, h(x, y)\} = H(x, y) , \tag{A4.3}$$

say. Clearly for extremum values, $\partial H/\partial x = \partial H/\partial y = 0$. In practice it is not always possible tc solve (A4.1) as above, but we have the following result:

In order to find extremum of f, consider the function

$$F(x, y, z) = f(x, y, z) + \lambda g(x, y, z) \tag{A4.4}$$

where λ is unknown. Suppose F has an extremum at (x_0, y_0, z_0). Then at (x_0, y_0, z_0) we must have

$$\frac{\partial F}{\partial x} = \frac{\partial F}{\partial y} = \frac{\partial F}{\partial z} = 0 . \tag{A4.5}$$

Solving (A4.5) together with the constraint equation (A4.1) determines x_0, y_0, z_0 and λ, and it can be proved that $f(x_0, y_0, z_0)$ *is an extremum value of f.*

The parameter λ is called a *Lagrange multiplier* and we illustrate the above result in the following simple example.

EXAMPLE:

Find the minimum of $f(x, y, z) = x^2 + y^2 + z^2$

subject to the condition that

$$x + 3y - 2z = 4.$$

SOLUTION:

Introducing the Lagrange multiplier λ, we consider the function

$$F = x^2 + y^2 + z^2 + \lambda(x + 3y - 2z - 4).$$

For extremum values of F,

$$\frac{\partial F}{\partial x} = 2x + \lambda = 0$$

$$\frac{\partial F}{\partial y} = 2y + 3\lambda = 0$$

$$\frac{\partial F}{\partial z} = 2z - 2\lambda = 0$$

Hence $x = -\lambda/2$, $y = -3\lambda/2$, $z = \lambda$; and substituting into the constraint,

$$-\frac{\lambda}{2} - \frac{3\lambda}{2} - 2\lambda = 4.$$

Thus $\lambda = -4/7$ and $x = 2/7$, $y = 6/7$, $z = -4/7$. The extremum value of f is 8/7, and it can be shown that this is a minimum.

Note: In this particular example, we can use the direct method by eliminating one one of the variables.

The method of Lagrange multipliers can be extended to functions of any number of variables, subject to a number of constraints. For instance, if we wish to find extremum values of

$$f = f(x_1, x_2, \ldots, x_n) \tag{A4.6}$$

subject to the constraints

$$g(x_1, x_2, \ldots, x_n) = 0 \tag{A4.7}$$

$$h(x_1, x_2, \ldots, x_n) = 0 \tag{A4.8}$$

we consider extremum values of the function

$$F = f + \lambda g + \mu h \tag{A4.9}$$

where λ and μ are Lagrange multipliers. Hence we obtain n equations

$$\frac{\partial F}{\partial x_i} = 0 \; (i = 1, 2, \ldots, n);$$

and with the two equations of constraint, (A4.7) and (A4.8), we have $(n + 2)$ equations for the $(n + 2)$ unknowns $x_{1_0}, x_{2_0}, \ldots, x_{n_0}$, λ and μ.

EXAMPLE:

Find the minimum of the function

$$f(x, y, z, t) = x^2 + y^2 + z^2 + t^2$$

subject to the constraints

$$x + y - z + 2t = 2,$$

$$2x - y + z + 3t = 3.$$

SOLUTION:

We consider extremum values of

$$F = x^2 + y^2 + z^2 + t^2 + \lambda(x + y - z + 2t - 2) + \mu(2x - y + z + 3t - 3).$$

Thus
$$\frac{\partial F}{\partial x} = 2x + \lambda + 2\mu = 0 \Rightarrow x = -\tfrac{1}{2}\lambda - \mu$$

$$\frac{\partial F}{\partial y} = 2y + \lambda - \mu = 0 \Rightarrow y = -\tfrac{1}{2}\lambda + \tfrac{1}{2}\mu$$

$$\frac{\partial F}{\partial z} = 2z - \lambda + \mu = 0 \Rightarrow z = \tfrac{1}{2}\lambda - \tfrac{1}{2}\mu$$

$$\frac{\partial F}{\partial t} = 2t + 2\lambda + 3\mu = 0 \Rightarrow t = -\lambda - \tfrac{3}{2}\mu$$

Substituting into the constraints gives

$$-\frac{7}{2}\lambda - 3\mu = 2$$

$$-3\lambda - \frac{15}{2}\mu = 3$$

which gives $\lambda = -8/23$, $\mu = -6/23$. Hence

$$x = \frac{10}{23}, \quad y = \frac{1}{23}, \quad z = -\frac{1}{23}, \quad t = \frac{17}{23}$$

and $f = 17/23$. (This is in fact a minimum.)

Answers to Problems

Chapter 2 Kinematics

2.2 Helix $x^2 + y^2 = a^2, \quad z = bt$;

$$\mathbf{v} = (a\omega \cos(\omega t), \ -a\omega \sin(\omega t), b) ;$$

$$\mathbf{f} = (-a\omega^2 \sin(\omega t), \ -a\omega^2 \cos(\omega t), 0) .$$

2.4

$$f = \frac{d^2x}{dt^2} = \frac{dv}{dt} = \frac{dv}{dx}\frac{dx}{dt} = v\frac{dv}{dx} .$$

2.5 Least maximum speed is 38·8 m s^{-1}.

2.6 0·23 mile.

2.7 10 mph, 30 mph, 45 mph; 10·96 s.

2.10

$$\mathbf{v} = a\dot{\theta}\,\mathbf{e}_\theta + a\sin\theta\,\dot{\phi}\,\mathbf{e}_\phi ;$$

$$\mathbf{f} = [-a\dot{\theta}^2 - a\sin^2\theta\,\dot{\phi}^2]\,\mathbf{e}_r + [a\ddot{\theta} - a\sin\theta\cos\theta\,\dot{\phi}^2]\,\mathbf{e}_\theta + [2a\cos\theta\,\dot{\theta}\dot{\phi} + a\sin\theta\,\ddot{\phi}]\,\mathbf{e}_\phi .$$

2.13 1°26′ West of South; 5·001 hours.

2.14 49 km per hour in direction N 34°56′ W.

2.17 Capture time $= 2b/3v_c$, where v_c is the speed of the cat.

2.18 (i) Yes (ii) No.

Chapter 3 Newtonian mechanics

3.1 $32{\cdot}4\ \mathrm{m\,s^{-1}}$.

3.2 $21{\cdot}4$ m.

3.3 $(g/k)^{1/2}$.

3.6
$$h = \int_0^u \frac{u\,du}{(g + 2au + b^2u^2)};$$
$$\tau = \int_0^u \frac{du}{(g + 2au + b^2u^2)}.$$

3.8
$$h = \frac{1}{2k}\log\left(1 + \frac{ku^2}{g}\right) \quad \text{and} \quad u = (g/k)^{1/2}\tan\{(gk)^{1/2}\tau\},$$
giving
$$h = \frac{1}{2}gk\,\tau^2\left(1 + \frac{1}{6}gk\,\tau^2 + \ldots\right)$$

3.9
(i) if $\dfrac{dm}{dt} = k\pi r^2, \; \dfrac{dr}{dt} = \dfrac{k}{4\rho};$

(ii) $r \approx \dfrac{kt}{4\rho}$ and $\dfrac{d}{dt}(mv) = mg;$ thus $v \approx tg/4$.

3.10
If $\dfrac{dm}{dt} = k(4\pi r^2)v, \; \dfrac{dr}{dt} = \dfrac{kv}{\rho} = \dfrac{k}{\rho}\dfrac{dx}{dt}$ and $r = r_0 + \dfrac{kx}{\rho}$.

Chapter 4 Rocket Flight Performance

4.3 Let $M(t) = P + M_f(t) + M_c(t)$, where M_c = mass of casing.

If $\dfrac{dM_o}{dt} = -k'$, we see that $M_c(t) = -k't + (1-\epsilon)M_o$.

But we require $M_c = 0$ when $\tau = \epsilon M_o/k$. Hence
$k' = (1-\epsilon)k/\epsilon$.

Equation of motion is
$$M\frac{dv}{dt} = -c\frac{dM_f}{dt}$$
so that
$$v = -c\,\epsilon\log\left\{1 - \frac{kt}{\epsilon(P + M_o)}\right\}.$$
Final velocity is $10{\cdot}86\ \mathrm{km\,s^{-1}}$.

4.4 $\sqrt{11} \approx 3{\cdot}32$.

4.6 (i) 77 ;
(ii) 107.

Chapter 5 Fields of Force

5.2 Equation of trajectory is $y = x \tan\alpha - gx^2 \sec^2\alpha/2u^2$;
Maximum range up inclined plane is $u^2/g(1 + \sin\beta)$.

5.5 Regarding y as a function of x, the boundary conditions are

$$y(0) = 0,$$

$$\frac{dy}{dx}(0) = \tan a\,,$$

$$\frac{d^2y}{dx^2}(0) = -\frac{g\sec^2\alpha}{u_o^2}$$

5.6 $\mathbf{r} = \{a + U/\omega - U\cos(\omega t)/\omega,\ b + U\sin(\omega t)/\omega,\ c + Vt\}$,
$\omega = qB/m$.

5.7 $z = Vt + qE_z t^2/2m$; uniform acceleration in z-direction, and projection of motion on xy-plane is a circle of radius $(1/\omega)(U + E_x/B)$, whose centre $\{a + (1/\omega)(U + E_x/B), b - E_x t/B\}$ is moving along negative y-axis.

Chapter 6 Conservation Theorems

6.3 (i) $\frac{88}{4}$, (ii) 4.

6.4 (a) (i) $4\frac{2}{3}$, (ii) 4;

(b) (i) $\frac{30}{7}$, (ii) 4 .

6.5 Yes; $\mathbf{F} = (y, x, 1)$ and curl $\mathbf{F} = \mathbf{0}$; hence $\mathbf{F}$ is conservative;
$\Omega = xy + z +$ constant.

6.6 Potential energy $= -xyz +$ constant.

6.7 Equation of motion $\frac{d\mathbf{v}}{dt} = q(\mathbf{v} \times \mathbf{B})$ — take scalar product with $\mathbf{v}$.

With electric field, $\frac{d\mathbf{v}}{dt} = q((\mathbf{v} \times \mathbf{B}) + \mathbf{E})$ and $\frac{d}{dt}(\frac{1}{2}mv^2) = q\mathbf{v}.\mathbf{E}$.

6.8 $V = \lambda a^2/2l$; $mv^2/2 + \lambda x^2/2l = \lambda a^2/2l$; $x = a\cos(\sqrt{\lambda/ml}\ t)$.

Chapter 7 Space Dynamics

7.1 Force $= mg(R_e/R_m)^2 (M_m/M_e)$, [$R_m$ = radius of Mass, M_m = mass of Mars] $= 1{\cdot}48mg$.

7.4 $m\sqrt{ak}$.

7.5 $17{\cdot}3\, a_e$, where a_e = semi-major axis of earth's orbit about the sun;

$R_a = 35{\cdot}4a_e$;

$R_p = 0{\cdot}5\, a_e$.

7.6 $\lambda = [2/(1+e)]^{1/2}$.

7.7 $\alpha = \sin^{-1}(1/4)$.

7.8 Three satellites are required to cover all points on the equator.

7.10 $1{\cdot}65\, R_E$; $6{\cdot}5$.

7.11 $\pm\ a \sin\alpha$.

7.15 $10{\cdot}2\ \mathrm{km\,s^{-1}}$; approximately 6 years; Saturn in front of Earth by $106°$.

7.16 $\beta_o = 28°$; $v_o = 6{\cdot}7\ \mathrm{km\,s^{-1}}$; $H = 1{\cdot}2 \times 10^3$ km.

Chapter 8 Stability

8.2 $x = -a$ is stable equilibrium position;

period of small oscillations $= 2\pi\sqrt{2ma^3/c}$.

8.3 If θ is the angle made with the downward vertical, positions of equilibrium are

(i) $\theta = 0$ (unstable); (ii) $\theta = \pi$ (unstable),
(iii) $\theta = \pm \cos^{-1}[b(mg + \lambda)/a\lambda]$ (stable).

8.4 (i) If $x = x_o + \xi, \xi = At^2$; unstable;

(ii) $\xi = Ae^{-t^2/2}$; stable;

(iii) $\xi = A/t(1 + t)^2$; stable.

8.5 Writing $x = x_o + \xi$, ξ satisfies.

$$\frac{d\xi}{dt} + t\xi + \frac{3t\xi}{x_o^4} = 0$$

and solving, $\xi = Ae^{-2t^2}/(1 + e^{-2t^2})^{3/4}$; stable.

8.6 Unstable ($\ddot{\xi} = 0 \rightarrow \xi = At + B$).

8.7 $3f(a) + af'(a) > 0$ for stable path.

Chapter 9 Oscillations and Vibrations

9.2 $\sqrt{3gb}$; $2\left(\frac{2\pi}{3} + \sqrt{3}\right)\sqrt{\frac{b}{g}}$

9.4 If x_1 and x_2 denote displacements of particles m, $\frac{2}{3}m$ respectively,

$$x_1 = -\frac{2}{5}d\cos\left(\sqrt{\frac{3\lambda}{am}}t\right) + \frac{2}{5}d\cos\left(\sqrt{\frac{\lambda}{2am}}t\right).$$

$$x_2 = \frac{2}{5}d\cos\left(\sqrt{\frac{3\lambda}{am}}t\right) + \frac{3}{5}d\cos\left(\sqrt{\frac{\lambda}{2am}}t\right).$$

9.6 If $n > \pi/\lambda$, the motion is overdamped, and no oscillations occur.

9.9 $39{\cdot}3\ \mathrm{N\,m^{-1}}$.

9.11 $\ddot{x} + k\dot{x} + \omega^2 x = \omega^2 ut.$

Chapter 10 Variational Calculus

10.1 (i) 4/3, (ii) 23/15, (iii) $(e^2 - 2e - 2)/(e-1)^2$.

10.2 $x^2 + (y-1)^2 = 1$.

10.3 $x = x_o \sinh\left(\frac{T-x}{\alpha}\right) \Big/ \sinh(T/\alpha)$ and $J_1 = \alpha x_o^2 \coth(T/\alpha)$;

$x = x_o \cosh\left(\frac{T-x}{\alpha}\right) \Big/ \cosh(T/\alpha)$ and $J_2 = \alpha x_o^2 \tanh(T/\alpha)$;

$J_2 < J_1$.

10.4 Path for minimum value of J is $x = \frac{1}{2} t^2 - 2t + 1$.

10.5 $L = \frac{1}{2} m(\dot{x}^2 + \dot{y}^2) - mgy$.

10.6 $P.E. = \frac{\lambda y^2}{2a}$; $L = \frac{1}{2} m\dot{x}^2 - \lambda x^2/2a$.

10.7 $L = \frac{1}{2}(a^2\dot{\theta}^2 + a^2 \sin^2\theta\, \dot{\phi}^2) - mga\cos\theta$.

10.8 $y = \frac{1}{2}x^2 - \frac{3}{2}x + 1$ and $J_1 = 23/24$;

$y = \frac{1}{2}x^2 - x + 1$ and $J_2 = 5/6$;

$J_2 < J_1$.

10.9 (i) $y = (b + 2a)x^3 - (b + 3a)x^2 + ax$;

(ii) $y = 0$.

Chapter 11 Optimal Control

11.1 Write x_1 in the form $x_1 = A\sin(\omega t + \epsilon) + kt/\omega^2$. Then system controllable when $k = a\omega^3/2(\pi - \epsilon)$ where ϵ satisfies $\tan\epsilon = -2(\pi - \epsilon)$ and $\pi/2 < \epsilon < \pi$.

11.2 Writing $x_2 = Ae^{-\sqrt{3}t/2} + Be^{-t/2}$, where A and B arbitrary constants, we see that

$$x_1 = -\frac{\sqrt{3}t}{2} A\, e^{-\sqrt{3}t/2} - \frac{B}{2} e^{-t/2},$$

$$u = \left(\frac{3}{4} - \frac{\sqrt{3}}{2}\right) Ae^{-\sqrt{3}\,t/2} - \frac{1}{4} Be^{-t/2}.$$

Without applying initial conditions, we can show that

$$u = \frac{1}{2}(1-\sqrt{3})\,x_1(t) - \frac{\sqrt{3}}{4} x_2(t)$$

$$\approx -\,0{\cdot}366\,x_1(t) - 0{\cdot}433\,x_2(t).$$

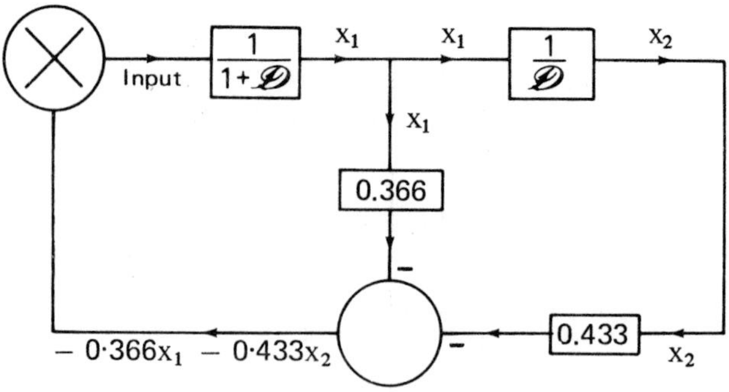

11.3 (i) $u = 18t - 10$ and $J = 28$;

(ii) $u = 6(t-1)$ and $J = 12$.

11.4 $$x = \frac{Q}{2}\left[1 + \frac{\sinh(\sqrt{2}t)}{\sinh(\sqrt{2}t_1)}\right] + \left(x_o - \frac{Q}{2}\right)\frac{\sinh[\sqrt{2}(t_1 - t)]}{\sinh(\sqrt{2}t_1)}$$

11.7 At any time t, $0 \leqslant t \leqslant \tau$,

$$u_1 = -\frac{F\tau}{2\alpha}(1-\alpha^2)\int \sec\phi\, d\phi + \text{constant}$$

giving $$v_1 = \frac{F\tau}{2\alpha}(1-a^2)\left[\log\left(\frac{1+\alpha}{1-\alpha}\right) - \log(\sec\phi + \tan\phi)\right].$$

Also

$$x = \frac{F\tau(1-\alpha^2)}{2\alpha}\left[t \log\left(\frac{1+\alpha}{1-\alpha}\right) + \frac{\tau(1-\alpha^2)}{2\alpha}\int \sec^2\phi \log(\sec\phi + \tan\phi)\,d\phi\right].$$

11.9

$$\frac{dk}{dt} - k^{\alpha} = -c,$$

$$\frac{dc}{dt} - 2(\alpha k^{\alpha-1} - \beta)c = 0.$$

General solutions and $c = A e^{\gamma t}$, $k = Be^t + \frac{A}{(1-\gamma)} e^{\gamma t}$, where

A and B are arbitrary, and $\gamma = 2(1-\beta)$. Applying the conditions gives

$$A = \frac{(k_T - k_o e^T)(1-\gamma)}{(e^{\gamma t} - e^T)}$$

and

$$B = \frac{(k_o e^{\gamma t} - k_T)}{(e^{\gamma T} - e^T)}.$$

Index

$$(P + M_0 - kt)\frac{dv}{dt} = ck \qquad \text{Eq}^n\ 4.12 \text{ page } 69$$

$$\therefore \int dv = \int \frac{ck}{(P+M_0-kt)}\,dt$$

$$v = \int \frac{ck}{(P+M_0)\left\{1 - \frac{kt}{P+M_0}\right\}}\,dt$$

$$V = \frac{ck}{P+M_0}\int \frac{dt}{\left\{1 - \frac{kt}{P+M_0}\right\}} = \frac{ck}{P+M_0}\left\{\frac{1}{-k/(P+M_0)}\right\}\ln\left(1 - \frac{kt}{P+M_0}\right) + c$$

$$= -c\ln\left\{1 - \frac{kt}{P+M_0}\right\} + V_0$$

$$(P+M_0-kt)\frac{dv}{dt} = ck - (P+M_0-kt)g \qquad \text{Eq}^n\ 418 \text{ page } 71$$

$$\therefore \int dv = \int \frac{ck}{(P+M_0-kt)}\,dt - \int g\,dt$$

$$= -c\ln\left\{1 - \frac{kt}{P+M_0}\right\}\Big|_0^T - gt\Big|_0^T + c \qquad \text{where } T = \frac{\varepsilon M_0}{k}$$

where $c = V_0$ at $t = t_0 = 0$